Research Reports ESPRIT

Project Group Microelectronics · Volume 1

Edited in cooperation with the European Commission

J. Mun A. A. M'baye (Eds.)

Gallium Arsenide Technology in Europe

Springer-Verlag

Berlin Heidelberg New York
London Paris Tokyo
Hong Kong Barcelona
Budapest

Volume Editors

Joseph Mun
BNR Europe Limited
London Road
Harlow, Essex CM 17 9NA, UK

Abdoul Aziz M'baye
European Commission
Rue de la Loi 200
B-1049 Brussels, Belgium

This report originates from ESPRIT Projects 2035, 5003, 5018, 5031, 5032, 5052, 6016, and 6050. These projects belong to the Work Area Compound Semiconductors and Other Non-Silicon Technologies within the Sunprogramme Microelectronics of ESPRIT, the European Specific Programme for Research and Development in Information Technology supported by the European Commission.

Collaboration is an essential element in European R&D. Collaboration has strengthened the European gallium arsenide (GaAs) community and helped to accelerate its growth against difficult times. This book illustrates some of the recent collaborative activities pursued by the aforementioned ESPRIT projects, including applications, MMICs, devices, fabrication technologies, modelling, as well as future manufacturing and technology issues. The book concludes with a section on nanoelectronics, which also includes some activities under various ESPRIT Basic Research Projects.

CR Subject Classification (1991): B.7

ISBN-13:978-3-540-57906-9   e-ISBN-13:978-3-642-78934-2
DOI: 10.1007/978-3-642-78934-2
CIP data applied for

Publication No. EUR 15677 EN of the European Commission,
Dissemination of Scientific and Technical Knowledge Unit, Directorate-General Information Technologies and Industries, and Telecommunications, Luxembourg.
Neither the European Commission nor any person acting on behalf of the Commission is responsible for the use which might be made of the following information.

SPIN: 10132168   45/3140 – 543210 – Printed on acid-free paper

Foreword

Some of the key milestones in GaAs technology were first demonstrated in Europe, for example, the first GaAs Field Effect Transistor (FET) with microwave performance and the first GaAs Microwave Monolithic Integrated Circuit (MMIC). The strategic nature of GaAs technology has attracted heavy investment from many vertically integrated companies in information technology, communication and defence, as well as from semiconductor manufacturers world wide. Europe always faced strong competition from the USA and Japan and until 1984, European GaAs activities were fragmented amongst various players, with some of the activities loosely grouped into national programmes.

In 1984, a number of collaborative projects were established under the European Specific Programme on Information Technology (ESPRIT) which crossed national boundaries. It has launched a new and exciting phase for GaAs in Europe and few of those involved at the time could have imagined where such collaboration may lead. In the beginning of those early projects, collaboration was approached with caution and suspicion because after all, many members of a newly formed project team were previously competitors. However, common technology problems soon became apparent and the opportunity to discuss these problems with engineers and scientists from different backgrounds had quickly broken down all barriers. Today, collaboration has become an essential element in European R & D. Collaboration has strengthened the GaAs community and helped to accelerate its growth against difficult times.

This book examines the importance of GaAs technology in Europe and illustrates some of the recent activities pursued under various ESPRIT projects. These activities include system applications, MMICs, devices, fabrication technologies and modelling, as well as future manufacturing and technology issues. The book is concluded with a section on the fascinating subject of nanoelectronics pursued under various ESPRIT Basic Research projects. We would like to take this opportunity to express our sincere thanks to all the editors and contributors who gave their valuable time and expertise to make this book possible.

Joseph Mun and Abdoul Aziz M'baye
February 1994

Contents

1. The Importance of GaAs Technology in Europe

J A Turner
GEC-Marconi Materials Technology Ltd., Caswell, Northants   NN12 8EQ, UK

1.1   Introduction

It is now (in 1993) over 27 years ago since the first paper demonstrating microwave performance from a Gallium Arsenide Field Effect Transistor was published and 17 years ago that the first microwave monolithic integrated circuit was produced (both European achievements) and yet it is only now that these devices and circuits are being produced commercially in reasonably large volumes. Why has it taken so long? The answer to this question is not a simple one. Perhaps it is because of the complex nature of the material itself which has taken some time to 'tame' and become reproducible enough to allow consistent device performance to be obtained, perhaps it is because the circuit and system designer took some time to realise the immense potential of the material or because manufacturers could not produce the components at the 'right' prices. Whatever the reason, the objections to its use have now been overcome and the market predictions for Gallium Arsenide based components show a steady increase to the year 2000 when merchant sales in excess of $2Bn are expected and thereafter to increase at 20% per annum.

There is no doubt that the commercial interest in GaAs was catalysed by the advent of satellite television. This growing consumer applications area began in Europe and is now about to spread into the US. At the high operating frequency of 10.7 - 12GHz the only viable technology for the signal down-converter was GaAs based and once the decisions were made to launch a television service via satellite this heralded the volume manufacture of GaAs parts, initially discrete devices and latterly GaAs MMICs. To date some 10 million receivers each using up to seven GaAs components have been produced. The ability of the manufacturers to produce such volumes has given the systems designer the confidence to look to other application areas where benefit is to be gained from the use of GaAs. These will be described later in this chapter.

As with all other semiconductor technologies, the European GaAs component manufacturers find themselves facing strong competition from the USA and Japan but thanks to support from the European Commission, National Governments and the Companies themselves, European technology is able to compete in performance with the best in the rest of the world.

Although funding for GaAs, mainly for military exploitation, began as early as 1964 this was relatively small and it was only in the early 1980s when the real potential of GaAs was realised that most of the major European companies began serious GaAs programmes. In 1984 the first ESPRIT programmes in GaAs were launched, and at first the consortium members were very wary of each other because, after all, they were competitors. However, as each other's strengths and weaknesses became known and social as well as working relationships were formed, collaboration increased to the point where very fruitful co-operation was possible. These collaborative projects have accelerated the growth of technological expertise across Europe and helped it keep abreast of the rest of the world in most of the presently available GaAs based technologies. The idea of collaborative programmes has also spread to National Programmes in many European Countries in which the large manufacturers and users have combined to accelerate the development of crucial GaAs components in order to improve their competitiveness in the world market place. It is seen that collaborative programmes are an essential ingredient for the survival of the European GaAs industry and that they continue to be supported by the EC and National Governments. Huge sums of Government money is being injected into our competitors, particularly in the US, and it is tipping the balance away from Europe. It would be disastrous if the current strengths in Europe could not be maintained through lack of available funding sources.

Such is the extent of European collaboration that seven of the leading GaAs manufacturers have formed a consortium. The EUROGaAs consortium consists of seven of the major GaAs component makers in Europe, Alcatel Telettra, Alenia, Daimler Benz, GEC Marconi Materials Technology, Philips Microwave, Siemens and Thomson and was formed to encourage co-operation and harmonisation across the whole of the European Community.

The many GaAs projects supported over the years by the Commission have involved Member States who have no indigenous sources of GaAs components but who never the less contribute to programmes through relevant theoretical expertise and through areas of application, Academia and small to medium sized enterprises (SMEs). All have played vital roles in maintaining the strength of GaAs in Europe. The EUROGaAs initiative plans to continue to utilise the European wide expertise in its ambitions to gain a greater market share of the world market place. It has a three phase plan to meet these high ambitions.

Phase 1   To assess the technological and manufacturing capability and market penetration of the European GaAs component makers, to determine the European user requirements for GaAs and to develop future technology plans.

Phase II   To seek support for aspects of the European infrastructure which is currently lacking and impeding penetration of products into the world market.

Phase III   To seek support for new technologies and techniques that will maintain the competitiveness of European component manufacturers and system builders.

Phase 1 is proceeding with zero Government funding under a EUROGaAs EUREKA umbrella project. Phase II is awaiting the outcome of financial discussions within the Commission. It is envisaged that a Phase III will be supported by the EC and National Governments through the Fourth Framework and EUREKA initiatives.

1.2   Why is GaAs important to Europe?

Europe's inventiveness as system designers and its ability to readily accept new technologies has meant that almost all of the current applications requiring GaAs components have been spawned in Europe. First it was satellite television, then hand held telephones, then road tolling and now Wireless LANs, and soon electronics in automobiles and, because the initial outlet for these products is, or will be in Europe, it is only natural that the European components industry should have a strong desire to service these markets.

Recent studies commissioned by the EC and national governments have highlighted the importance of key 'enabling' technologies to the future of European industrial competitiveness. Without access to European sources of these technologies, European manufacturers of added value systems, e.g. telecommunication networks, computers, automotive management systems, air traffic control networks, all of which will rely to some extent on the supply of GaAs based components will be dependent on non-European suppliers of this technology. Such reliance will almost certainly lead to European systems manufacturers being put at a competitive disadvantage and ultimately replaced by the component suppliers particularly as many are part of vertically integrated companies such as Fujitsu, Hitachi, NEC, TRW, Motorola, Texas Instruments, etc.

In general it is the wish of the European system builder to purchase from European sources and ease of communication, simple access to the manufacturer and a general 'built-in' desire to buy locally if at all possible are positive reasons for this.

So, because the initial markets for many GaAs components are in Europe and because Europe has potentially the total capability to satisfy these markets it is important that European industry capitalises in this situation. Opportunities are being created for all sections of the industry from SME service providers and component suppliers through to large systems builders and exploiters to benefit from this massively expanding microwave market and these opportunities must not be missed. Active and co-ordinated steps to satisfy them will mean that Europe not only benefits from the employment created in its various manufacturing industries, but also through the usage of the systems by businesses, emergency services, the general public, etc. for the much broader benefit of the whole community.

1.3   European Community Supported GaAs Programmes

1.3.1   Introduction

The European Commission has over the past four years supported four major programmes in GaAs aimed at specific end point objectives. The technologies and demonstrator circuits in these projects were required to meet specifications such that after further development and production engineering they could be used in 'real life' applications. These programmes, AIMS, CLASSIC, COSMIC and MANPOWER address receiver and transmitter applications from 1GHz (L-band) through to 60GHz. Brief outlines of the objectives of these projects are given in this section and show the diversity of applications in which GaAs based components have a key role to play.

The technological objectives of the ESPRIT GaAs programmes have been very clear : to establish a European core technology base for microwave and millimeter wave integrated circuits. For this to be achieved low cost, reliable, high performance and, in a word, competitive products are needed. Such an objective was, and still is, very ambitious. Recent years have seen an enormous growth in potential applications including personal communications, DBS television, satellite communications, automotive and navigation aids. As frequency bands become ever more congested there is increasing pressure to move to higher frequency bands. The ESPRIT GaAs programmes are, and have been, addressing the technologies required to realise MMICs for these applications. The circuits, designed by the major European Systems Companies, have used a range of enabling technologies, MESFET, HEMT, pseudomorphic HEMT and HBT and have been fabricated in the GaAs facilities of GMMT, Siemens, Daimler Benz, Thomson and Philips. Where possible the technology has been based on existing established commercial processes with performance enhancements developed as required to meet application needs.

As the technology matures, programmes are looking at second sourcing arrangements. This is seen as a key requirement in the acceptance of GaAs components into volume commercial applications.

1.3.2   ESPRIT - 5018 - COSMIC

GaAs Monolithic Analog Circuits for Microwave Communications Systems up to 23GHz.

This program was based on the MESFET technologies of GMMT and Siemens and the SAG HEMT technology of Siemens. The main applications areas addressed by this program were

- Point to multipoint (PTMP) Systems in L- and K-band.
- Mobile and Navigation Communication Systems, L-band.
- Optical communications to 2.5GBits.
- Direct Broadcast TV, X-band.

Table 1 COSMIC Technology Demonstrators

Circuit	Design	Fabrication	Technology
L-Band PTMP MMICs	Alcatel	Siemens	MESFET
Transimpedance amplifier	Telefonica	GMMT	MESFET
Transimpedance amplifier			
Transimpedance amplifier	PT Madrid	GMMT	MESFET
Mobile Communications MMICs	Forth	GMMT	MESFET
DBS low noise converter			
DBS low noise converter	Siemens	Siemens	MESFET
DBS low noise amplifier			
SAR receiver MMIC	GMMT	GMMT	MESFET
K-band PTMP MMIC	Siemens	Siemens	MESFET
	Siemens	Siemens	HEMT
	Siemens	Siemens	MESFET
	Alcatel	Siemens	HEMT
	Alcatel	GMMT	MESFET

1.3.3   ESPRIT-5032-AIMS

Advanced Integrated mm-wave Sub Assemblies

The AIMS project extends the work of the COSMIC program to higher frequencies based on pseudomorphic HEMT and HBT devices which have cut off frequencies above 100GHz. An industrial base for manufacturing circuits has been established with confirmed second sourcing features following very close co-operation between the two MMIC manufacturers, Daimler Benz and Thomson.

The technology being developed in AIMS can be applied to frequencies up to 100GHz but initially demonstrators were restricted to the 30GHz area. Two sectors were chosen with a primary interest in developing mm-wave technology, satellite communication and small hop land links in difficult terrain. These were necessarily generic functions as the actual systems did not exist at that time. Since then actual applications have indeed appeared including an Airport Ground Surveillance System and PCN Beacon Interconnect System.

1.3.4   ESPRIT-6050-MANPOWER

Manufacturable Power MMICs for Microwave System Applications -

The MANPOWER project which commenced in April of 1992 is extending the capabilities of the Siemens, GMMT and Philips MMIC technologies to higher output power levels, circa 2W, necessary to meet the requirements of PTMP and mobile communication systems.

The MANPOWER program is not restricted to MESFET technology, Siemens are also developing HBT and HEMT technologies for power requirements.

1.3.5   ESPRIT-6016-CLASSIC

Components for Large Signal Sixty GHz GaAs Integrated Circuits.

The CLASSIC project represents a further technology extension to higher mm-wave frequencies around 60GHz on the basis of PMHFET. PMHFET technology was chosen because it combines very high frequency operation with a low noise figure and has demonstrated good power densities for large signal devices.

The aim of the CLASSIC project is to develop a 0.15µm gate length T-gate process based on delta doped MBE material supplied by PicoGiga to the two fabrication facilities, Thomson-TCS and Daimler Benz. The technology developed will allow the integration of Schottky diodes (for mixers) and PMHFETs (for low noise or power amplifiers) in monolithic form using the pseudomorphic HFET layer structure and HFET technology. To date, cut off frequencies of the Schottky diodes of between 500 and 600GHz have been achieved.

1.3.6   Summary

The ESPRIT projects described have firmly established Europe in the forefront of GaAs for the three major technologies MESFET, PHEMT and HBT. The current programmes are aimed at developing a mm-wave technology which will be essential as systems are forced to higher frequencies as microwave frequency bands become more an more congested. European industry is in an excellent position to meet the requirements of the emerging volume applications. This potential to be a world technology leader must be fully exploited after these programmes so that Europe can be the leader and not the follower in the intercontinental market place.

1.4   Market for GaAs Components

Market analysts from across the world are predicting a steady expansion in the use of GaAs components for many systems applications many of which will be first introduced in Europe. Table 2 below shows some of the major opportunities being created by the use of GaAs components. These applications areas have hitherto been unaddressable as the performance required could not be reached by silicon technology alone. Now a very close working relationship between the silicon and gallium arsenide industry exists as each understands the importance of both technologies to the creation of these new markets. In many of the applications given in the table, the percentage of the semiconductor content represented by GaAs is relatively low being in most cases only 10 - 20% but GaAs is key to the system performance being met. If we look at each of these market sectors in turn it will be come obvious that many market analysts and

Table 2 World GaAs Merchant Markets : 1992 - 1997

	Small Signal						Power		Digital ICs	
	Microwave				mm-wave		All GaAs		All GaAs	
	Discrete		MMIC		All GaAs					
	1992	1997	1992	1997	1992	1997	1992	1997	1992	1997
Consumer										
Terrestial TV	18	33	6	19	-	-	-	-	-	-
DBS	42	64	14	36	-	-	-	-	-	-
Transport										
Two way communications	0	5	0	7	0	0	0	0	-	-
Road Tolling	0	1	0	6	0	0	0	0	-	-
Traffic Control (inc AVL)	0	<1	0	1	0	0	0	3	-	-
Car Electronics*	0	1	0	3	0	2	0	0	-	-
Wireless Communications										
Cellular, Cordless and Pagers	35	63	12	69	0	0	3	8	-	-
Terrestrial Links	5	10	1	7	1	8	<1	6	-	-
Wireless LAN	<1	2	3	18	-	2	-	-	-	-
VSAT	6	14	7	15	-	-	2	4	0	<1
Civil Radar	2	6	3	6	-	-	5	11	-	-
Space Systems										
Total	16	23	3	12	1	4	8	14	<1	3
Computers	-	-	-	-	-	-	-	-	31	101
Other	-	-	-	-	-	-	-	-	-	-
Instrumentation	5	10	3	7	2	5	-	-	5	13

* Includes Collision Warning/Avoidance Radar, Intelligent Cruise Control, Digital Audio Broadcast
Figures are in $millions

indeed industry itself sees a continuing expansion in the use of GaAs in a board range of applications.

1.4.1 Consumer

Satellite television reception is only possible through the use of components fabricated in GaAs. This was the first major application of GaAs and its successful usage in this application was instrumental in the adoption of GaAs in many other important system areas. GaAs technology, in the form of discrete MESFETs and HEMTs (high electron mobility transistors) and microwave monolithic integrated circuits will continue to be used at least until the end of the century. New television broadcast standards involving the use of digital transmission pose new problems for the systems designers and component manufacturers and the present thinking is that the new requirements may only be met using discrete GaAs devices rather than a MMIC solution on both performance and cost grounds.

1.4.2 Transport

Europe manufactures and uses many millions of motorcars per year and is at the forefront of development that will reduce automobile accidents and help to reduce traffic congestion on our European roads. In areas such as road tolling, roadside communication, automatic cruise control, anti collision radar systems and AVL European industry is active in developing GaAs based electronics to perform the necessary control and communication functions

Roadside communication techniques at 5.8 and 18GHz are being studied to give the automobile driver up to date information on weather conditions for his journey, availability of car parking space at his destination and identifying congested roadways that he should avoid. Automatic cruise control and anti collision radar systems operating in the 70GHz frequency range are presently being demonstrated that sense the movement of the vehicle in front and maintain a safe distance from it either by visual and/or warnings, or through a sophisticated throttle and brake control system.

1.4.3 Wireless Communications

Ever since computers became an everyday piece of office equipment there have been problems in making the interconnects between them. This is soon to be a thing of the past as silicon and GaAs technologists and circuit designers have been working together to remove the need for interconnecting wires between computers by introducing microwave transmit/receive modules to each computer allowing them to 'talk' to each other over the 'air-waves'. This is perhaps the largest growing market world wide for GaAs components and one which will almost certainly lead to 'spin off' markets for point of sale cash points and shelf edge labelling in retail outlets such as supermarkets.

Another communication application area where silicon and GaAs technology is already being utilised together is in the hand-held telephone market. Without doubt this will be the largest single application for GaAs components and already European GaAs device and circuit makers are supplying large volumes of parts to the hand-set makers. Portable phones are already revolutionising our working practices, our domestic arrangements and our emergency services and will continue to do so far into the 21st century.

Very small aperture terminals (VSAT) are increasing the ability of major companies to communicate. Operating at around 14GHz these systems allow inter-company communication to be carried out without recourse to the public telephone. Already companies, such as motor car selling agencies, credit card organisations and 'trucking' concerns are utilising the private communication systems available through VSAT to reduce their communication changes and improve their links across their network of outlets.

The potential to produce electronically scanable radar systems utilising GaAs based transmit receive modules in a phased array radar configuration at affordable prices has opened up the possibility to produce compact equipments for civil applications. This technology will be based on that currently being developed for the military and has the advantage that it does not require a rotating aerial and has the ability to switch from one 'target' to another very quickly.

1.4.5   Computers

The high operating speed and low power consumption of GaAs digital circuits has finally been harnessed and super computers utilising this technolgy have been built. Digital GaAs is also finding its way into the office computer where plug in replacements for the slower silicon technolgy are already available.

1.5   Conclusions

We are now experiencing a period in time when the potential of GaAs is being realised in volume requirements for the world wide market place. Systems designers have gained the confidence they need to specify GaAs parts and at last the performance advantages of GaAs are being recognised by companies steeped in the silicon technology. GaAs and silicon are now living happily together in many high frequency systems available today - this can only be a good thing for GaAs and is certain to lead to ever increasing opportunities for the technology. Support from the National Governments and the European Commission has helped to position the European technology competitively with the rest of the world. Manufacturing expertise is now the key to successful market penetration. Further support for European Industry is necessary to develop the lacking infrastructure and to increase its technology base.

2.1 Satellite Communication

C Tronche
Alcatel Espace, Toulouse, France

Ka Band Satellite

The use of Ka-band has become necessary in order to prevent radio-frequency interference because C and Ku band are already widely used for commercial operations. Moreover, Ka band satellite allocation offers larger transmission bandwidth in each direction (up to 2.5 GHz) and the possibility of further reductions in the size of the terminal antenna. The attenuation experienced at this frequency band is, however, significantly higher than at Ku and C band. Rain fade is also a critical problem which requires due attention when designing a Ka band system and specifying an application.

Because of the development of fibre optic cables for terrestrial integrated services networks and trunk interconnections, a reorientation of the role of satellites seems to be necessary. Communication satellites can be used to interconnect a great number of widely spread, low traffic ISDN users as well as a small number of high traffic users such as a TV-van studio or a TV production centre. To take into account actual communications requirements, two specific applications have been investigated.

Point to point communication between pico-terminals (also called V^2SAT in this book) and **central terminals** at a low rate: 32 Kbps users for data transmission, voice communication, fax as well as medium quality picture (low scanning TV) with an objective of link availability close to 98%, and a BER close to 10^{-6}.

Point to point communication for HDTV: 63 Mbps for high quality TV transmission from TV vans to TV production centres with a global link availability close to 99% and a BER of 10^{-6}.

Those applications, described in the next pages, have led to the following technical specifications for proposed Ka-band satellite system payloads:

1. Multibeam antennas for the up-link and both multibeam and beam hopping for the down-link.
2. FDMA/SPCPC access techniques for a large number of small users (700 V^2SAT users) and a small numbers of large users (12 HDTV channels)

3. Incorporation of on-board processing to cope with low cost earth terminals V^{2}SAT (small antenna size, low power) as well as by-pass transparent HDTV services.
4. Coding of information in order to minimize bit error rates.
5. Reconfigurability of the capacity per beam beacause of traffic disparity between spots.

System definition

Frequency plan constraints
Ka band has not yet been used in Europe for fixed (terrestrial) communications systems. However a new Ka band satellite system designed to serve micro and pico terminals could avoid future terrestrial interference by using 30 to 31 GHz up-link and 20.2 to 21.2 GHz down-link bands. Those bands are exclusive for both the fixed and the mobile satellite communications systems.

Up-link coverage
The satellites receiving antenna produces a number of fixed regional beams that provide complete coverage. Europe is covered by 14 fixed spot beams that re-use a set of 3 frequency sub-bands (see Fig. 1 below).

Sub-band : 30.00 to 30.26 GHz : central frequency Fu1 = 30.13 GHz
Sub-band : 30.26 to 30.52 GHz : central frequency Fu1 = 30.39 GHz
Sub-band : 30.52 to 30.78 GHz : central frequency Fu1 = 30.65 GHz

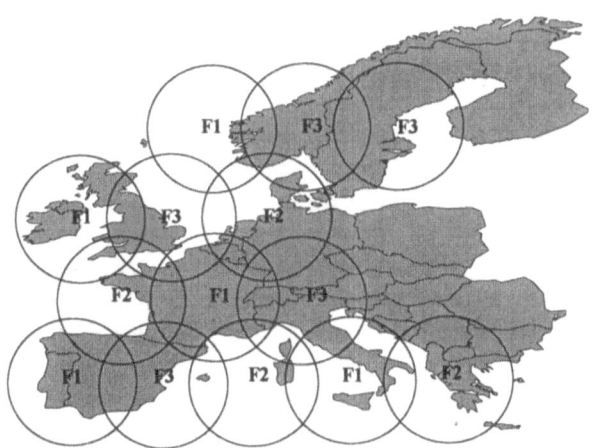

Fig. 1 Coverage of Europe by 14 spot beams

For each beam, the 3 dB beamwidth is equivalent to 1°. The beams are shaped simultaneously and one frequency is permanently assigned to each beam. Adjacent beams use two different frequencies but non-adjacent beams can use the same frequency.

Down-link coverage

The satellites transmit antenna is of the DRA type (Direct Radiating Array) and produces four hopping beams for small user's traffic as well as five re-configurable regional beams anywhere in Europe, which provide a complete coverage for the TV transmission system. The hopping beams use the same frequency band. However, the regional TV beams, as for the up-link case, re-use a set of three frequency sub-bands. Each of the five re-configurable regional beams is fixed during the transmission and independent of the others.

HDTV :   Sub-band 1 : 20.200 to 20.435 GHz
              Sub-band 2 : 20.435 to 20.670 GHz
              Sub-band 3 : 20.670 to 21.000 GHz

V^2SAT :   Sub-band 4 : 21.000 to 21.045 GHz

The four hopping beams can hop to 14 discrete locations providing a complete coverage of Europe. In a period equal to a TDMA (Time Division Multiple Access) frame, a beam can hop to many locations, dwelling long enough at each location for the specific traffic. By adaptatively varying each dwell time, the system's capacity is efficiency matched to a non-uniform demand for the traffic.

Modulation and coding

To maximise power and bandwidth efficiency and simplify the system, QPSK transmission is used on both the up and down-links. Differential detection of QPSK could be employed to simplify the demodulation process on the satellite but, as opposed to coherent detection, it would result in a power penalty of 2.3 dB. So, even if coherent detection is more complex and expensive than differential detection, the coherent detection technique is employed.

Concerning the coding scheme, a range of options is available, from no coding to rate 1/2 Forward Error Correction Coding (FEC). On the one hand, the uncoded approach is attractive to minimize the bandwidth, but that is not relevant for Ka-band. No coding scheme would also allow simpler demodulation on-board the satellite and on the terminals and consequently a demodulation cost reduction. On the other hand, low rate FEC schemes such as 1/2 convolutional encoding/Viterbi decoding allows terminal and satellite EIRP reductions. The following table compares 3 different coding schemes, as regards required Ed/No and bandwidth for BER = 10^{-6}, a bit rate of 32 Kbps, and QPSK modulation. It assumes a spectral efficiency equal to 2 to calculate the required bandwidth.

Coding scheme	Required Ed/No	Occupied bandwidth
Uncoded	10.5 dB	32 KHz
Viterbi 3/4	6.1 dB	42.75 KHz
Viterbi 1/2	5.2 dB	64 KHz

The main disadvantages of on-board regeneration are cost and payload complexity. However, on-board regeneration does offer many advantages over conventional transparent repeating communications satellites. These advantages are summarised below:

1. The bit error rate of the up-link and the down-link can be improved separately.
2. A link budget advantage: up-link thermal noise does not degrade the C/N ratio on the down-link. On-board demodulation and remodulation make it possible to separate up-link and down-link signal degradations, resulting in an overall improvement of the link budgets.
3. The data transmitted from one up-link user can be separated into smaller messages which are transmitted separately to down-link users. Similarly, messages from several up-link users can be combined into one burst. A regenerative transponder can then provide **full interconnectivity between earth stations.** Management of the traffic is carried out according to a time plan, stored on-board through the telecommand channel. The time plan is generated and transmitted by a control station.

Up-link scheme
The terminal is desired to be portable and not permanently connected to the network. Before establishing communication, the terminal is at an unknown location. As a result, the V^2SAT should always be the first to try to communicate with a larger permanent central station or other, already connected, V^2SATs. For instance, the central station could be located inside a plant, a laboratory or a main office and terminals in an isolated area. Many sub-networks can then be constituted using one central station and a number of widely spread low rate terminals per sub-network.

Because of OBP complexity **the number of simultaneous small users per spots is limited to 700 over Europe** (0 to 75 per spot). Satellite channels are frequency-shared by the users using an SCPC (Single Channel Per Carrier) access technique. This technique is selected to minimise V^2SAT EIRP requirements. It permits the use of easily transportable and maintainable small ground stations for low volume users. To establish a satellite connection, a request is made to the control station (CS) over an orderwire channel. This request will contain an indication of the data type, amount, location and channel required. The CS assigns a channel to the pertinent user pair. **It requires a frequency agile terminal.** Each user terminal is provided with the capability to transmit a single narrow band (less than 32 Kbps per terminal) digital carrier. A no coding scheme with QPSK modulation has been chosen. The carriers transmitted by terminals located within the same spot are distinguished by frequency. The up-link signal set is comprised of a large number of SCPC carriers spread over a specified bandwidth. Dual polarisation is not used to simplify the user terminals. The traffic in each spot beam is selected around its own frequency. For each beam, the receiver down-converts the SCPC signals at 30 GHz to the unique IF at 4 GHz. The signals are then down-converted to 70 MHz where a coherent demodulation is carried out. A total of 14 receivers are required.

The payload must be able to handle various traffic to spot assignments for V^2SAT during its lifetime. An ability to re-allocate capacity between spot beams within the European coverage can be provided by sub-dividing the up-link carriers into sets of 25 users. The sets are separated by "1 to 3" multiplexers and routed by a 42x28 IF switch matrix. Each output of the matrix is then presented at the input of an MCD (Multi Carriers Demodulator). This method allows the MCDs to be shared between the spots and improves the overall flexibility. MCDs accept as input a composite FDMA signal and provide a TDM pulse stream. A single MCD might simultaneously process 25 FDMA carriers, thereby providing a 1:25 reduction in the required number of demodulators. Acceptable MCD complexity will be be a key factor determining the V^2SAT mission capacity.

Baseband Processing
The MCDs outputs are fed to a BaseBand Processor (BBP) that buffers, sorts, formats, and route adresses the data bits for down-link transmission. The BBP stores, in parallel, data frames from all the inputs channels, that is 7000 32 Kbps data frames. Each time stage operates by storing a complete time-frame and reading it out in a different order controlled by a pre-determined programme. The programme is generated and up-linked from a Control Station (CS) to an on-board control module and is continuously changing in response to varying demand-assignment of the traffic pattern within the networks. **This architecture allows the up-link and down-link time plans to be independent of each other.**

Down-link scheme
Down-link transmission is accomplished by means of a small number of scanning beams, each of which supports a wideband TDM carrier. The number of down-link beams is smaller than the number of spot areas. Four down-link beams are programmed to scan over a portion of the European coverage in a repetitive pattern. In this configuration, TDM channels are time shared among several beams. The proportion of time spent in any one beam is equal to the total traffic in that beam. in such a case, on board DC power is well exploited. The selection of the transmit section architecture is not independent from the choice of the receive section. the receive and transmit sides of the Baseband Processor (BBP) should work with equal input and output throughput.

Transmit antenna

The transmitter antenna is of the DRA type (Direct Radiating Array), composed of 112 rectangular radiating elements. That DRA is divided into two equal parts, each consisting of 56 radiating elements: one used by the HDTV application , and the other one used by the V^{2}SAT application. The area allocated to the V^{2}SAT system is fed by 4 BFNs (Beam Forming Networks) dedicated to 4 V^{2}SAT hopping spot beams. The area allocated to the HDTV system is fed by 5 BFNs dedicated to the 5 HDTV re-configurable regional fixed spot beams. The global dimensions of the DRA are 1.47m x 1.08m. The arrangement of the radiating elements is based on the geometry of the global rectangular European coverage. The dimensions of the radiating elements $7\lambda x 9\lambda$ (with $\lambda = 1.5$cm at 20 GHz), has been chosen to avoid the grating lobes over the European coverage.

A DRA separated into two parts offers the following advantages:

- Each SSPA is operated with equal amplitude carriers resulting in better SSPA efficiency, and better (C/I) ratio).
- Lower required output power per SSPA set of modules; so a smaller number of SSPA modules per set and lower thermal control difficulties result, giving simpler connections and enabling separate optimisation of the SSPAs.

Satellite capacity

Global system architecture

The figure below (Fig. 2) schematically describes the two applications studied.

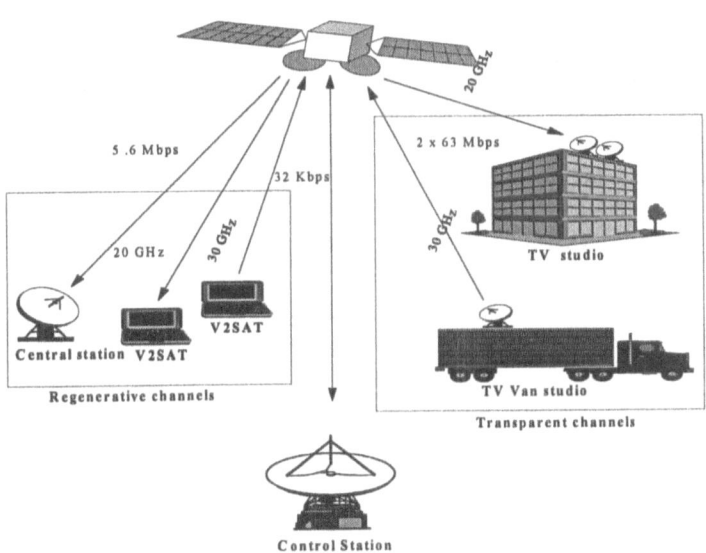

Fig. 2   Global Ka Band system architecture

Inter-satellite links

An inter-satellites link is a connection between two geostationary satellites (Inter-satellite-Link: ISL) or one geostationary and a low orbit satellite (Inter-Orbital Link: IOL). Within the context of the telecommunications, the commercial applications have been defined for satellites in geostationary orbit. In this chapter, only the ISL will be examined.

As demand increases for orbital slots within prime regions of the geostationary arc, ISLs can circumvent orbit saturation by increasing the orbital arc. The interconnection between two stations located above two different coverage areas, achieved by using ISL, avoids double-hopping, thus providing reduced signal propagation delay, and consequently improving system quality. Increasing geographical coverage and traffic interconnectivity can be achieved with ISLs. A new satellite location can be selected to provide coverage for high-traffic areas. An ISL allows considerable flexibility in the management of a satellite system network. While multiple earth station are needed for interconnectivity between satellites in conventional multi-hop systems, only one earth station antenna per traffic mode is required in ISL systems. The up and down-link bandwidths of the relay station are replaced by the inter-satellite frequency bands, so the space segment spectrum utilisation is improved.

At the World Administration Radio Conference, the 60 GHz frequency (i.e. 59 to 64 GHz band) was chosen for inter-satellite transmissions because the earth's atmosphere is opaque to transmission in this frequency range. This makes crosslink transmission virtually invulnerable to interception from the ground, thus improving the link confidentiality, and protecting the link against interference due to terrestrial noise.

The ISL application is highly dependent on the supporting communication system. One of the main candidate systems for the implementation of an operational ISL service is the INTELSAT system. The INTELSAT organisation is the main satellite communication operator, already operating more than 15 satellites to provide its 121 nation members with international and intercontinental communications for voice, video, data transmission and distribution services from any location over the three ocean areas.

The figure on the next page (Fig. 3) presents a view of the INTELSAT satellites constellation.

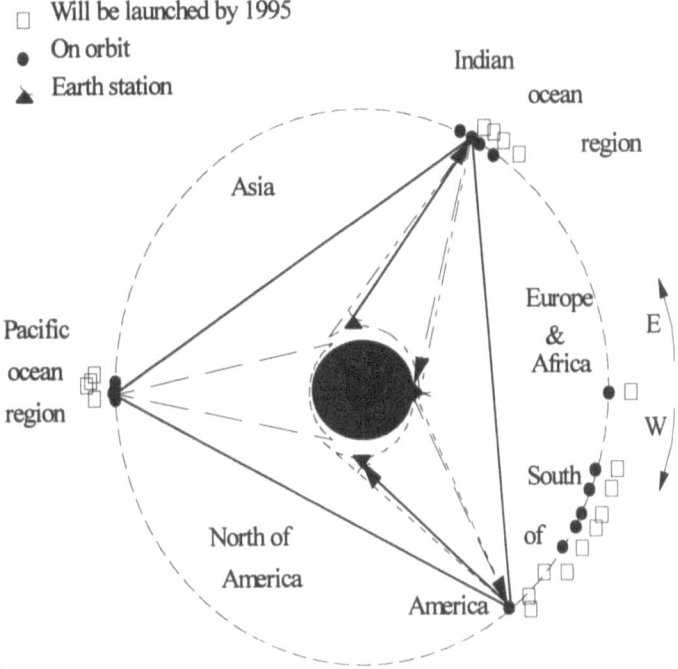

Fig. 3 INTELSAT satellite constellation

We can see that the implementation of ISLs between satellites would permit the definition of a world satellite network, whereas the present system is composed of three distinct regional zones. Connections between the three regions can only be done by means of double-hop schemes :

$$\text{Ground} \longrightarrow \text{satellite 1} \longrightarrow \text{ground} \rightarrow \text{satellite 2} \longrightarrow \text{ground}$$

The use of ISL avoids this double hopping :

$$\text{Ground} \longrightarrow \text{satellite 1} \longrightarrow \text{satellite 2} \longrightarrow \text{ground}$$

The INTELSAT system has been selected as an example for the use of an ISL payload architecture. The ISL payload has been designed in order to be implemented as an additional payload for an INTELSAT VII spacecraft.

Fig. 4 presents a simplified block diagram of the INTELSAT VII payload including an ISL transponder. The INTELSAT VII payload nominally includes two transponder groups operating in C Band (6/4 GHz) and Ku Band (14/12 GHz). Various coverages are available, including global earth coverage, zone beams and narrow spot beams. The different transponders may be interconnected with flexibility one to the other by means of a common C Band switch matrix. The same switch matrix interface has been chosen for ISL transponders. This allows us to connect, on demand, the ISL with almost any C and Ku band transponders. The present switch matrix design is based on the use of static switches, which means that the matrix configuration is changed only a few times (per day) by a new set of group commands. Hence, considering that the ISL capacity between two satellites is limited to two channels in each direction, only two different coverages may access the ISL simultaneously. However, the implementation of a dynamic switch matrix can be contemplated for future INTELSAT satellites. Then, assuming a satellite switched time-division multiple access scheme, all coverages could access sequentially the ISL transponders during a common time frame. The capacity per beam could be changed because of traffic disparity of the ISL. Such a configuration would allow us to derive maximum benefit from this new ISL capability. The ISL transponder interface specifications, including frequency band and level requirements, have been derived from the present INTELSAT VII switch matrix design.

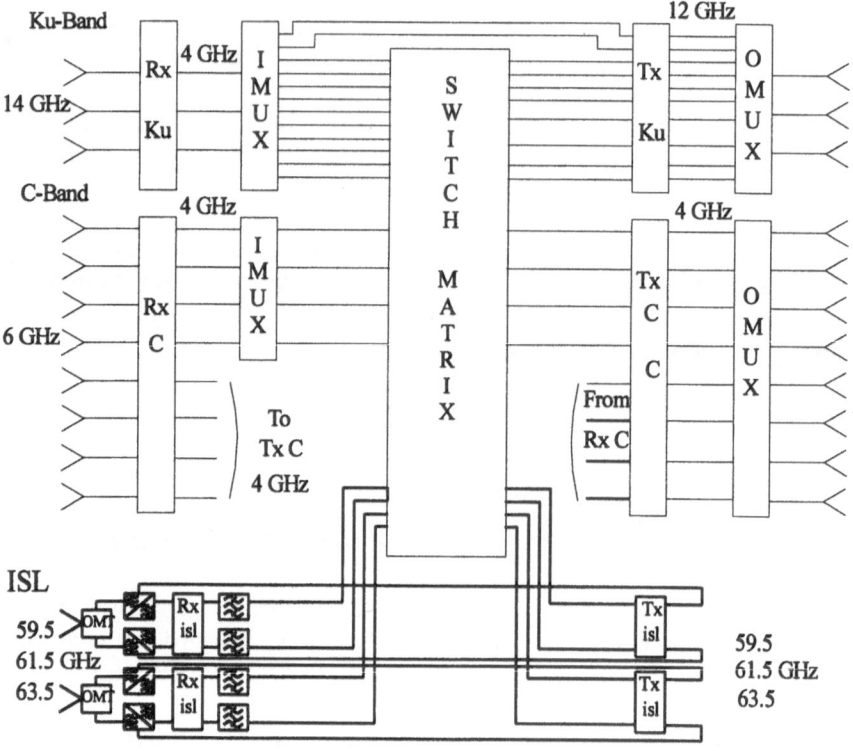

Fig. 4 INTELSAT VII block diagram

System definition

ISL distances

The maximum theoretical ISL distance corresponds to the case where the geostationary satellite's line of sight is almost tangential to the earth surface, as indicated in Fig. 5. A 150 km altitude margin is kept to avoid atmospheric disturbance. The maximum ISL distance obtained is 83 311 km, and this has been used for the link budget evaluation. However, we can note that if geostationary satellites offer a complete earth coverage (for latitudes below ± 70 °), the angular spacing between two satellites cannot exceed 120° to 140° . Hence, the maximum operational ISL distance will be limited to about 750 00 km.

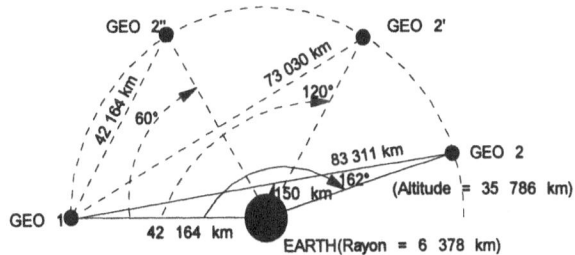

Figure 3 : ISL DISTANCES

Fig. 5 ISL distances

Signal characteristics

The proposed ISL payload architecture is fully transparent, which means that it can accommodate any kind of analogue or digital modulated signal. The only limitations are due to the channel frequency bandwidth and the C/N_0. The channel bandwidth selected is 72 MHz which permits the connection of the ISL transponders with most of the INTELSAT VII transponders. In order to propose satellite EIRP and G/T specifications, specific link budgets are established in the next section. In the frame, each ISL channel supports a 70 Mbps digital transmission. To minimise the bit error rate, the information is coded. Signals are QPSK modulated, throughout the whole transmission (from ground to ground via ISL). Taking advantage of the wide available channel bandwidth, rate 1/2 convolution coding is implemented. Together with a Viterbi decision decoding algorithm, this offers a 6 dB improvement in the link budget.

Link budget

A classical bit error rate (BER=10^{-6}) and a lower bit error rate (BER=10^{-8}) for better quality of the link have been chosen. The C/N_0 up-link and the C/N_0 down-link have been fixed respectively at 115 and 100 dBHz. These C/N_0. related to ground/satellite transmission are kept consistent with fixed service transmission performances. The budgets exhibit comfortable margins: around 3 dB for 10^{-6} bit error rate. By decreasing the ISL distance, the margins become more and more comfortable. At 42000 km, an antenna diameter of 0.7 m is enough to provide a margin higher than 1.5 dB. The ISL transmission is the limiting factor and thus, the overall C/N_0 obtained is very close to the ISL C/N_0. We can note that the operational availability is not a driving parameter for such an ISL budget, since no atmospheric perturbations can occur. The only possible source of performance degradation would come from an increase in the receivers thermal noise incurred by the sun. Such an event is considered to be rare enough to be neglected. Hence, global service availability is only dependent on ground/satellite transmission availability.

Frequency plan constraints

The frequency plan was chosen to reduce both the interferences and parasitic antenna coupling to be a minimum. The 59-64 GHz frequency band is used since it prevents terrestrial interference. For one antenna, the transmit frequency must be different from the receive frequency in order to separate them from each other. At this centre frequency, a diplexer can separate two 1 GHz bands spaced by more than 2 GHz. For one satellite, the transmission of one antenna must not disturb the reception of the other antenna, as shown in Fig. 6. Although antenna decoupling may provide sufficient isolation, the co-existence of low level Rx and high level Tx signals in the same satellite at the same frequency is a critical point. So the emission frequency (Fe) must not equal to the reception frequency (Fr).

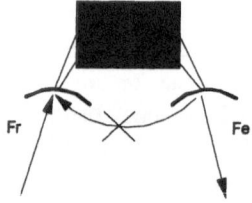

Fig. 6   Parasitic antenna coupling

On the other hand, a satellite can receive (or transmit) at the same frequency if the angle between antenna pointing direction is greater than three times the 3 dB aperture angle. We can also reuse the same frequency on different polarisations. The circular polarisation is retained because it is less restrictive for the pointing than linear polarisation. Fig. 7 shows that three frequencies are required to connect all the satellites.

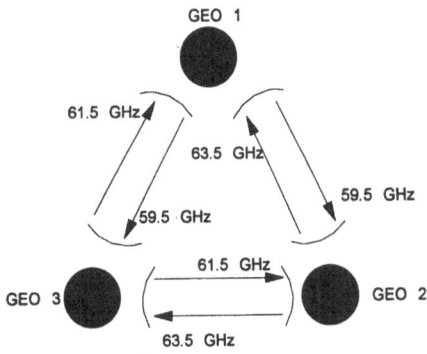

Fig. 6   Frequency scheme for total links

Payload description

Antennas

For a diameter of one meter, the antenna has a gain of approximately 54 dBi. The gain variation is limited to 0.25 dB per antenna by assumption. If the 3 dB beamwidth is equal to 0.35 degrees, the depointing angle θ is calculated to be roughly 0.05°. As the satellites accuracy is of the order of 0.15°, a pointing system will be required. The pointing range could be low if the satellite to link were clearly identified. However, it would be useful to have a movable pointing system in case of a satellite breakdown, or for the addition of a new satellite in the constellation. Since all satellites are in the equatorial plane, the ISL case is much simpler than the IOL case because the pointing is mainly restricted to axis pointing. Moreover, the pointing subsystem must deal only with a very slow variation as the geostationary positions are stabilised relatively, one in relation to the others. By using only the east INTELSAT satellites, the angle between the tangent of the first satellite trajectory and the direction of the second satellite can vary from 0° (co-located satellites) to 81° (maximum distance between two geostationary satellites). To minimise the pointing range, the initial pointing is set up at half of the geostationary arc.

We can think of several mechanically adjustable antenna configurations. At millimeter wavelengths, lenses or arrays cannot be used as the gain is not large enough and the beamwidth too wide. It could be preferable to analyse configurations based on reflector design (single or dual, centred or offset, shaped or not) that minimise the overall losses and the system complexity. The dual reflector beam waveguide system has been identified as a promising solution in terms of compactness, lightness and simplicity.

The feed is fixed. The first mechanism provides azimuth rotation by moving both sub and main reflector. A second mechanism ensures an elevation rotation by moving only the main reflector. At 90° angle between the two axes is greatly preferred in order to simplify the pointing data treatment.

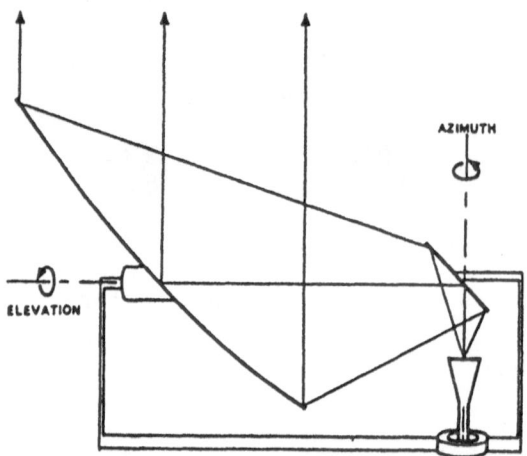

Fig. 7   Dual reflector beam waveguide scanning principle

The design of this of antenna makes it very well adapted to wide angle scanning. If necessary, its compactness and lightness allow it to be mounted at the top of a boom to avoid any interception of the beam by the platform. An accurate tracking system is necessary to cope with antenna pointing requirements. This tracking system is a critical point of such a system and should be carefully analysed. However this is out of the scope of this study and shall not be addressed.

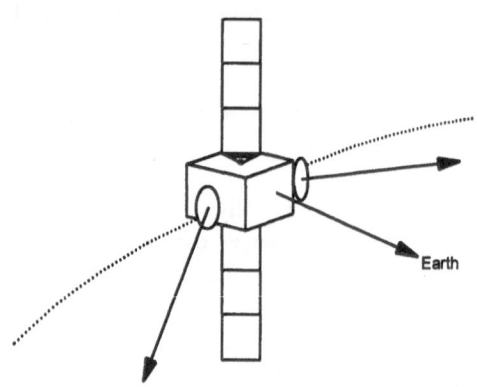

Fig. 8   Antenna implementation

Repeater

The payload simply uses a by-pass repeater with an intermediate frequency switch matrix. The terminals are made of a common antenna for reception and transmission. This antenna type requires a diplexer. The receivers provide preamplification at 59.5, 61.5, or 63.5 GHz, down conversion to the 4 GHz band, and further broadband amplification at 4 GHz. The receivers are in a 3 for 2 redundancy configuration. The receiver outputs are routed to the channel filter input. Through the switch matrix, the signals reach the transmitter. They have the same redundancy. Each transmitter includes an attenuator to compensate the gain variation. After amplification and up-conversion into the 60 GHz band, the signals are routed to TWTAs for high power amplification before transmission via the diplexer and antenna system. Each amplifier is operated with equal amplitude carriers. This results in a better efficiency, and a better C/I ratio.

The ISL block diagram is shown figure below:

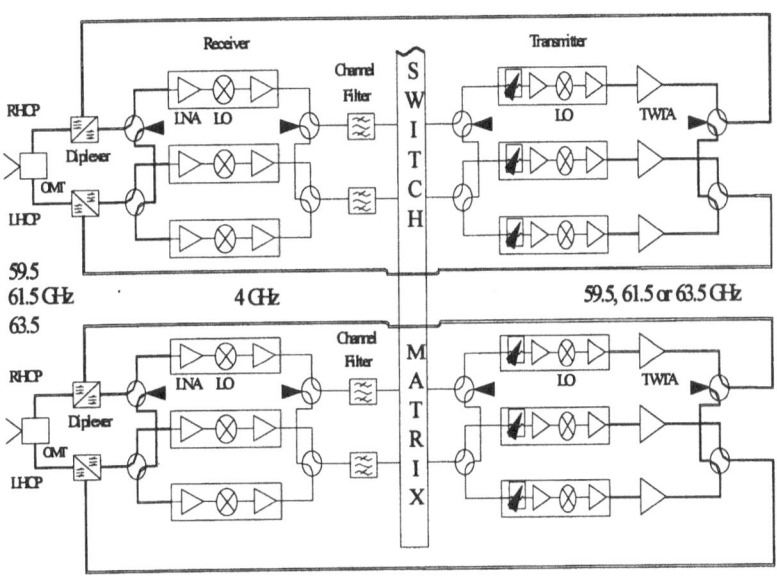

Fig. 9 ISL block diagram

The repeater noise figure is about 8 dB and the C/I_3 is better than 44 dBc.

Mass/power consumption budget

The mass/power consumption budget is presented below.

Equipment list	Equipment number	Mass (Kg)	Consumption (W)	Total mass (Kg)	Total Consumption
Antenna + Pointing system	2	20	17	40	34
Orthomode transducer	2	0.05	0	0.1	0
Diplexer	4	0.03	0	0.12	0
WG Switch	8	0.04	0	0.32	0
Receiver	4	0.32	2.5	1.28	10
Coax, Switch	8	0.02	0	0.16	0
Channel Filter	4	0.20	0	0.80	0
Switch Matrix 6x6	1	6	0	6	0
Transmitter	4	0.18	2.2	0.72	8.8
TWTA + EPC	4	1.5	50	6	200
Harness	1	5	0	5	0
Total				60.5	252.8

Conclusion

The feasibility of two satellite communications systems operating in the mm wave frequency range has been studied. Both systems offer substantial advantages over competing systems. As yet, these systems have not been implemented. However, the increasing availbility of monolithic GaAs MMICs as being developed in the ESPRIT AIMS and CLASSIC projects will bring forward the point in time where these systems are both technically and economically feasible.

2.2   Mobile Millimetre-wave Communication

A Plattner
Deutsche Aerospace, Ulm, Germany

Introduction

Mobile telecommunications is one of the markets with the largest growth potential all over the world. The increasing number of subscribers for modern digital mobile telephone services as well as requirements for new services such as interconnection of PCs and workstations, operation control of trains, route guidance and co-operative driving are some of the reasons for this development.

The radio spectrum resources, however, remain the same and most of the spectrum from VLF up to the microwave range is occupied by a large number of users. This necessarily leads to the extension of the usable radio spectrum up to the mm wave range where sufficient bandwidth for high data rate services is available.

Systems and services

Several applications dealing with mobile mm wave communications are currently in the system definition phase, most of them partially sponsored by the European Community. The first group of applications concerns the vehicle environment. To increase the safety of driving, to prevent the individual transport from breaking down and to charge for the use of the infrastructure are key objectives of the programs which are comprised of vehicle-beacon and vehicle-vehicle communication. In addition to a frequency band at 5.8 GHz, which is used for the first beacon-vehicle applications with limited bandwidth requirements, the 60 GHz range is foreseen, offering larger transmission capacity and the advantage of good discrimination between lanes using small beacon antennas. For vehicle - vehicle communication (co-operative driving) the 60 GHz frequency band is well suited because the high atmospheric attenuation reduces the transmission range. One cluster of vehicles can communicate with each other without disturbing other clusters which are further away and exchange of data with them is not required /1/.

The realisation of the B-ISDN network based on optical fibres, the increasing demand for high data rate transmission and the requirement for mobility leads to the concept of a Mobile Broadband System (MBS) which is one of the projects of the RACE program. The MBS approach allows mobile communication to be fully integrated into the IBC network /2/. MBS will support a very wide range of services like speech, data and video transmission with bit rates of up to 155 Mbit/s. Some of the possible applications as a function of mobility and data rate are depicted in Fig. 1. The high bit rate requirements leads to a natural choice of carrier frequencies in the mm wave range as the required bandwidth is not available at lower frequencies. From the evolutionary point of view, MBS could start with stand alone configurations supporting private networks and end up with a fully integrated public broadband mobile network.

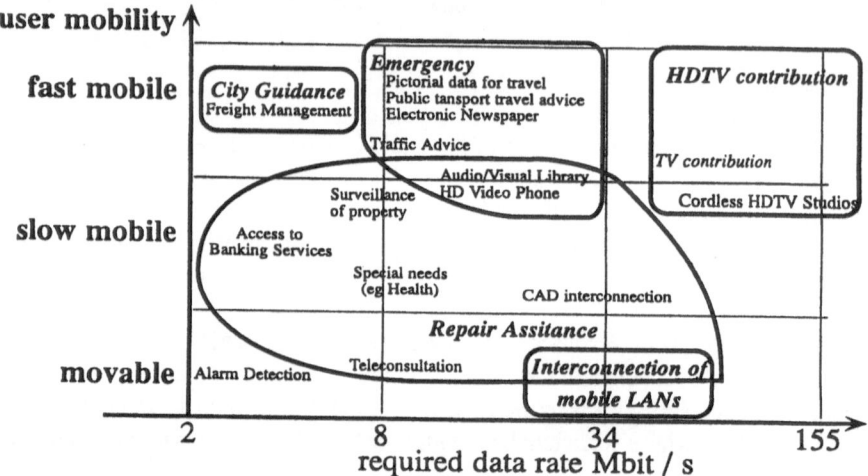

Fig. 1 Proposed MBS Applications

For high speed trains mm wave radio is an attractive candidate for the communication between the train and a fixed network. The communication may include operation control data and user services which are expected in a modern train like telephone, telefax, ISDN and user information. The special advantages of mm wave communications for this application are:

- Small base- and mobile station antennas will focus the electromagnetic energy along the track leading to a lower required transmitter power and lower environmental pollution with electromagnetic energy.

- Due the high directivity of the antenna beam and the moderate transmission range, disturbance from external mm wave transmitters is reduced.

- The available bandwidth is sufficient for the required services.

- The discrimination against multipath fading due to the small directive antennas enables high data rate transmission without the requirement for an expensive adaptive equaliser.

- A very low bit error rate can be achieved along the track, even without the implementation of forward error correction

A mm wave communication system which was realised in 1989 for the German MAGLEV research centre /3/ has demonstrated reliable operation up to vehicle speeds of 450 km/h.

Radio LAN will be a rapidly growing market in the near future . Although several microwave bands are foreseen for this application (2.4 to 2.483 GHz, 5.15 to 5.3 GHz and 17.1 to 17.3 GHz) additional requirements for the allocation of mm wave frequencies could be necessary when one considers the increasing complexity of computer programmes especially in the CAD area. As Radio LAN is mainly a low range indoor application, frequencies around 60 GHz would be well suited.

Frequency band allocation

CEPT has the long term objective to harmonise the use of mm wave spectrum particularly between 30 and 300 GHz in Europe. and to establish detailed regulations for the various applications. In March 1993 the Radio Communication Office, a permanent body of the CEPT/ERC made detailed spectrum investigations in the frequency range 3.4 to 105 GHz /4/. Some mm wave frequency ranges are provisionally allocated for mobile applications (see table below).

Frequency in GHz	Bandwidth in MHz	System
39.5 to 40.5	1000	Possible for broadband mobile system shared with civil and non civil users
42.5 to 43.5	1000	Future fixed and mobile users. Possible for broadband mobile system. May be paired with 39.5 to 40.5 GHz band
59 to 62	3000	Cordless local area networks
61 to 61.5	500	ISM applications with special authorisation
62 to 63	1000	Broadband mobile system for connection to IBCN, paired with 65 to 66 GHz
63 to 64	1000	Road Transport Telematics (RTT) vehicle to vehicle, vehicle to road
65 to 66	1000	Broadband mobile system for connection to IBCN, paired with 62 to 63 GHz

From the table above it becomes obvious that the provisionally allocated frequency spectrum will support the bandwidth requirements of future high data rate communication systems. Furthermore the frequency reuse distance at mm wavelengths is small (especially around 60 GHz). This feature also increases the total channel capacity.

Mm wave propagation

At mm wave frequencies radio propagation is mainly restricted to line of sight (LOS) paths, except for low data rate or extremely short range applications. The reason is the low transmissivity through building materials (0 to 15 %) and the high attenuation of natural obstacles like bushes or trees. Furthermore, as free space propagation is related to the wavelength (e.g. for 100m range -109 dBm at 60 GHz compared to -78 at 2 GHz) only a limited range can be achieved using mm wave communication unless high gain, highly directive antennas can be used, which is unlikely for mobile applications. Mm wave propagation also suffers from additional attenuation mainly due to rain. At a rainfall rate of 25 mm/h, the additional attenuation amounts to 7 dB/km at 40 GHz and 10 dB/km at 60 GHz /5/. In Europe this event occurs approx. 0.03% of the time. Depending on the required reliability of the link this additional attenuation must be taken into consideration. At 60 GHz the well known oxygen absorption causes an attenuation which amounts to 16 dB/km at 60.2 GHz and 2 dB/km at 54 and 66 GHz /6/. For low range application (less than 100 m) this additional attenuation is beneficial as the same carrier frequency can be re-used in short distances. Mm wave propagation normally suffers from severe fading due to ground reflections. As there is only a small delay between the direct path and the reflected path this fading is flat as a function of frequency and follows, in principle, the two-way propagation model. If omnidirectional antennas are used, the reflections from the surrounding environment causes additional frequency selective fading. In a measurement campaign which was carried out at 60 GHz in different rooms, a delay amounting to 200 ns (-15 dB level) for large assembly halls was observed /7/. Fig. 2 shows an example of the measured power - delay profile. However, in smaller rooms the significant impulse responses are normally delayed by less than 100 ns.

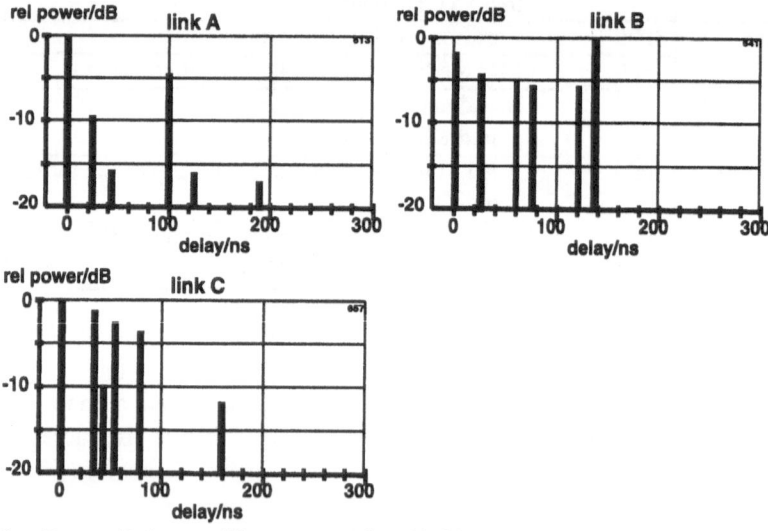

Fig. 2   Power-Delay profiles measured at 60 GHz

It is worth mentioning that the delay spread at 60 GHz is much less compared to that at 5 GHz as comparative measurements in the same environment have shown. (All measurements use omnidirectional antennas in the horizontal plane). The delay spread observed causes intersymbol interference which limits the data rate to approximately 1 to 2 Mbit/s. For higher transmission rates, measures to overcome this problem must be implemented.

Modulation and multipath cancellation

For the selection of the most suitable modulation scheme for a system the following issues have to be considered

- data rate
- delay spread
- carrier to interference ratio
- spectrum occupancy
- transmitter linearity requirements
- receiver sensitivity
- modem complexity
- equaliser complexity
- implementation cost

At present, when selecting a modulation scheme for a millimetrewave communication system, the transmitter output power and linearity is one of the major constraints. With the Gunn element or Impatt diode based oscillators used in the past, only constant envelope schemes could be used. With the advent of linear mm wave amplifier MMICs one can consider modulation schemes requiring linear amplifiers. However in this case, the output power is reduced and, having the present technology constraints in mind, limited to approx. 10 to 20 mW depending on the mode of operation unless linearisation methods are implemented. For low to medium data rate applications (below 2 Mbit/s) and reduced performance requirements, FSK modulation with limiter discriminator demodulation is a good choice. This modulation scheme is easy to implement and robust against carrier frequency offset and rapid Doppler variation. For high bit rates more sophisticated modulation schemes are favoured. For MBS, offset QPSK is foreseen for bitrates up to 20 or 40 Mbit depending on the environment and on the residual delay spread. This scheme requires no linear amplifier and will be used in the initial phase of implementation. It is expected that with the progress of GaAs technology, linear transmitter amplifiers will become available and 16 QAM can be used. So the bitrate can be doubled while maintaining the symbol rate. COFDM was also considered as this scheme could solve the delay spread problem. However the linearity requirements especially when taking into account adjacent channel emissions are beyond that what is expected to be available in the near future.

For high data rate transmission and 4 or 16 QAM schemes, measures against the delay spread are required. The adaptive equaliser is one of the possibilities to enable the transmission. The high data rate and the high vehicle speed are challenges for this component. At low bitrates, digital signal processors (DSPs) may be used for the application. However at 34 Mbit/s and more, the development of ASICs are the only way to achieve a cost effective equaliser with moderate power consumption.

Another method to reduce the intersymbol interference is the utilisation of highly sophisticated antennas. As the different delayed beams normally have different angles of incidence at the receiving antenna, the increase of the antenna directivity reduces the delay spread and thus the requirements for the equaliser are reduced or the equaliser may even be omitted. However, depending on the cell coverage, a number of lobes must be used for each antenna. An additional benefit is the increase in antenna gain which improves the system margin or the transmission range.

The enhancement of the antenna directivity can be performed by two methods:

-       Lobe switching antennas

This approach simply switches the transmitter or the receiver to different narrow beam antennas. The disadvantage is the occurence of switching losses which then partly compensates the gain improvement. (Fig. 3a), unless preamplifiers and power amplifiers are used for each antenna (Fig. 3b). Lobe switching antennas are relatively simple to control, however, a handover algorithm must be developed, in order to switch the relevant antenna lobe for the best transmission.

-       Phased array antennas

The phased array antenna, well known from radar technology, uses one transmitter and/or receiver module for each antenna element. (Fig. 3c). At the transmitter side the unique advantage of this approach is the power combining feature leading to an output power of n-times the element power (n: number of antenna elements). Another benefit is the multipath cancellation capability /8/. However complicated steering algorithms are required and the cost of this type of antenna is high. For a high data rate mobile mm wave communication system, a trade-off between antenna and equaliser complexity taking into account cost, size (MS), weight (MS) and power consumption (MS) could be the best approach.

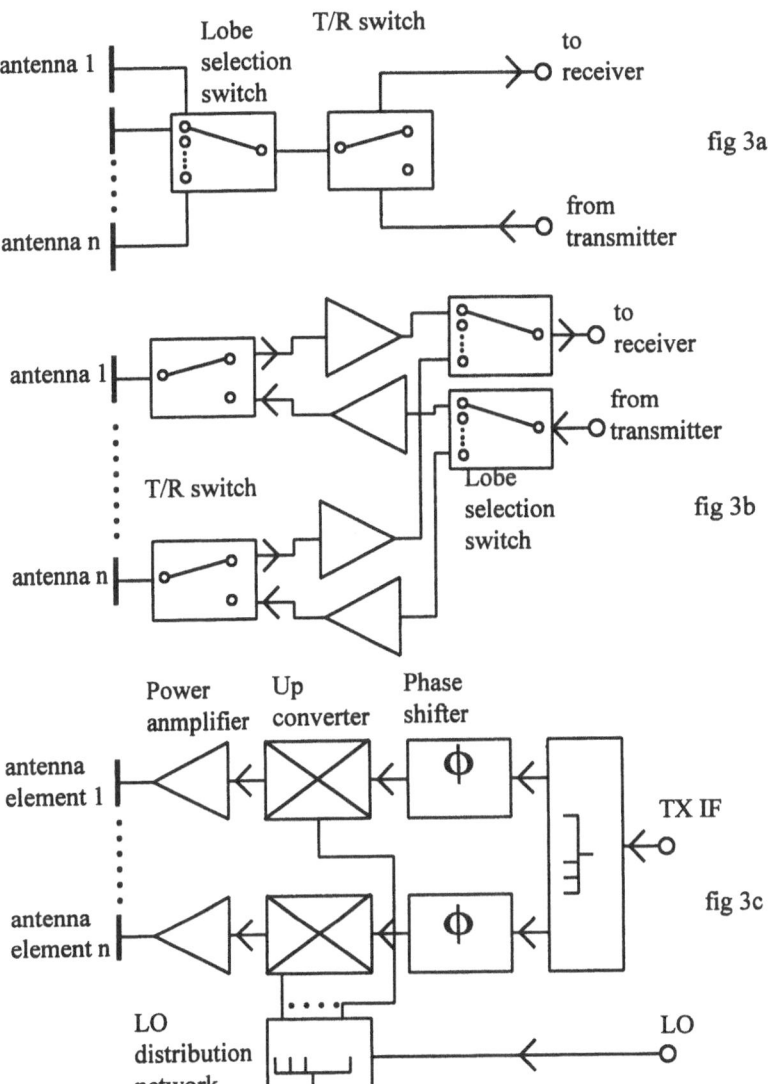

Fig. 3 "intelligent" mm wave antennas

Mm wave transceiver design

In the following chapter some of the possible transceiver architectures are discussed.

Fig. 4a shows the block diagram of one of the simplest solutions. The mm wave oscillator (VCO) operates directly at the transmit frequency. Modulation is performed by applying a tuning voltage to the VCO and causes a frequency modulation of the oscillator. Thus FSK and (G)MSK schemes are possible.

As the frequency drift of free-running mm wave oscillators is too high (except for very low performance applications), phase locking to a reference oscillator is mandatory. For the receive mode the oscillator is shifted in frequency. The unmodulated oscillator acts as the LO and a low IF frequency is generated. In order to eliminate the image noise, the use of an image rejection mixer is advantageous. Image rejection is only 20 dB maximum, which is not sufficient if the radio spectrum is densely occupied. This approach is only applicable for semi-duplex operation.

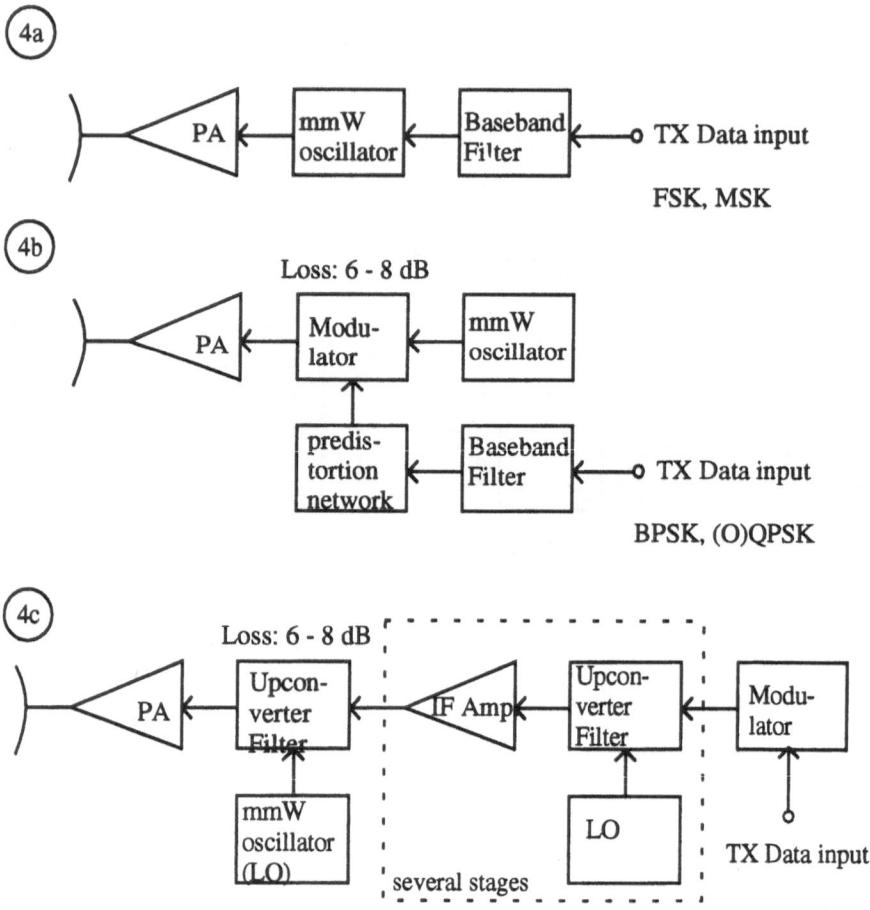

Fig. 4 Mm wave transmitter architectures

The second approach (Fig. 4b) also uses an oscillator operating at the output frequency. However, modulation is performed by a phase modulator which may be a BPSK or QPSK modulator. If non-linear modulators are used, the output spectrum is of the sin x/x type causing adjacent channel distortions which usually cannot be accepted. A linear modulator in combination with an appropriate pre-distortion network enables the generation of a cleaner output spectrum, however, the insertion loss of a linear modulator is considerably higher. The cost of the pre-distortion network at data rates of 34 Mbit/s and above should also not be underestimated. For the receiver the same problem as above occur if only one oscillator is used.

Probably the best approach in terms of performance is depicted in Fig. 4c. This architecture converts an IF signal, generated and modulated at low IF frequencies, up to the required transmit frequency. Several IF frequencies are selected in order to keep spurious emissions to the required low level and to enable the use of low cost filters. The last transmit IF (and first receive IF respectively) must be high enough to enable the use of low cost planar microstrip filters at the transmitter side (for spurious rejection) and at the receiver side (for image rejection) as this type of filter is the only one which is compatible to MMIC technology. Any type of modulation scheme can be used with this approach which is compatible to full duplex or semi-duplex mode of operation. Image rejection is high due to the use of filters, however, the generation of spurious frequencies at the transmitter must be carefully examined. Another advantage of this architecture is that the mm wave local oscillator which contributes most to the phase noise of the system need not be tuned or modulated as modulation is performed at the IF and channels can be selected by the second local oscillator. Therefore, fixed frequency, high Q oscillators can be used at this stage offering the lowest possible phase noise.

For the receiver part of all three architectures, diversity reception giving significant improvement of the quality of transmission is foreseen and is achieved by using two receiving antennas separated by some wavelengths in height. In the MBS demonstrator, different types of diversity selections, based on bit error rate or based on power level will be investigated in order to find the best trade-off between performance and cost.

In the following table an attempt to compare the three transmitter architectures described above was made. The selection of the optimum architecture depends on the system application and the required performance.

TRANSCEIVER ARCHITECTURE

	1	2	3
Complexity	o	o	-
Flexibility	-	o	+
Data rate	-	o	+
Power efficiency	+	o	-
Spurious rejection	+	+	o
Size	+	+	o
IF/BB processing complexity	o	o	o
Oscillator phase noise	-	o	+
Band switching	-	o	+
Modulation schemes	FSK, MSK	BPSK, OQPSK	all
Out of band radiation	o	o	o
Transmit/receive switching	-	o	+
Sum	3*(+), 3*(o), 5*(-)	2*(+), 9*(o), 0*(-)	5*(+), 4*(o), 2*(-)

+ good
o fair
- poor

Requirements for millimetrewave components

In the past, millimetrewave transceivers were equipped with waveguide components machined from solid metal and screwed together with waveguide bends and twists. This approach would never be a cost effective solution for commercial application of mm wave technology. However for several years, integrated millimetrewave technology based on GaAs substrates has shown tremendous progress and new components are available now. This MMIC technology, being potentially low cost, is the only way towards the widespread use of mm wave communication enabling the systems described above to be implemented. In addition to the fabrication of the MMICs, the interconnection and packaging technology is important. As the MMIC fabrication process becomes mature, the assembly cost may become predominant. The following section indicates the requirements for some of the building blocks of modern mm wave transceivers.

Upconverter

Assuming the transceiver architecture No. 3 described above, an upconverter is required from an IF signal in the vicinity of 5 to 10 GHz to the output frequency at e.g. 62 to 53 GHz and 65 to 66 GHz for MBS. This component is comprised of an IF and an LO buffer amplifier, the upconverter mixer and an RF output amplifier.

A bandpass filter is required to reduce the level of the sideband and the LO in order to fulfil the spurious emission specification. At present the upconverter consists of several MMICs. However, for applicable systems, a reduction of the numbers of MMICs is required and the upconverter may consist of a single MMIC with RF, LO and IF buffer amplifiers, the upconverter mixer and a moderate LO and sideband filter.

Power amplifier

This is one of the most challenging devices, especially if linear modulation schemes are considered. The required output power shall be 100 mW minimum which means a 1 dB compression point of more than 400 mW if 16QAM schemes are considered and even more if the adjacent channel emission specification is stringent. As upconverter mixers operating in the linear mode will deliver approximately -5 dBm output power, the total amplification must be 25 to 30 dB depending on filter loss. This gain is shared between preamplifier and power amplifier. So the gain of the power amplifier must be around 15 dB. Ideally, in the future, the mm wave part of the transmitter will consist of maximum 2 MMICs: the upconverter and the power amplifier with an alumina microstrip filter in between. Alternatively, if the isolation can be achieved, a single MMIC containing all the required functions including on-chip microstrip bandpass filters (Fig. 5) could be implemented.

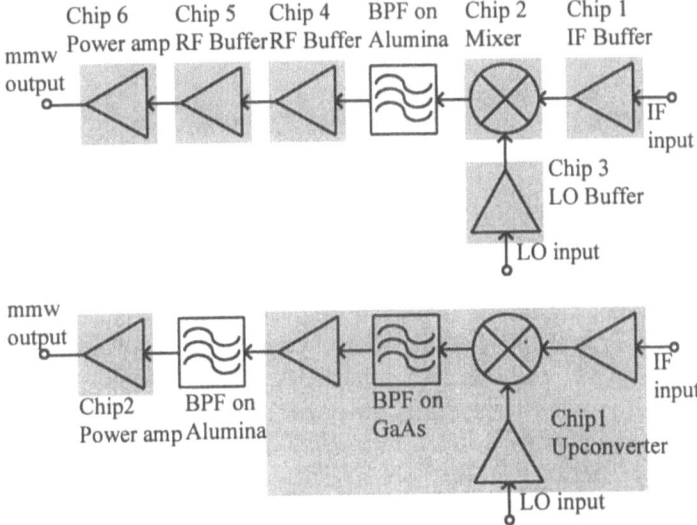

Fig. 5 Present and future mm wave transmitters based on GaAs MMICs

T/R duplexing and antennas

Depending on the mode of operation (semi-duplex or full duplex) different T/R duplexing methods must be implemented. A simple semi-duplex transceiver requires a T/R switch which could be an MMIC based on pin diode or FET technology. Low insertion loss is the key parameter as the switch is both in the TX and RX paths. The isolation in the off state must be sufficient to prevent the preamplifier from heavy saturation, which could result in long recovery times if switched to the receive mode. Switching time is not a major constraint as GaAs FET switches with commutation times of a few ns are available at lower frequencies. For full duplex operation either a diplexer or separate transmit and receive antennas are required. The insertion loss of the diplexer must be low (< 1dB) Microstrip filters are probably not the best choice as the insertion loss requirements is hard to fulfil with low Q filters. The optimum performance is achieved with a waveguide diplexer which could be cost-effectively fabricated using the plastic moulding technique. If separate transmit and receive antennas shall be used, the isolation between the antennas must be high enough to prevent the receiver from degradation while transmitting. This isolation must be maintained if the radome is polluted e.g. with dust, rain or snow.

The antennas to be used are dependent on the system requirements. For mobiles, an omnidirectional radiation pattern is usually required, whereas the base station may adopt the antenna beam pattern of the radio cell to be illuminated. Planar antennas on a suitable substrate material are favoured because this technology matches best to MMIC technology. Another approach, especially if a higher gain is required, is to focus the beam to a small patch /9/, which could be part of a MMIC, by using inexpensive dielectric lenses. Even waveguide antennas can be used if suitable manufacturing techniques are used. The benefits of intelligent antennas have already been described above.

Receiver front-end

A low noise figure is the most important feature for the receiver front-end which consists of the low noise preamplifier, an image rejection filter, a downconverter and the IF amplifier stages. As stated above, most applications will rely on diversity reception requiring a dual channel front-end including the LO distribution network or, for the simple switched diversity approach, an antenna switch which could be located after the preamplifier in order to avoid system degradation caused by switching loss. The large signal behaviour of the receiver front-end is not a large constraint unless very high gain antennas are used. At 10 m distance and 15 dB antenna gain for both TX and RX antenna the received signal level at 62 GHz is only -38 dBm (assuming 100mW transmit power). So the preamplifer must have enough gain to mask out the influence of filter and mixer on the overall noise figure.

Mm wave Oscillators

Depending on the transceiver architecture, mm wave oscillators are either used as the transmitter oscillator or as the first local oscillators for up- and/or down conversion. Unlike waveguide oscillators based on Gunn elements, which offer a tunability of several hundred MHz and good phase noise performance, the trade-off between these parameters is more difficult to achieve using planar MMIC-based oscillators. Without the implementation of a dielectric resonator, phase noise is poor and not compatible to 4 or 16 QAM modulation schemes as the residual phase noise causes a high bit error rate. Hence, a dielectric resonator must be used in order to meet the phase noise requirements. This however reduces the tuning range to a few MHz and channel switching cannot be performed. As mentioned above, this problem is eliminated in transceiver concept 3.

Dielectric resonators at carrier frequencies around 60 GHz are small and difficult to machine with the required accuracy. An oscillator concept which generates the required frequency by doubling or even quadrupling the output of the DRO solves this problem and the multiplier acts as a buffer which prevents the oscillator from pulling. Drawbacks of this solution is the generation of sub harmonic signals, which must be filtered, and a lower efficiency.

Technology requirements

We expect that GaAs pseudomorphic high electron mobility transistors will be used for all active mm wave devices. A gate length of $0.15\mu m$ or less is required to obtain a reasonable gain per stage at 60 GHz. The description of the technology which meets the requirements for mm wave communication systems is the scope of chapter 3 in this book and is also described in /10/. From the applications point of view, the effort on GaAs technology shall focus on both performance enhancement and cost reduction. The cost contribution of all the mm wave parts of a transceiver shall be less than 20 to 25 %. Several measures to decrease the cost of the mm wave parts must be taken i.e.:

- Increase in process maturity and yield improvement.
  It is obvious, that yield has a main influence on cost.

- Increase of circuit density.
  The reduction of GaAs area per circuit function increases the number of chips per wafer and cost is reduced.

- Multifunction chips.
  The interconnection of chips may contribute to the cost even more than the chip itself. Therefore multifunction chips with enhanced circuit complexity must be developed. A good example is a chip developed for

MBS by one of the partners which contains an input buffer amplifier, a frequency doubler, bandpass filter, power splitter and two output buffer amplifiers on a single chip (Fig. 6).

-    Improved assembly techniques.
As the tolerances of circuits and interconnections like wire or ribbon bonds must be extremely small at mm wavelengths, new low cost circuit assembly techniques must be developed. This may include precision pick and place and bonding machines or flip chip techniques /10/; a technique where bond connections are avoided, but grounding and heat dissipation problems are not yet solved.

-    Package technology.
MMICs must be protected from the environment in an appropriate housing. For prototypes and small quantities, these housings are currently milled from solid metal. A low cost solution for the assembly enclosure, taking into account moisture protection, heat dissipation and the mm wave interconnection to other assemblies and antennas must be found.

-    Reliability.
Commercial mm wave communication systems require a reliable operation over a long period of time. The MTBF (mean time between failure) of mm wave MMICs must be investigated before they can be implemented in such systems.

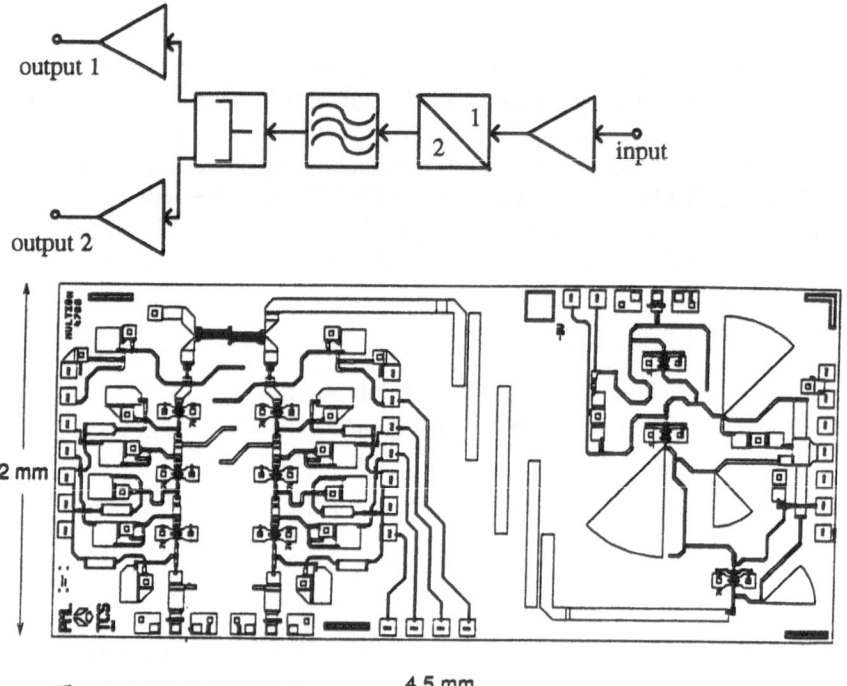

Fig. 6 Multifunction MMIC

Conclusion

Millimetrewave communication systems are required in order to fulfil all the present and future communications needs. One of the prerequisites for this development is mm wave MMIC technology based on GaAs. Although tremendous progress has been made in recent years, the cost requirements for commercial or even private systems cannot yet be achieved.

In addition to development of the chip technology leading to lower prices, engineers must focus on the integration and packaging techniques. Various research programmes of the European Community like ESPRIT and RACE give the European industry a good chance to combine their efforts in order to meet the objectives of low cost mobile mm wave communication.

References

/1/   H.J.Fischer: "Transparent Universal Communication System for RTI-/ATT Applications Beacon-Vehicle / Vehicle-Vehicle Link" DRIVE Technical Days Brussels 3/93

/2/   L. Fernandes: "Progress towards a Mobile Broadband System" Microwave Engineering Europe Nov. 1993 pp 17 - 25

/3/   A. Platter: "A Millimetrewave Communication System for MAGLEV Application" 1994 IEEE MTT - S Digest

/4/   European Radiocommunications Office: "Results of the Detailed Spectrum Investigation (DS); First Phase 3.4 to 105 GHz."

/5/   CCIR Report 721-2

/6/   CCIR Report 719-3

/7/   A. Plattner et al: "Indoor and Outdoor Propagation Measurements at 5 and 60 GHz for Radio LAN Application 1993 IEEE - MTT-S Digest Vol. 2 pp 853 to 856.

/8/   Thomas S. Fong: "An Interferene Nulling Algorithm for a Single Receiver Phased Array" IEEE Transactions on Antennas and Propagation June 1990 pp 951-953

/9/   J.C.Alder et.al. "Microwave and Millimeterwave Staring Array Technology." Proc. EUMC 1990 Budapest, p 454ff

/10/   M. Chelouche, A. Plattner: "Mobile Broadband System (MBS): Trends and Impact on 60 GHz Band MMIC Development" Electronics & Communications Engineering Journal June 1993 pp 187 -197.

2.3   Airport Surveillance Radar

J. M. Dieudonné
Deutsche Aerospace, Ulm, Germany

Introduction

For the improvement of airport safety a multisensor system has been proposed by the radar and radio division of Deutsche Aerospace. Spatially distributed radar sensors are the conceptual key of the envisaged system. Such a configuration of networked radars offers the potential for accurate, timely and reliable target location and identification under all weather and visibility conditions. A compact, low cost and high performance millimetre-wave front-end is required. The most promising way to fulfil this objective is the use of GaAs MMICs with a high level of integration such that a complete front-end can be realised on a single GaAs chip.

The objective of AIMS was to develop a basis technology for millimetre-wave integrated circuits in Europe. Active devices with state of the art characteristics have been developed and are now reproducibly obtained. These devices include Heterobipolar Transistors (HBT) and Pseudomorphic Hetero-Field-Effect Transistors (PMHFET). All the key components for the airport radar millimetre-wave fronted have been fabricated by Daimler-Benz and the results have demonstrated, that the GaAs technology is now mature enough for the realisation of the proposed system.

System concept

Dependent on the actual airport structure, a number of radar-sensors are positioned at different airport locations (Fig. 2). Individual sensor data are transfered via data link to a central station where a fused radar image is generated for monitoring of the airport surface traffic.

The main advantages of this approach using networked radar sensors are:

- The spatial sensor distribution ensures complete and redundant surveillance coverage of runways, taxiways and aprons. Shadowed areas can be excluded by appropriate sensor distribution.

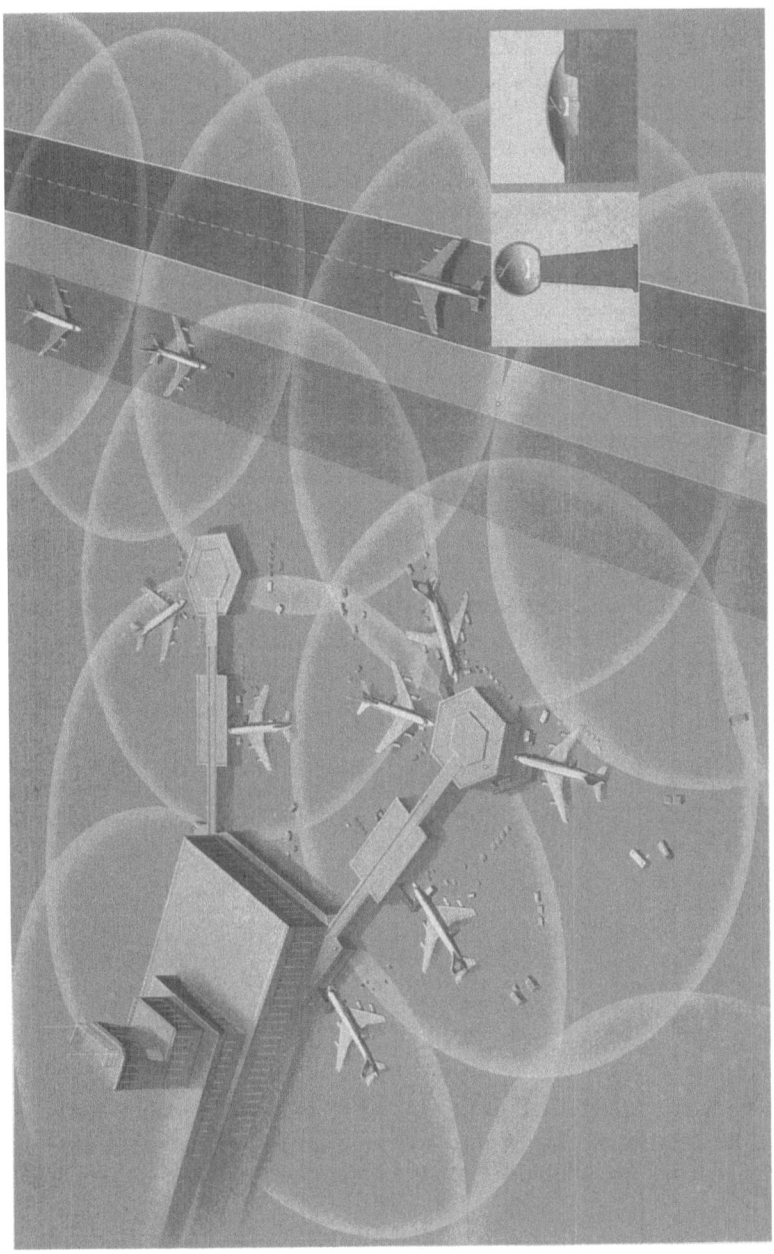

Fig. 1   Artist's view of RADAR controlled airport ground traffic

- The modular structure allows a stepwise implementation matched to the airport topology.
- Each target is generally illuminated by several (typical 3) sensors from different viewing angles.
- Stationary and moving targets (aircraft, ground vehicles and obstacles) are detected and tracked during good and poor visibility conditions.
- Excellent operational reliability and availability are the obvious consequences of the redundant coverage concept.

The network approach relies on the concept of applying compact low cost mm-wave sensors of limited range but high resolution. A 35 GHz sensor is chosen in accordance with these requirements.

The radar head comprises transmitter, receiver, and A/D-converter. A low power solid state transmitter with linear continuous wave frequency modulation (FMCW) is used for economic power generation and high range resolution. The transmitted waveform from a linearly swept oscillator is returned back from a target and mixed with a sample of the transmitted signal. The resulting beat frequency is composed of a range and a doppler dependent component. FMCW modulation has been successfully demonstrated with 0.25 m range resolution. The transmitted waveform is circulary polarised. A dual channel receiver separates copolar and crosspolar returns which are subsequently processed by spectral analysis. A FFT processor executes the range/doppler gating function.

The main characteristic of the millimetre-wave sensor are given below.

Operating frequency	34.7 GHz
Output power	24 dBm
Bandwidth	1 GHz
Radar mode	FMCW
Range	60 - 1000 m
S/N	13 dB
False alarm rate	10^{-6}
Rain Rate	16mm/h

Plots of detected targets are extracted and reported via data link to the central station for data fusion. These plots contain a list of relevant parameters such as position, doppler and attributes.

In Fig. 2 several mm-wave sensors are distributed over the airport surface. The hatched triangular area can be defined as the generic cell of a cellular architecture. Inside the "area of authority" of such a network, cell targets are advantageously illuminated by 3 sensors from different viewing angles ("angle diversity"). The conceptuel idea of clearly defined cell boundaries is the key for accurate,

43

unshadowed, redundant and reliable surveillance. The cellular architecture allows the network to be adapted to different topologies (runways, taxiways, apron areas) and modular extensions.

In Fig. 3 the fusion process with respect to the millimetre-wave network is described. Plots reported from different sensors are transfered to the central station for data fusion. These plots concentrate the relevant target parameters as provided by the spatially distributed sensors. Due to the network architecture and the all-weather performance of each sensor, redundant target detection and parameter estimation is ensured inside the area of each generic cell.

In the first stage plots are aligned in time to account for non-synchronous target illumination. Subsequently the polar coordinate system inherent to sensor plots is converted to the common reference coordinate system. The spatial alignment procedure associates plots from different sensors to a common source origin using spatial and attribute criteria. Finally the data fusion process results in a target track and classification file.

Undoubtedly the multi-sensor/multi-target problem is considerably alleviated by the availablility of target doppler and attribute data that are reported from different viewing angles. Expected benefits are:

• Support of multiple target association and state estimation

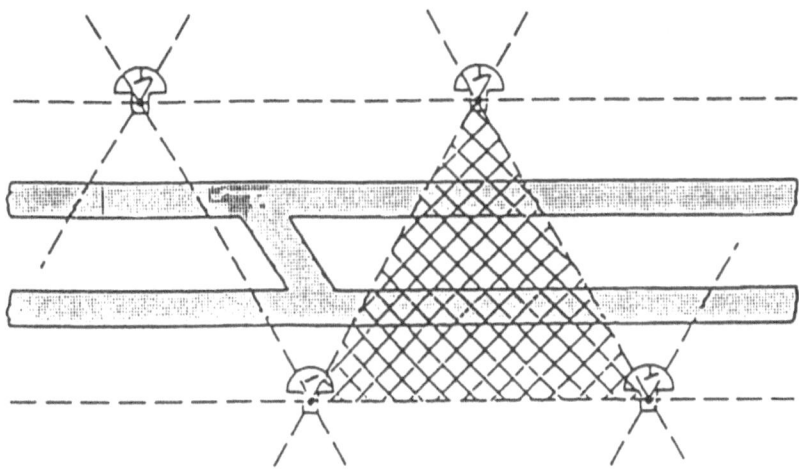

Fig. 2   Positions of sensors w.r.t. taxi way and run way

mmW-Sensors

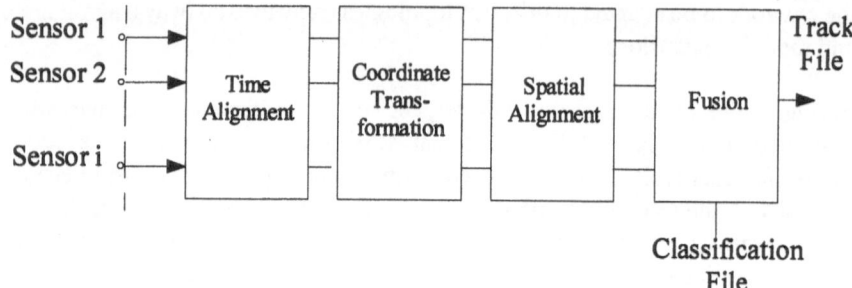

Fig. 3   Architecture of plot file level data fusion

- Provision of the velocity vector based on plot fusion. This vectorial information is helpful for timely indication of target manoeuvres.

- Automatic target recognition: Different target classes (large or small aircraft, vehicle or aircraft) can be discriminated.

Millimetre-wave front-end

The HBT and PMHFET MMIC technologies developed in the scope of AIMS by Daimler Benz have allowed the realisation of all the key components for the millimetre-wave front end.

The block diagram of the radar front-end demonstrator is shown in Fig. 4. The specifications are given on the figure. It consists of 5 monolithic components:

- Voltage Controlled Oscillator (VCO)
- Buffer Amplifier (BA)
- Solid State Power Amplifier (SSPA)
- Low Noise Amplifier (LNA)
- Mixer.

The center frequency is 34.7 GHz with a bandwidth of 1 GHz. Good phase modulation noise and amplitude modulation noise are needed for a high resolution FMCW system.

The HBT technology has been chosen for the realisation of the VCO because of the expected better noise performance. Moreover, this technology allows the realisation of good varactor diodes for wide frequency tuning. A photo of the VCO chip is shown in Fig. 5. As active device a self-aligned HBT with an emitter area of $2 \times 1.5\ \mu m \times 20\ \mu m$ is used. The realised free runing VCO exhibits a phase noise of -80 dBc/Hz at 100 kHz off carrier. The tuning bandwidth is greater than 1 GHz with an output power about +3 dBm.

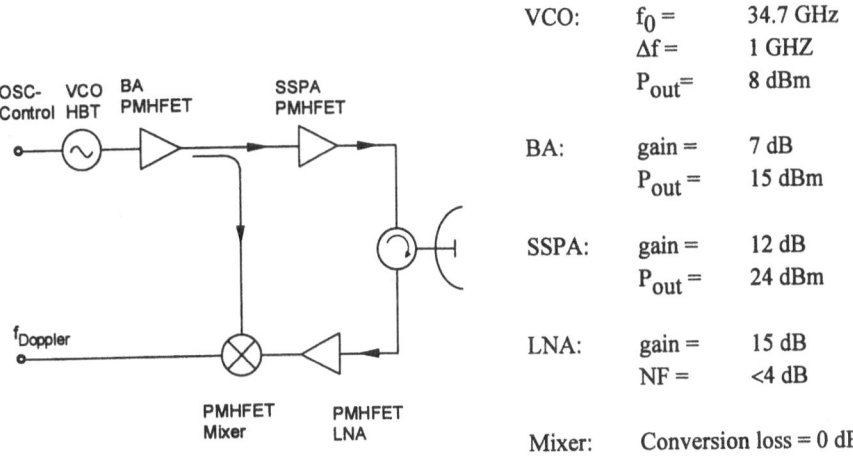

VCO:	$f_0 =$	34.7 GHz
	$\Delta f =$	1 GHZ
	$P_{out} =$	8 dBm
BA:	gain =	7 dB
	$P_{out} =$	15 dBm
SSPA:	gain =	12 dB
	$P_{out} =$	24 dBm
LNA:	gain =	15 dB
	NF =	<4 dB
Mixer:	Conversion loss = 0 dB	

Fig. 4   Block diagram of the millimetre-wave front-end for the airport surveillance radar demonstrator

The developed 0.25 µm PMHFET technology is used for the low noise and power MMIC circuits. A photo of the realised LNA is shown in figure 2.5.6. The amplifier is a 3 stage design. The device used in the circuit is a 6 finger transistor. Each finger has a width of 20 µm. The amplifier chip has a gain of 17 dB associated with a noise figure of 3.7 dB at 35 GHz.

The power amplifier is based on a balanced configuration using Lange couplers at the input and output as shown in Fig. 7. The transistor used in the output stage has a total gate width of 480 µm. The measurement results have shown a gain of 8 dB at 35 GHz with an output power of 21 dBm at 1 dB compression. An optimized version of the amplifier which is under fabrication will achieve a gain of 15 dB with an output power greater than 23 dBm at 1 dB compression.

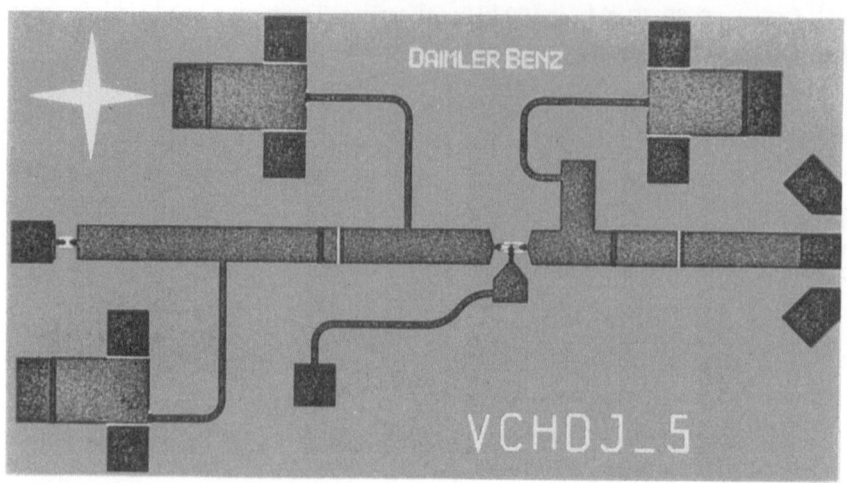

Fig. 5 Photo of the HBT VCO

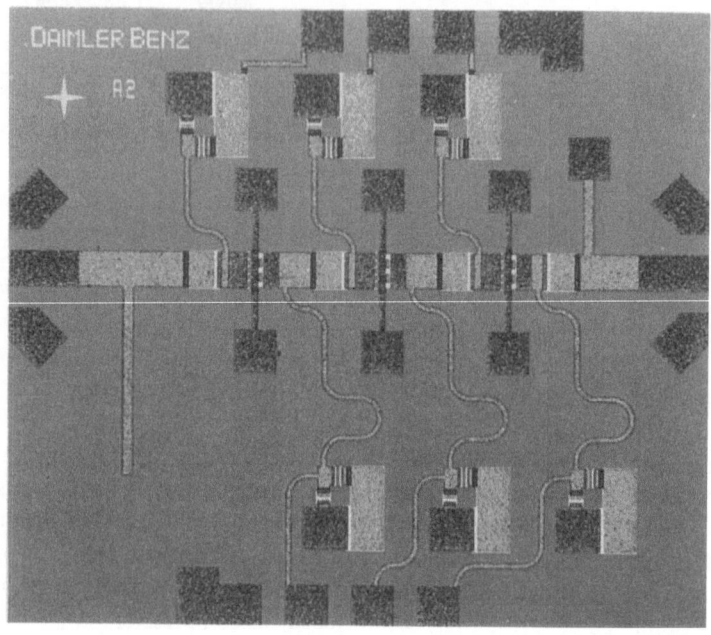

Fig. 6 Photo of the 35 GHz PMHFET LNA

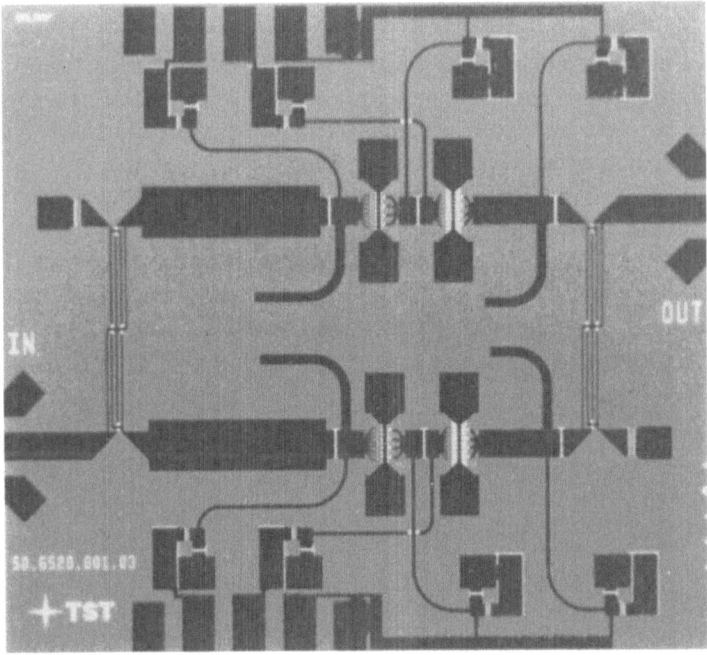

Fig. 7 Photo of the 35 GHz PMHFET SSPA

On the mixer side 3 different types of mixer are under fabrication on PMHFET material:

- a single balanced Schottky diode mixer
- a single ended single gate transistor mixer
- a single ended dual gate transistor mixer.

For the Schottky diode mixer a conversion loss of about 7 dB is expected. The active mixers will achieve conversion gain of more than 5 dB. The best candidate for the radar system is probably the balanced configuration because of the cancellation of the AM noise at the mixer output.

Conclusion

It is expected that a network of properly sited radar sensors is capable of meeting the essential requirements of the airport surveillance system. The system will provide an accurate, timely, complete and reliable presentation of the actual airport situation even under adverse conditions.

In the current activities of the AIMS project, a PMHFET technology and an HBT technology have been developed for the millimetre-wave front-end. The MMIC circuits have demonstrated compatibility with the system requirements. These technologies can be used for the fabrication of low size and low cost millimetre-wave front ends.

2.4   Mobile Communication

E Pettenpaul, L Scharf and K J Schopf
Siemens Semiconductor Group, D-81617 Munich, Germany

1. Introduction

The mobile communication market is the current driving force of the European telecommunication industry with remarkable growth rates. In March 1990 about 2.5 million mobile communication systems have found application in Europe with a forecast of more than 90 million DECT-GSM-PCN systems up to the year 2000.

This situation is of great benefit for the semiconductor industry in general and the GaAs components suppliers especially. The reason is that GaAs components maintain their sophisticated performance - low NF and high PAE - under the stringent low power consumption conditions of a handheld. Moreover, the established low-loss GaAs MMIC technology allows in the meantime the production of small SMDs at a reasonable price.

In the following, we describe a complete set of GaAs and Si devices suitable for the new digital communication systems.

2. System Description

The second generation of mobile communication services is characterized by the change from analog to new digital transmission techniques. The main digital systems to be considered worldwide are summarized in Table 1.

The new system standard is fixed by the GSM (Group Special Mobile) telephone system. GSM is a full featured telephone network which allows communication (receiving and calling) with automatic allocation of all users. The system can be accessed with various types of user equipment: from lightweight microportables up to high power mobile transceivers. The system will automatically transfer users from ons base station to another if the car is moving between zones. GSM is a cellular system for mobile and handheld communication. It will be the first mobile communication standard to be introduced all over Europe. A first version at 900 MHz is being introduced and will be followed by a second generation at

System	Description	Frequencies MS - BS BS - MS channel	Max. Power
GSM	Cellular European net with int.roaming	890-915 MHz 935-960 MHz 200 kHz	20 W
DCS 1800 (PCN)	Cellular European net with int.roaming	1710-1785 MHz 1805-1880 MHz 200 kHz	1 W
ADC (D-AMPS)	Cellular radio (North America)	824-850 MHz 869-895 MHz 30 kHz	6 W
JDC	Cellular radio (Japan)	940-960 MHz 810-830 MHz 25/50 kHz	?
CT2	European cordless phone and and telepoint system	864-868 MHz 864-868 MHz 100 kHz	10 mW
DECT	European cordless telephone	1880.92-1898.208 1880.928-1898.208 1.728 MHz	250 mW (450 mW)

Table 1:   Mobile communication system

1800 MHz in 1994 developed to increase the number of channels from 992 to 2992. This latter system is called PCN (Personal Communication Network) or DCS 1800. The expected user range is 1-8 km compared to about 35 km for the 900 MHz GSM. DECT (Digital European Cordless Telephone) is the digital counterpart of the CT1-CT2 analog telephone working at 900 MHz. In addition, DECT can be used as private wireless PABX. This system will improve todays paging systems to full intercompany telephone networks. Operating frequency is about 1.9 GHz and typical ujser range is up to 250 m. The introduction of DECT in Europe is planned for 1992 - 1993.

3. General Components Concept

Fig. 1 shows a schematic block diagram of a complete receiver - transmitter unit which consists of an RF part, a signal processor, a multifunction interface, a microprocessor, and a frequency synthesiser. Those areas where GaAs components can be used have been identified. The RF part with the receiver components preamplifier and mixer, and transmitter branch with the power amplifier chain, should be fabricated in GaAs technology.

The specifications of the systems are such, that many features are best fulfilled by GaAs components giving the following benefits: low current consumption, low battery voltage requirement, high linearity, low noise figure (receiver), high power added efficiency, PAE, (transmitter). Si technology is highly competing with GaAs in terms of price and maturity but especially at 1.8 GHz operation frequency clear performance advantages with comparable price levels can be expected from GaAs devices. This is not obvious for the VCO and T/R switch where the Si bipolar transistor and diode technology is still dominant.

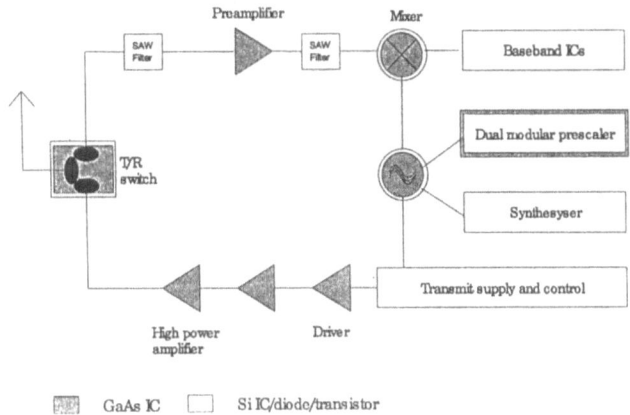

Fig.1 : Schematic block diagram of the complete GSM, PCN DECT receeive/transmit unit

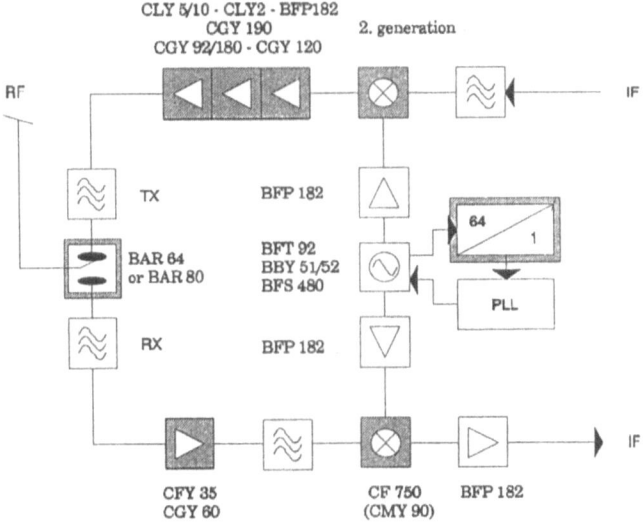

Fig. 2 : Devices for the GSM, PCN, DECT RF part.
(By courtesy of Siemens)

In any case todays state-of-the-art RF components (discretes and MMICs) in GaAs and Si have to demonstrate 3V-low current capability, high performance up to 2.5 GHz, surface mount packaging and attractive large-volume prices.

A product line-up suitable for 900 MHz up to 2.5 GHz is shown in Fig. 2. Based on this component set, which represents a mixture of GaAs and Si parts, the current and future component strategy will be discussed in the following.

4. Receiver Part

The requirements for all systems discussed in Table 1 are similar, i.e. low NF and intermodulation distortion are important. The current generation fabricated is based on a 0.5 μm MESFET as preamplifier followed by an MMIC mixer with the described performance:

Function	Type	Device	G dB	NF dB	IP3 dBm	Package
Preamplifier	CFY 35	MESFET	17	0.5	0	SOT 143
	CGY 60	MMIC	12	1.5		MW-6
Mixer	CF 750	MMIC	14	5.0	- 1	SOT 143

Table 2: Performance of Receiver Components at 1.8 GHz (I = 2.5 mA, 3V)

The GaAs mixer MMIC shows a simple but very effective circuit topology. It is a DG FET configuration with source LO allowing the desired low current - low voltage - low LO level operation. At 1.8 GHz, a conversion gain of 14 dB with a SSB-NF of about 5 dB and 15 dB isolation has been measured with an LO power of 0 dBm.

The next generation of preamplifier, a 1-stage 50 Ω-MMIC, is already available in first samples. Fig 3 contains a chip photograph of the LNA MMIC based on a reactively matched feedback configuration with active load and unipolar voltage supply. The performance revealed 12 (15) dB of gain and 1.5 (1.3) dB NF at 2.5 (5) mA between 1.7 and 2.0 GHz. A simple adaption to all band of interest beween 900 and 2500 GHz is possible.

For the VCO we currently recommend a Si bipolar transistor and a 3 V optimised varactor diode with sufficient capacity shift and Q factor.

A longterm development strategy requires a permanent device improvement. As a result, a one-chip receiver unit containing a preamplifier, mixer, postamplifier and LO is already in the design phase.

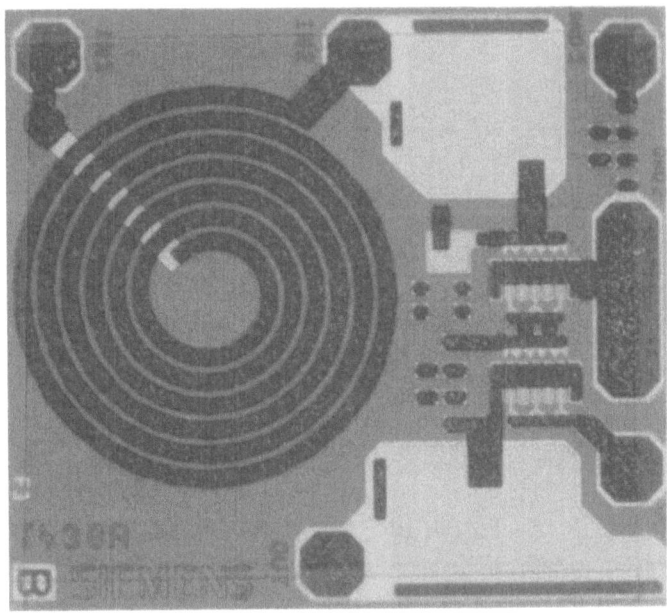

Fig. 3 : GSM, DECT, PCN GaAs LNA MMIC

5. Transmittter Part

For the transmitter part, a less uniform picture is visible since the systems have different output power and gain control requirements. In any case this part shows the highest power consumption and as a result the questions of the PAE and 3V-supply are of major importance.

The system related device specifications and selection are summarized in Table 3. The currently used devices for the 3-stage High Power Amplifiers (HPAs) are the GaAs Power MESFETs CLY 2, 5, 10 with up to 1000 mW output power. Very high drain efficiencies of 50 - 60 % have been measured at only 3 V supply voltage. A short summary of the P-MESFET test results are shown in Table 4. The devices are assembled in a standard or modified SOT 223 SMD package.

The next important step is the monolithic integration of the GaAs power amplifiers for DECT, GSM and PCN systems. The first MMIC available is a 3-stage DECT HPA. The device is specified with 27 dBm output power, PAE of > 35% and IP3 > 30 dB again with 3 V supply voltage and 50 Ω terminations. Fig 4 contains a chip photograph of the DECT HPA realized on 2 mm² chip area. The device is mounted in a modified 12-pin SOT 223 package called MW-12.

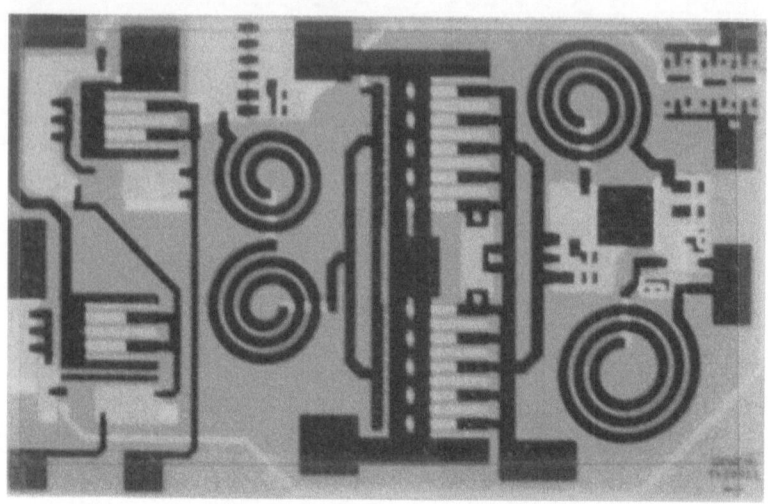

Fig. 4 : DECT GaAs HPA MMIC

Applica-tion	Performance					Device Selection			Device Generation
	Pi dBm	PO dBm	PAE %	VGA	F MHz	1. Stage	2. Stage	3. Stage	
1. GSM ADC JDC	>-11	31-32	>30	Yes	890 - 915	BFP 183 -	CLY 2 CGY 120	CLY 10 CGY 92	SiBip; P-MESF. P-MMICs
2. DECT	>-5	27	35	No	1880- 1900	CLY 2 -	CLY 5 CGY 190	- -	P-MESFETs P-MMIC
3. PCN	>-18	31-32	>30	Yes	1710- 1785	CF 739 -	CLY 2 CGY 120	CLY 5 CGY 180	DGFET,P-MESF. P-MMICs
4. w-LAN	0	20	>20	Yes	2400- 2500	CLY 2	CLY 5 CGY 250	-	P-MESFETs P-MMIC

Table 3 : Power Amplifier Generation and Specification

In comparison to DECT, the GSM and PCN HPAs need higher output and gain accompanied by a gain regulation stage. As a result, a 2-chip MMIC solution is foreseen for both systems. A GSM/PCN Variable Gain Amplifier (VGA) based on a 2-stage DG FET feedback configuration with 50 Ω terminations is also available in first samples (Fig. 5). The VGA MMIC is specified with 12 dBm output power at 1 dB compression and 50 dB gain control range. This device is ideally suited to drive the GSM/PCN final P-HPA MMIC stages requiring 31 to 32 dBm output power. In this case, the design work has been completed (GSM) or first samples under test (PCN).

Device Type	$P_{(-1dB)}$ [dBm]	G_p [dB]	η_D [%]	f [GHz]	V_{DS} [V]	I_D [mA]	Package
CLY 2	23	15	> 60	1.8	3	175	MW 6
CLY 5	27	9.5	> 55	1.8	3	350	SOT 223
CLY 10	29	9.0	> 55	1.8	3	700	SOT 223

Table 4 :   Test results of GaAs power MESFETs

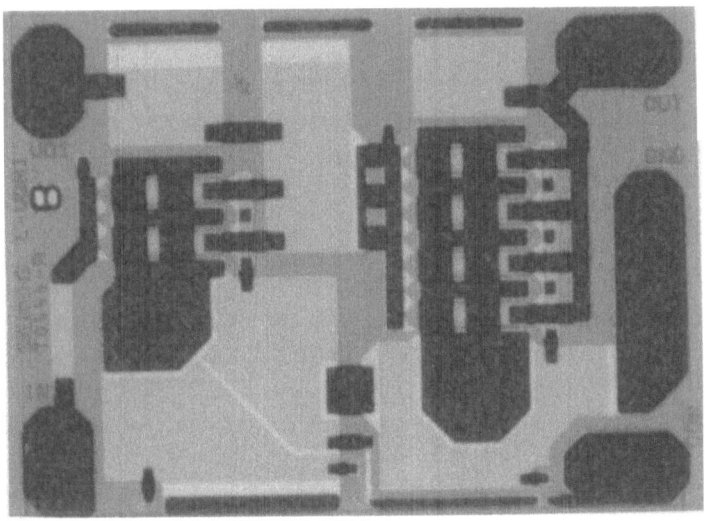

Fig. 5 : GSM, PCN GaAs VGA MMIC

6. Device Technology

The above described GaAs devices have been fabricated using the well-known 3"-wafer process DIOM 15 (1-3). It comprises a planar process up to the gate deposition with localized ion implantation, selfaligned gate technology realized by i-line wafer stepper, and airbridge crossovers. The only option for the different low noise and power devices is the adaption of the channel profile. Based on this process, more than 20 million MESFETs and 500 000 MMICs have been fabricated.

The often adressed cost advantages of Si compared to GaAs are not as high as expected for the mobile communication devices. The reason is that processing, testing and packaging can be realized on a comparable cost level due to the fact that identical or similar production resources can be used. As a result, higher GaAs costs are coming from the pure material part, i.e. this contribution is small compared to the overall costs.

7. Conclusions

A set of GaAS/Si SMD devices for the new GSM, DECT and PCN mobile communication systems has been described. The advantages of GaAs for the RF part in terms of LNA and mixer NF and IP3, and especially PAE of the HPAs have been demonstrated. The GaAs MMICs available are LNAs with 1.5 dB NF and mixers with a SSB-NF 5.0 dB. In first samples available are a DECT HPA MMIC (P = 27 dBm, PAE = 35%) and a GSM/PCN VGA MMIC. The GSM/PCN HPA MMICs are in a phase of design and test completion, respectively. All devices have been optimized for 3 V supply voltage and will be available as SMD parts.

Acknowledgement

The autor would like to thank his colleagues for their contributions to this work and acknowledge financial support by the EC through ESPRIT projects 5018 COSMIC and 6050 MANPOWER.

References

1.   E.Pettenpaul, W.Heidenreich, J.Huber, W.Flossmann,
        A High-Temperature Sensor Based on Monolithic GaAs Hall IC,
        GaAs IC Symp. Digest, 1985, pp. 169-172.

2.   E.Pettenpaul,
      State-of-the-Art of MMIC Technology and Design in West Germany,
      IEEE MTT Symp. Digest, 1987, pp. 763-766.

3.   C.R.Green, E.Pettenpaul, H.Müller, T.Meier, J.E.Müller, W.Kellner,
      GaAs MMIC Power Amplifiers for Radar and Communication Systems,
      Military Microwave Conference, Brighton, 1992.

2.5   Direct Broadcasting TV

T Meier and E Pettenpaul
Siemens Semiconductor Group, D-81617 Munich, Germany

1. Introduction

The direct broadcasting and reception of television programs via satellite (DBS) has been attracting growing interest since end 1988 due to the variety of programs, including foreign programs, as well as providing better coverage of service areas. This will demand the introduction of systems implemented in a GaAs microwave technology of higher quality than hitherto been customary in consumer products. GaAs multifunction MMICs in MESFET and HEMT technology have been developed for these satellite TV systems by GEC-Marconi and Siemens.

2. System Description

Several European DBS satellites are today broadcasting from their geostationary orbit position in a height of about 36000 km. The location of the most important satellites is illustrated in Figure 1. High sophisticated antenna techniques provide an optimized footprint, allowing the reception of the corresponding programs all over western Europe. As an example, the footprint of the very popular satellite ASTRA 1B is shown in Figure 2.

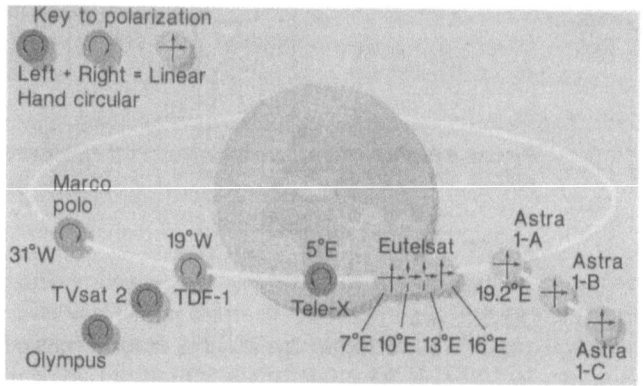

Fig. 1: Orbit Positions of European DBS Satellites

The original plans implied a reservation for the frequency band 11.7 - 12.5 GHz used by the German TV SAT and the French TDF satellites. But in the meantime, the most popular frequency band is 10.95 - 11.7 GHz used by the ASTRA satellite family (ASTRA 1A : 11.2 - 11.45 GHz, ASTRA 1B : 11.45 - 11.7 GHz, ASTRA 1C : 10.95 - 11.2 GHz). The main reason is the provision of in total 48 channels.

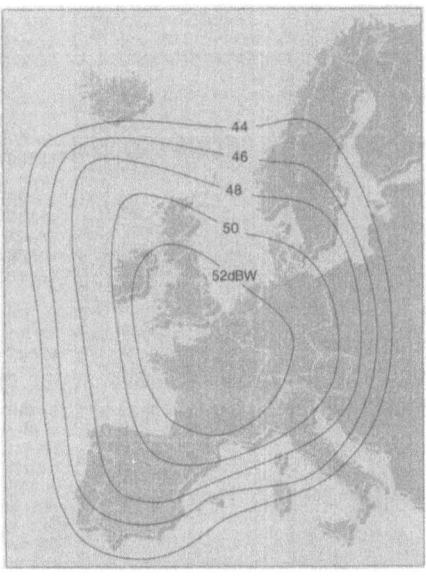

Fig. 2 : Footprint of ASTRA 1B BDS Satellite

High power satellites like TV SAT and TDF, showing about 60 - 65 dBW EIRP (effective isotropic radiated power) per channel, have been developed to allow reception with reasonable antenna sizes (80cm diameter) assuming receiver noise figures of about 1.5 dB. Due to the rapid improvements of the applied semiconductors, todays DBS receivers show a noise figure of about 1.2 dB allowing the use of medium power satellites like ASTRA (52 dBW EIRP) and antenna diameters of only 60 cm.

A DBS receiver unit (Figure 3) requires a frequency down converter (LNC) mounted on a small parabolic dish antenna (outdoor unit) and an indoor unit comprising a channel selector, a demodulator and other signal processing devices. Essentially, a DBS front end performs five basic functions: low-noise amplification at x-band frequencies (11 - 12.5 GHz), stable oscillation at a somewhat lower x-band frequency, down-conversion to an IF of 0.95 - 1.75 GHz, IF amplification, and finally image rejection filtering.

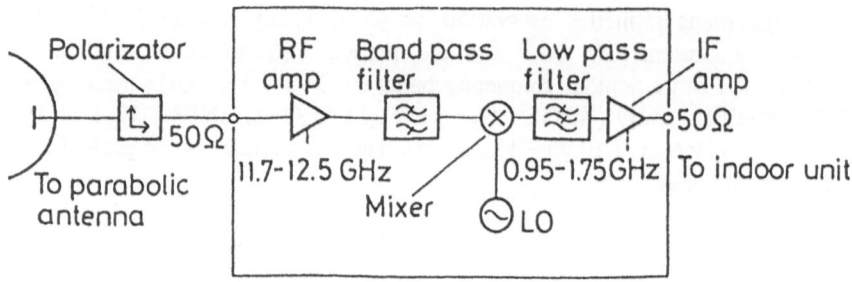

Fig. 3 : Schematic Diagram of DBS Low Noise Converter (LNC)

The usual approach for the LNC is up to now based on discrete devices (GaAs HEMTs, MESFETs, Si diode mixers, Si bipolar IF transistors) mounted on a microstrip RF-board. Although very good performance has been achieved with the hybrid frontends, integrated circuits (MMICs) will prepare the way for further technical and economic progress by greatly accelerating assembling and tuning routines as well as reducing the need for external circuitry.

3. DBS MESFET LNC MMIC by Siemens

Fig. 4 shows the realized chip. The DBS MESFET LNC contains a 2-stage Low Noise Amplifier (LNA) using reactive matching and inductive series feedback to achieve good S11 and low NF simultaneously. A 7-element high pass filter has been placed between the LNA and the following mixer to achieve sufficient image rejection. Perhaps the most critical part is the active mixer. A DG FET multiplicative mixer configuration in the low noise mode has been selected. The mixer is driven by an on-chip SG FET local oscillator MMIC using S/D series feedback principles and an off-chip dielectric resonator (DR). A 3-stage IF amplifier (IFA) based on a parallel resistive feedback circuit principle has to deliver the highest contribution to the overall LNC gain. A 5-element high pass filter has been inserted between the mixer and IFA.

The simulation results of the LNC and its subcomponents are summarized below:

		LNA	IRF	MIX	IFF	IFA	LNC
G	dB	14	-2.5	1.5	-2.0	24	35
NF	dB	3.0		11		3.5	5.5

Table 1 : LNC Subcomponent Simulation

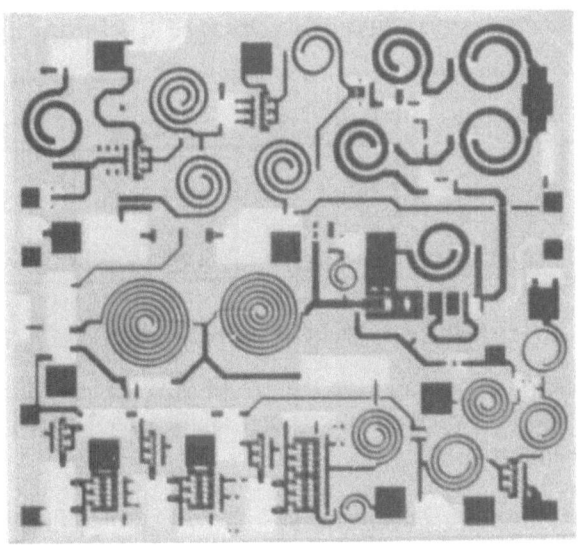

Fig. 4a :   Chip Photograph of DBS MESFET LNC MMIC
(Siemens, Chip Size 4mm²)

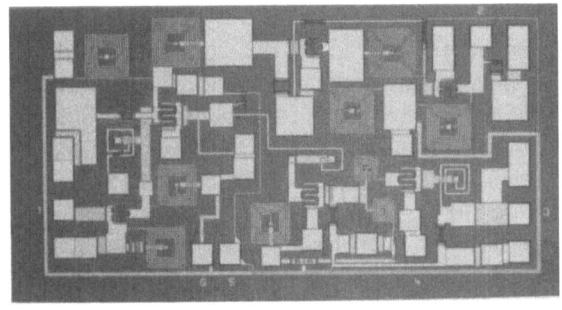

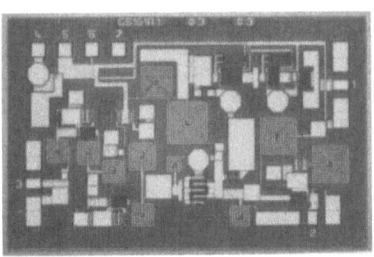

Fig. 4b :   Chip Photograph of DBS MESFET LNC MMIC
(GEC-Marconi, Chipsize 4 and 2 mm²)

A design strategy aspect is the realization of a broadband MMIC covering the most important satellite reception bands which can be described as follows:

Satellite	RF/GHz	LO/GHz	IR/GHz	IF/GHz
TV-SAT	11.70 - 12.50	10.75	9.00 - 9.80	0.95 - 1.75
ASTRA	10.95 - 11.75	10.00	8.25 - 9.05	0.95 - 1.75

Table 2 : LNC System Frequencies

The wafer process used is based on the DIOM 20P 0.5 µm-MESFET for all active devices, airbridge interconnections and source viaholes after the substrate thinning to 100 µm.

To meet all the stringent requirements of a broadband multifunction MMIC, a structured array design approach has been introduced which includes on-chip tuning elements consisting of lines with variable lengths and variable MIM-C top plates. To make use of this approach, optional reticles are available for the airbridge interconnections to be selected after the in-process OW test of the MESFETs.

The initial tests of the DBS LNC MMIC have been performed after chip assembling on a suitable ceramic test carrier. One of the attractive features of the realized LNC is that the LO frequency can be fixed to frequencies between 9.8 and 10.8 GHz by a mechanical tuning of the DRO which allows the reception of several subbands and TV satellites between 10.95 and 12.5 GHz.

The most important test results are summarized in Table 3 for the case of the ASTRA reception, a frequency response of the conversion gain and SSB-NF is shown in Figure 5. The LNC noise figure of 5.5 dB is impressive. The conversion gain shows values between 30 and 32 dB with an operating current of 110 and 150 mA. The LO-RF and LO-IF leakages are in the range of -40 and -20 dBm, respectively, which is important with respect to expected package losses. The on-chip image rejection filter delivers a contribution of at least 15dB as an add-on to the external filter placed between the MESFET LNC and a HEMT preamplifier.

A critical comparison to published results shows that only the already available commercial products of Anadigics (AKD 12000, 12010 and 12011) fulfill in the respective subbands all requirements of a DBS converter. A test comparison of the Anadigics and Siemens devices revealed advantages of our DBS MESFET LNC with respect to NF, isolation and filtering.

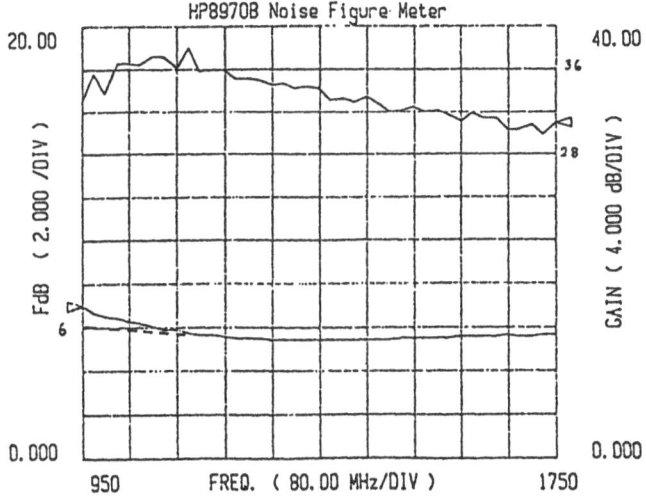

Fig. 5 : Test Results of Siemens DBS MESFET LNC MMIC

In a following phase, preliminary test of the DBS MESFET LNC in a 8 pin TO-package and new SMD package have been carried out. Both packaged devices delivered encouraging results. To compensate the visible gain losses of the SMD package, a minor redesign of the LNCs LNA and IFA will be realized for the following production phase.

Parameters		Target	Test 1. Run
RF	GHz	10.95 - 11.75	10.95 - 11.75
LO	GHz	10.0	10.0
IF	GHz	0.95 - 1.75	0.95 - 1.75
Gc	dB	35	> 32
P (-1dB)	dBm	+ 5	
NF	dB	< 5.5	5.5 - 6.0
LO-RF Leak	dBm	- 10	- 41
LO-IF Leak	dBm	- 10	- 22
IM Rej	dB	> 15	15
LO PN	dBc/Hz	- 70 - 97	- 70 - 95
at Offset	KHz	10 100	10 100
IP 3	dBm	+ 10	+ 13
O/P	VSWR	2: 1	2 : 1
U_d	V	+5	+ 6
U_σ	V	- 5	-5
Iop	mA	< 120	110

Table 3 : Siemens DBS MESFET LNC MMIC Specification & Test Results

4. DBS MESFET LNC MMIC by GMMT

Initial miniature components for high packing density (HPD) techniques were fabricated on mask sets GS135 and GS138. Included were 2µm through polyimide vias (formerly 4µm minimum), and reduction of M3 minimum width from 10µm to 4µm. Sub-circuits for use in downconverter MMICs were fabricated on mask sets GS146 and GS 148. The first multifunction circuit, included on GS148, was purposely narrow-band and utilised no HPD components except 12µm stacked spirals. Consequently the circuit occupied some 8mm². Predicted and measured conversion gain and noise figure performance are shown in Figure 6.

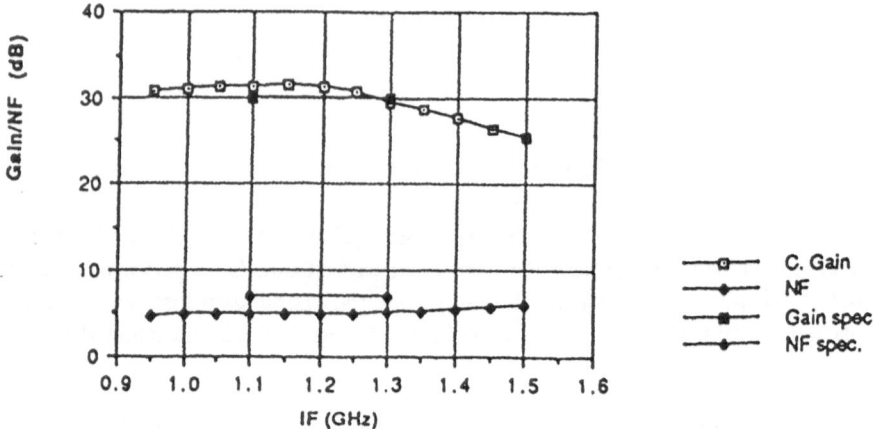

Fig. 6 : Test Results of GMMT DBS MESFET LNC MMIC
8 mm²-LNC, Jig-measurement with 10 GHz DRO

Following the success of this circuit a HPD design was produced on the next mask set, GS154. After consultation with potential customers a full FSS-band specification was determined. Tthe resulting chip included HPD techniques such as 6µm spirals, FETs with 2µm poly vias, and Nichrome resistors (Figure 4). Close packing density techniques were also used, following simulations on LINMIC. The circuit was designed to be fully RFOW testable, by injection of an external LO signal into the oscillator port. Figure 7 shows a typical RFOW measurement. Figure 8 illustrates the performance of the chip when measured with the true dielectric resonator ocsillator, in a suitable test jig. Table 4 details the excellent correlation between specification and measured results, for a wide variety of measurements. Chip size was 4mm².

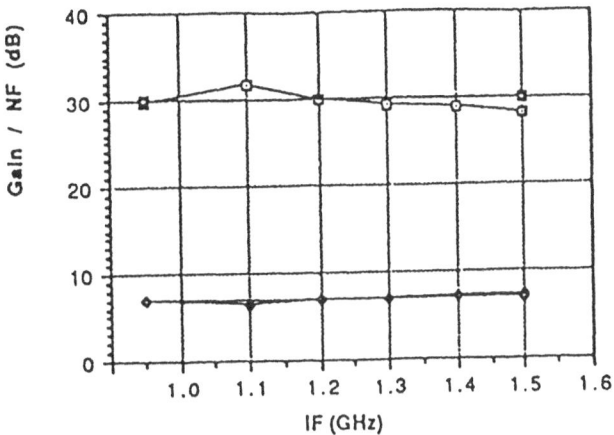

Fig. 7 : Test Results of GMMT DBS MESFET LNC MMIC
4 mm²-LNC, RFOW-measurement with 10 GHz LO

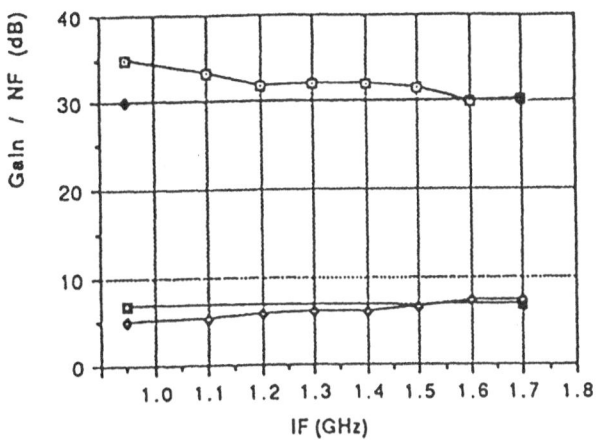

Fig. 8 : Test Results of GMMT DBS MESFET LNC MMIC
4 mm²-LNC, Jig measurement with 10 GHz DRO

In order to minimise production costs a further size reduction to less than 2mm²
was deemed necessary. Further additional HPD test components were thus
included on a further mask set, namely GS157. The primary components were
6µm stacked spirals, miniature active devices, and custom miniature active
devices incorporating self-bias and active loads. Miniature self-biased LNAs,
active loaded IFAs, close-packed mixers and oscillators were included on mask
set GS158.

After measurement and modelling of GS157 and GS158 components a complete HPD database was established. The final sub-2mm² LNC was designed on mask-set GS161. The chip is again illustrated in Figure 4. Preliminary results from this circuit are very encouraging. Figure 9 shows conversion gain and noise figure measured RFOW. Although performance appears slightly low in gain and high in noise figure, this is expected to improve significantly when jigtested with the true dielectric resonator oscillator. Further measurements of this circuit will be performed beyond the end of the COSMIC programme.

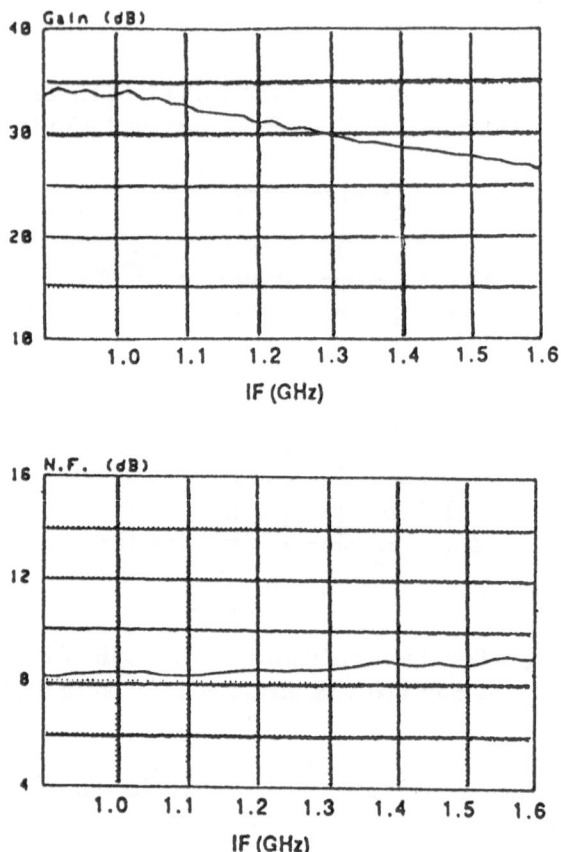

Fig. 9 : Test Results of GMMT DBS MESFET LNC MMIC
2 mm²-LNC, RFOW-measurement with 10 GHz LO

Parameter	Specification	Measured
RF range	10.95 - 11.7 GHz	10.95 - 11.7 GHz
IF range	0.95 - 1.7 GHz	0.95 - 1.7 GHz
LO frequency	10 GHz	10 GHz
Conversion gain	> 30 dB	> 30 dB
Noise figure (SSB)	< 7 dB	< 7.5 dB
LO-RF leakage	< - 20 dBm	- 42 dBm
LO-IF leakage	< 0 dBm	- 19 dBm
Image rejection	> 0 dB	12 dB
3rd order intercept point	> 10 dBm	16 dBm
LO phase noise @ 10 KHz offset	≤ - 70 dBc/Hz	-79 dBc/Hz
Current	≤ 150 mA	150 mA

Table 4 : Specification vs Measured Results for GS154 4mm² LNC

5. DBS HEMT LNA MMIC by Siemens

A HEMT low-noise preamplifier MMIC is necessary to fulfill the DBS reception requirements. A broadband HEMT LNA configuration has been developed which allows with help of the structured array design technique to cover a band of more than 3 GHz.

The circuit topology is the well-known combination of reactive matching and series inductive feedback principles. The 2-stage circuit is based on 4 x 45 µm x 0.25 µm-HEMTs. We use lumped element matching for the interstage and output circuits to reduce the chip area, but strip lines and passive component arrays at the input.

The HEMT LNA MMIC has been fabricated using the 0.25 µm SAG HEMT 40 process, airbridge interconnections and source viaholes. Figure 10 shows a chip photograph of the LNA, chip size is 1.55 x 0.9 mm².

The rf characterization has been done on-wafer. The test results together with the specification and simulation are summarized in Table 5.

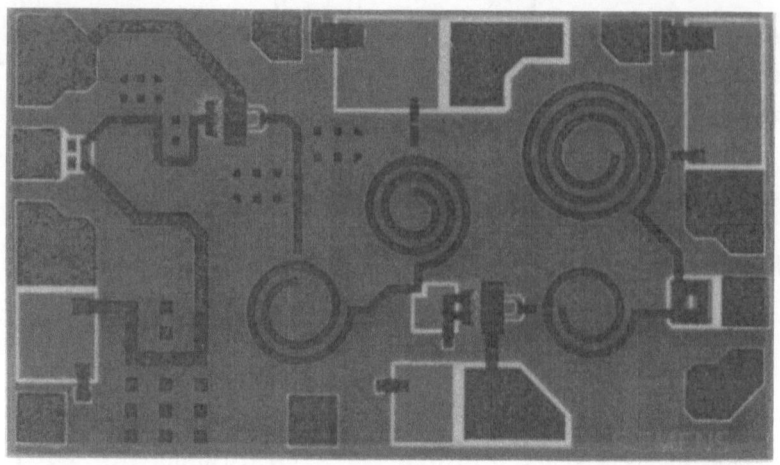

Fig. 10 : Chip Photograph of Siemens DBS HEMT LNA MMIC

	F	S11	S22	S21	F_{50}	U, I
	GHz	dB	dB	dB	dB	V, mA
Spec	11 - 12.5	10	10	15	1.5	3, 30
Sim.	11 - 12.5	14	12	17	1.2	3, 30
Test	11 - 12.5	14	15	19	1.25	2.5, 30

Table 5 : DBS HEMT LNA MMIC Results

The results show that the device is clearly in target and that the accordance between simulation and test is good. The achieved performance, NF = 1.25 dB and G = 19 dB, is remarkable compared to state of the art. The frequency dependend test results of Figure 11 show the expected broadband behaviour and that a good noise matching has been achieved.

The potential of the structured array approach is obvious by considering the results of this HEMT LNA optimized for the three subbands of Table 6.

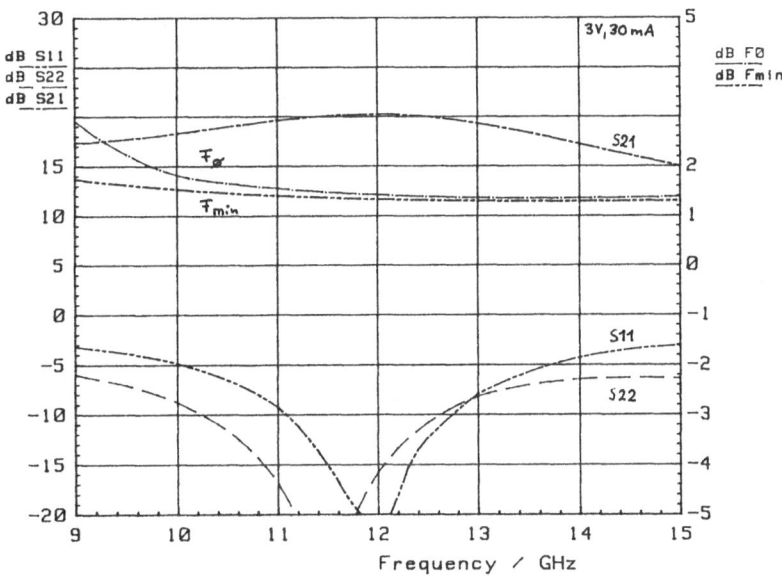

Fig. 11 : Test Results of Siemens DBS HEMT LNA MMIC

Circuit		1	2	3
Frequency	GHz	11 - 12.5	8.5 - 10.5	12.5 - 15
Gain	dB	19	20	18
NF	dB	1.25	1.2	1.3
S11	dB	> 10	> 10	> 10
S22	dB	> 10	> 10	> 10

Table 6 : DBS HEMT LNA MMIC Broadband Results

6. DBS LNC System Test by Siemens

In the end, a DBS converter system test has been performed by combining the two MMICs (HEMT LNA, MESFET LNC) together with an IR-filter according to the schematic block circuit of Figure 12. The system requirements (NF) mainly depending on satellite transmitter power and antenna diameter are summarized in Fig. 13. It shows for example that for the ASTRA reception (EIRP = 50 dBW, antenna $\varnothing$ = 0.8 m) a DBS system NF of 1.5 dB is sufficient to receive a high quality TV in central Europe. Table 7 contains the total receiver

NF as a function of the HEMT LNA MMIC performance assuming that the realized MESFET LNC delivers NF = 5.5 dB and G = 33 dB. As a result, using the fabricated HEMT LNA MMIC (NF = 1.25 dB, G = 19 dB), a converter NF of 1.4 dB can be realized which meets the target.

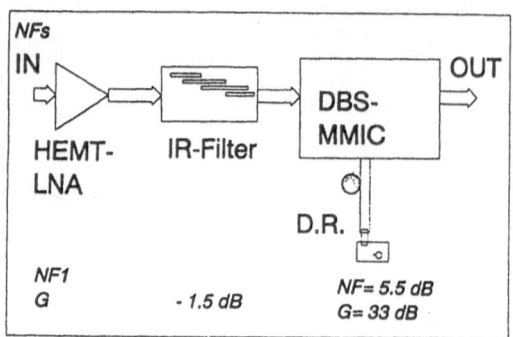

Fig. 12 : Schematic Diagram of DBS LNC MMIC Configuration

On the other hand, a request of TV receivers with NF = 1.2 dB max is a new target of some antenna producers to increase the reception coverage. This requires a HEMT LNA MMIC with NF = 1.1 dB max and G = 20 dB min, which is achievable using the available pseudomorphic SAG HEMT technology.

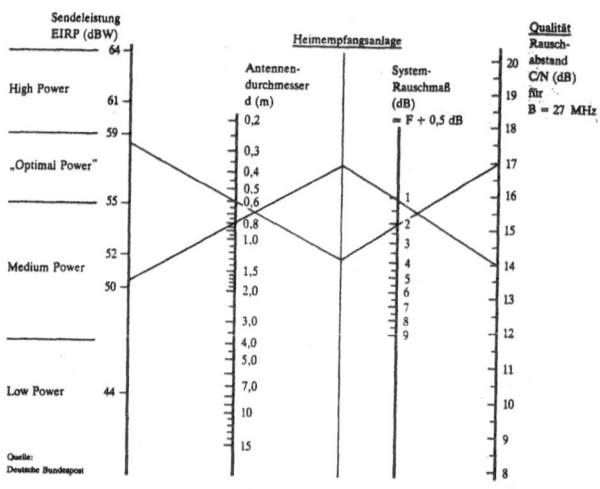

Fig. 13 : DBS Receiver System Requirements

HEMT LNA				LNC
NF1		G/Gain		NFs
0.8 dB		18 dB		1.024 dB
		19 dB		0.978 dB
		20 dB		0.942 dB
1.0 dB		18 dB		1.214 dB
		19 dB		1.170 dB
		20 dB		1.136 dB
1.1 dB		18 dB		1.309 dB
		19 dB		1.267 dB
		20 dB		1.233 dB
1.2 dB		18 dB		1.404 dB
	Test R.	19 dB		1.363 dB
		20 dB		1.330 dB
1.35 dB		18 dB		1.547 dB
		19 dB		1.507 dB
		20 dB		1.475 dB
1.5 dB		18 dB		1.691 dB
		19 dB		1.652 dB
		20 dB		1.621 dB

Tab. 7 :   Siemens DBS 2-Chip LNC
(HEMT LNA + MESFET LNC)
Specification and Test Results

2.6 Point to Multipoint Communication

F G Pedraja
Alcatel SESA, Madrid, Spain

1. Introduction

The main field of application of point to multipoint (PTMP) systems is to provide access to both public and private networks (PSTN, PDN, ..) particularly for remote subscribers. By means of PTMP systems the network service area may be extended to cover both distant and scatterd subscribers locations.

These remote subscribers, in a similar way to the city subscribers, are offered the full range of services by the particular public or private network. Subscribers have access to these services by means of the various standardised user network interfaces (2-wire loop, data, ...).

PTMP applications in the Metropolitan and urban environment are mainly for the provision of new data services for business subscribers and for the extension of ISDN services to local subscribers.

2. System Description

2.1 PTMP Systems at L-Band

Radio is often the ideal way of obtaining communications at a low cost and almost independently of distance, and difficult topography. Moreover, a small number of sites are required for these installations, thus facilitating rapid implementation and minimising maintenance requirements of the systems.

The frequency bands below 3 GHz (L-Band) are particularly suitable for the extension of telecommunications service to distant rural and suburban subscribers.

PTMP systems provide standard network interfaces and transparently connect subscribers to the appropriate network node (local, switch, ...). These systems allow a service to be connected to a number of subscribers ranging from a few users to several hundred, and over a wide range of distances.

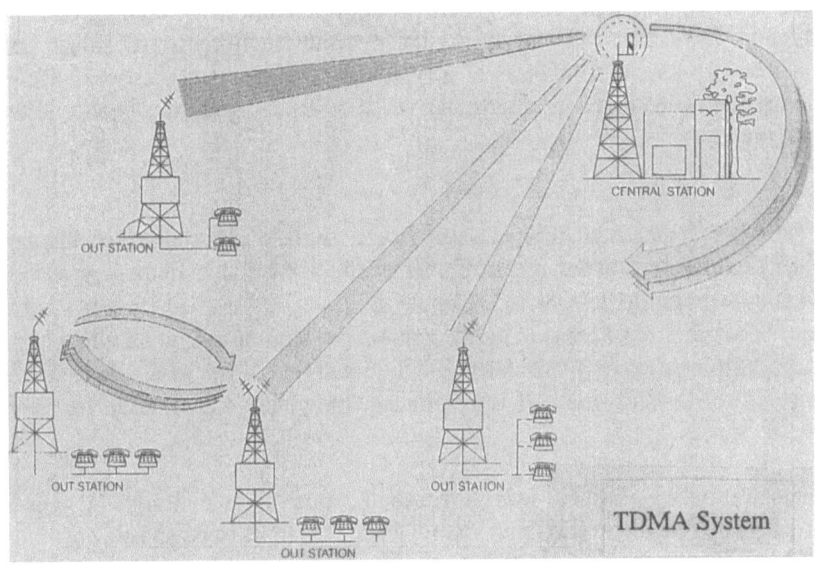

Fig. 1 : PTMP L-Band TDMA System

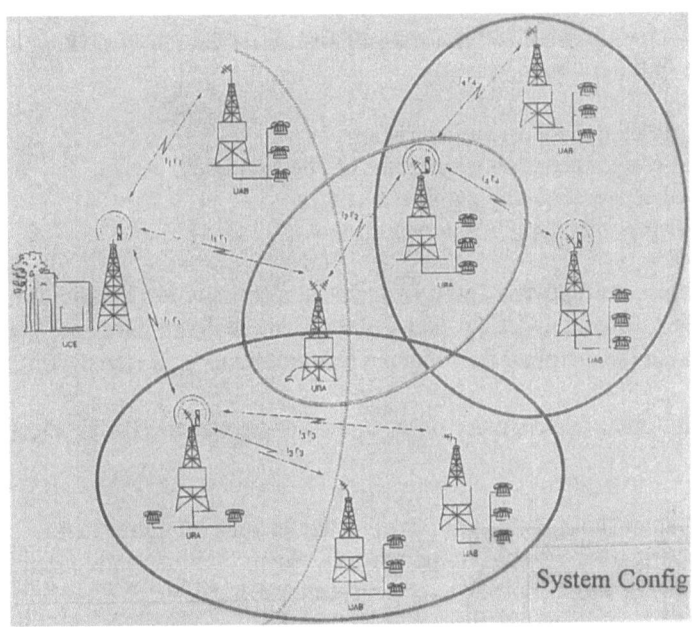

Fig. 2 : PTMP L-Band System Configuration

Point to multipoint systems are generally, but not necessarily, configured as Pre-Assigned Systems as Demand Assigned Multiple Access (DAMA) Radio Systems. The essential features of a typical PTMP Demand Assigned Multiple Access Radio Systems are: efficient use of radio spectrum, concentration, transparency.

Concentration means that N subscribers can share n channels (N being larger than n), allowing a better use to be made of the available frequency spectrum and at a lower equipment cost. The term "multi-access" derives from the fact that every subscriber has access to every channel (instead of a rigid assignment as in most multiplex systems). When a call is initiated one of the available channels is allocated to it. When the call is terminated, the channel is released for another call.

Concentration requires the use of distributed intelligent control which in turn allows many other operation and maintenance functions to be added.

Transparency means that the exchange and the telephone communicate with each other without being aware of the radio link.

Alcatel SESA has developed a point to multipoint digital radio system, basically designed to serve peripherals layers of the different existing communications networks in the following cases:

- Lacking of a basic infrastrucutre.
- Laying unfeasibility by means of physical lines.
- Absence of nearby public exchanges.
- Temporary supply of services.

The system uses time-division multiple access techniques (TDMA). In this type of systems, communication between a Central Station and a number of peripheral units, distributed throughout a particular area, is effected via radio.

It is a full-duplex system using only one pair of frequencies for the connection of stations.

The Exchange Unit is fitted with one antenna located on a risen surface to permit visibility with the associated peripherals. Each remote station sends to the antenna one or more short signals (synchronized traffic bursts); so when a burst reaches the Exchange, it is the only signal present, and the collision with signals coming from other remote stations is thus avoided (see figures 1 and 2).

Generally the structure of this type of system is based on three units:

- Exchange Unit (UCE)
- Subscriber Unit (UAB)
- Repeater Unit (URA)

The UCE handles the interconnection with the Exchange, and the accurate concentration required for assignment of the channels available in the system. The UAB is responsible for connection of subscribers loops, while the URA acts a radio-signal distributor in both senses and can handle its own subscribers. In the UCE-UAB sense, the transmission is continuous in a TDM (Time Division Multiplexing) format; in the UAB-UCE sense, the transmission is in the form of TDMA burst.

The SMD-30/1.5-2.4 developed by Alcatel-SESA is a point to multipoint digital radio system with TDM-TDMA link, with concentration between subscribers and channels, and with low demand (DAMA) or permanent assignment, perfectly suitable for the above mentioned scenarios. It uses two frequency bands: 1.5/2.4 GHz, which provide it with greater versatility under band-congestion conditions.

The radiolectric and system characteristics have been summarized in the following tables.

	SMD-30/1.5	SMD-30/2.4
Frequency Band	1427-1535 MHz	2300-2500 MHz
TX-RX Separation	≥ 40 MHz	≥ 65 MHz
Channelling	2 MHz	2 MHz
*NFD at 2 MHz	25 dB	25 dB
Channelling step	250 KHz	500 KHz
TX output power	+ 27 dBm	+ 30 dBm
	+ 18 dBm	+ 18 dBm
Modulation type	4 QAM	4 QAM
Demodulation type	Central: Differential	Central: Differential
	Remote: Coherent	Remote: Coherent
Receiver threshold	Central: - 92 dBm	Central: - 92 dBm
BER = 10^{-3}	Remote: - 94 dBm	Remote: - 94 dBm

Table 1 : PTMP L-Band System Description

Number of channels	30
Max number of subscribers	512
Max number of UABs	64
Max number of UABs + URAs	64
Max number of subscribers per URA/UAB	64
Max number of subscribers in one UAB 16 (this includes battery charger in the same container)	16
From central to remote	TDM
From remote to central	TDMA

Table 2: L-Band System Data and Access Method

2.2   PTMP System at K-Band

For urban systems normally the frequency band is higher (band Ku, K for instance) and the radio part of the equipment has to be near the antenna and in an outdoor container.

It is widely recognized that the future path for public communication networks is an evolution towards an universal ISDN network. The ISDN services will be introduced first in the urban areas and for business customers.

A good solution for providing ISDN services quickly, especially where provisional installations are needed, is multiaccess systems. The system architecture is similar to that of L band systems, (normally repeater stations are not required).

The radioelectric and system characteristics can be seen in the following tables.

Frequency Band	18 GHz
TX-RX Separation	1010 MHz
Channelling	7 MHz
TX output power	+ 17 dBm
Modulation type	4 QAM
Demodulation type	Central: Differential
	Remote: Coherent
Receiver threshold	Central: - 87 dBm
BER = 10^{-3}	Remote: - 85 dBm

Table 3 : PTMP K-Band System Description

Number of channels	30
Max number of subscribers (in alternative)	156
ISDN (DAMA)	117
Kbps (dedicated access)	30
Max number of outstations	
From central to remote	TDM
From remote to central	TDMA

Table 4 : K-Band System Data and Access Method

3. L-Band PTMP MMICs and System Test (Alcatel SESA, Alcatel-Telettra, Siemens)

For this kind of application several circuits with different approaches have been realized in a very close collaboration between Alcatel SESA, Alcatel Telettra and Siemens. The objective was the design of a Low Noise Converter working at the 1.5 + 2.5 GHz freqency band. This Low Noise Converter should meet the required performances needed for the peripheral and central stations of the PTMP systems. The targets for the design were: 20dB of gain, 2dB of NF for central station devices and 4dB of NF for peripheral station devices. These devices were specified as follows:

Parameter	LNC
Input Frequency	1.5 + 2.5 GHz
Output Freqeuncy	≤ 100 MHz
Zin	50 Ω
Zout	≤ 2 KΩ
Gain	≥ 15 dB
NF	≤ 2 dB (4 dB)

Table 5 : Specification of L-Band LNC

Among the designed circuits there are three different versions of LNA (600μ, 900μ, and shrinked version), three different versions of mixer circuits (MIX 1, MIX 2, and SSB MIX) and also combinations of LNA and mixer circuits on a single chip to obtain a complete front-end module.

The standard technology DIOM 15 and the P Buried Layer Channel (DIOM 15P) have been employed. The performances of the second type devices are slightly better than the former ones. The devices have been designed by Alcatel SESA (Telettra Spain), have been produced by Siemens and mounted and assembled by Alcatel Telettra.

Typical results of the Low Noise Converter are summarized in the following table:

	LNA-MIX 1
LNA Gain	≥ 20 dB
LNA NF	≤ 1.7 dB
LNA S11	≤ -10 dB
Mixer1 Gain	≥ 7 dB
Mixer 1 NF	≤ 14 dB
LNC Gain	≥ 26 dB
LNC NF	≤ 2.5 dB
LO-IF Leakage	≥ 30 dB

Table 6 : L-Band LNC Test Results

These results are in good agreement with the previously simulated results and also with the project goals. The deviaton of the NF (2.5 dB versus 2 dB) is due mainly to the NF of the mixer circuits, although its performances are in good agreement with the expected values for these kind of active mixers, accordingly to literature.

For the LNA + MIXER 2 device, two external RF chokes are necessary for biasing. Its characteristics are similar to LNA + MIXER 1. These results are also well in accordance with the foreseen values and comply with the system necessities. For the LNA + SSB mixer device the conversion gain is greater than 23 dB and the image rejection better than 15 dB.

In Figures 3 and 4 a photograph of the LNC Type 1 and the measurement results of the device are shown. Typical values of 26 dB of gain and 2.5 dB of NF have been achieved.

The system assessment of the devices has also been done, the key parameters of the system that can be affected by the LNC performances have been checked. For this purpose the device under test have been included in a real operating system and all the ordinary test done in the production line have been realized. Among these parameters the following ones were considered the most relevant:

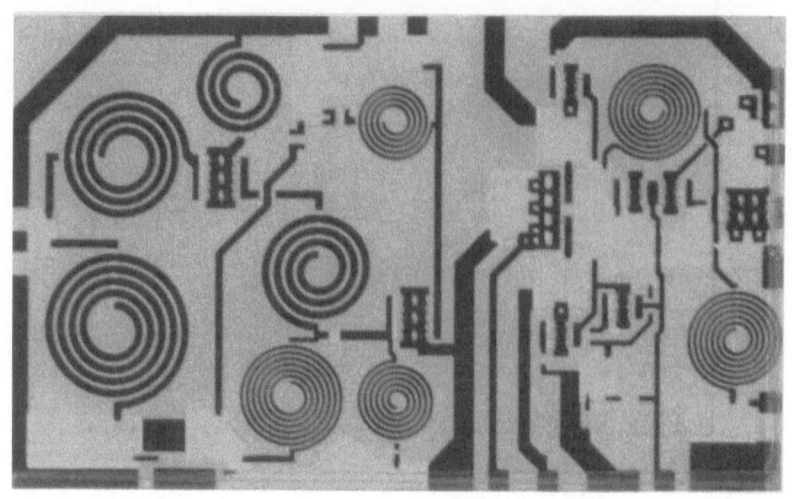

Fig. 3 : L-Band PTMP LNC MMIC Type 1

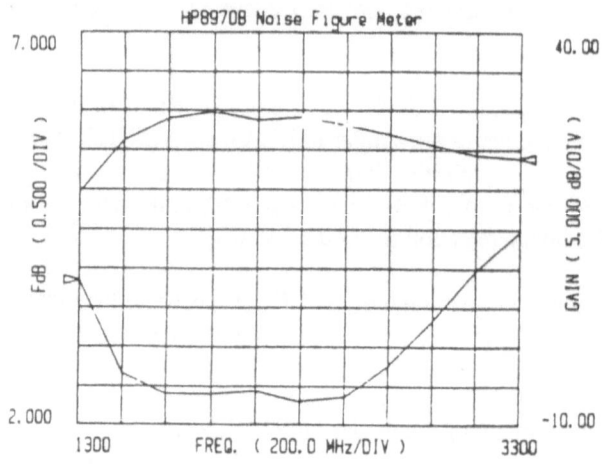

Fig. 4 : L-Band PTMP LNC MMIC Test Results

- BER performance versus received signal power level
- Receiver input level range
- Interference sensitivity for adjacent channel case
- VSWR
- Environmental conditions.

All the measurements performed on the above parameters indicate that the circuits designed fulfil the application requirements; the main improvements obtained, when the results are compared with the results of our standard equipment, are in BER versus received level, with 2 dB of threshold (BER = 10^{-3}) improvement. Additional advantages are size and weight.

Some of the designed chips in ESPRIT 5018 are already being considered to be employed in a new PTMP system that is currently under development in Alcatel.

4. K-Band PTMP MMICs and System Test (Alcatel SESA, Alcatel Telettra, GEC-Marconi, Siemens)

For this application two different devices have been considered. Both of them are LNC, one for the peripheral stations of the system with low cost and relatively moderate performance and another one with higher requirements for the central stations. The most important design goals for these devices were noise figure and gain.

Two different technologie have been employed: A HEMT LNA, produced at the Siemens Foundry, for the central device and a MESFET LNA and MESFET mixer, produced at the GEC Marconi Foundry, for the peripheral device.

The main LNA devices specification parameters for this application were:

Parameter	Peripheral LNC	Central LNC
Input Frequency	17.7 ÷ 19.7 GHz	17.7 ÷ 19.7 GHz
Output Frequency	≤ 500 MHz	≤ 500 MHz
Zin	50 Ω	50 Ω
Zout	≤2 kΩ	≤2 kΩ
Gain	≥ 10 dB	≥ 15 dB
NF	≤ 8 dB	≤ 4 dB
Iop	≤ 50 mA	≤ 50 mA

Table 7 : Specification of K-Band LNCs

Two chip solutions were considered: HEMT or MESFET LNA and MESFET mixer.

Several HEMT LNA prototypes have been fabricated, mounted on suitable test fixtures and characterized. Results are in good accordance with simulations for the three different types of LNAs (types A, B designed by Alcatel SESA and type C designed by Siemens). The process was 0.25 μm SAG HEMT 40. Fig. 5 shows one of the HEMT LNA MMICs.

In the following table a summary of the results obtained in the design type B and in the more risky design C is shown.

	f (GHz)	S11	S22	S21	NF (dB)
Type B	17.7 ± 19.7	6	10	13	3
Type C	17.7 ± 19.7	8	10	17	2.4

Table 8 : Test Results of HEMT LNA MMICs

In comparison to that the realized MESFET LNA MMIC showed a gain of ≥ 10 dB and NF ≤ 8 dB. Again Fig. 6 contains a chip photograph.

The passive mixer implemented with MESFET technology showed the following performance at the considered K-band frequencies: conversion loss of 13 dB, LO matching of -10 dB, RF matching of - 5 dB, LO-IF isolation of ≥ 40 dB.

In the following, the HEMT LNA MMIC and passive MESFET mixer MMIC have been configured to an LNC with an NF lower than 4 dB and the desired gain, i.e. complying with the objectives for the central PTMP station.

The selected combination has been considered for use in a system with the following characteristics:

Frequency Band	17.7 + 19.7 GHz
System Capacity	8 Mbps
Modulation Type	4 QAM
Demodulation Type	Coherent
Channeling (IF Bandwidth)	7 MHz

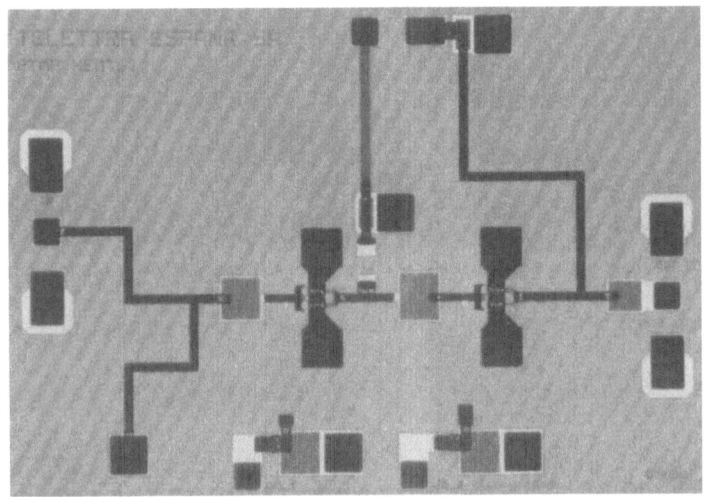

Fig. 5 : K-Band PTMP HEMT LNA MMIC
(Design Alcatel SESA, Foundry Siemens)

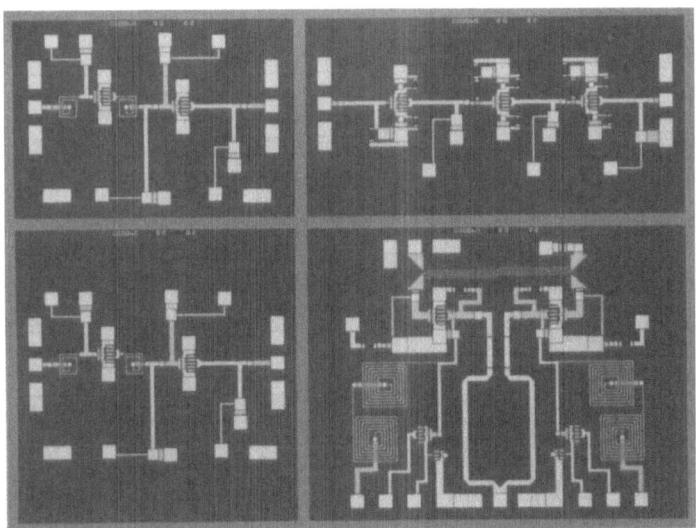

Fig. 6 : K-Band PTMP MESFET LNA & Mixer MMIC
(Design Alcatel SESA, Foundry GMMT)

With the above assumptions, the key parameters that are affected by the LNC performances, together with the related values that can be achieved using these device, are:

-      BER versus received level: - 87 dBm for BER = 10^{-3}

-      Input level range:
       From -35 dBm to the BER 10^{-3} threshold. This value permits coverages of up to 10 Km in type k regions (CCIR) for unavailability time due to rain lower than 30 minutes/year when typical antennae and transmission power levels are chosen.

2.7 Optical Communication

J L Conesa and J M Hernandez, Telefonica Madrid, Spain
M Salazar-Palma, Polytechnic Madrid, Spain

1. Introduction

High speed optical transmission has enormous bandwith capability, low losses and high immunity to EMI. A number of different systems have been developed so far for optical communications. Until very recently, the only system used was the intensity modulation direct detection known as IM-DD. This system lead to simple and robust designs that are relatively inexpensive. But due to the limitation of the electronic components it only uses an insignificant part of the bandwidth of the fibre.

Recent advances in semiconductor laser have demonstrated the coherent systems to be feasible, at reasonable cost and space. Although the idea is not new, the outstanding performances required for the lasers make them impractical until now. Coherent systems are more complicated than direct detection systems, but offer two very important advantages, a significant increase in sensitivity and the possibility of high density multichannel.

This section describes the development of different GaAs based transimpedance amplifiers used in optical transmission receivers, in direct detection systems (155 MBit/s to 2.5 Gbit/s) or coherent detection systems (155 MBit/s or 622 MBit/s), as well as for a high data rate LAN (Local Area Network) system.

The work includes different contributions from Telefonica, FORTH, PT Madrid, U.Rome and GEC-Marconi carried out under the ESPRIT project 5018.

2. System Description

2.1 Direct Detection System

In direct detection systems, the information is carried on the power of the optical signal. A typical system consists of a laser diode transmitter that is turn on and off depending whether a one or a zero is transmitted, and some equalizer and temperature control circuitry, and a receiver where a photodetector senses the power from the fibre and amplifies and decodes that signal in the electrical domain.

The actual transfer rate of these systems is limited by the electronic circuits involved, and far below the possibilities of the fibre. The theoretical sensitivity of those systems is very high, but it deteriorates rapidly with the noise from the photodetector and the amplifier.

Possibility also exists for building multichannel systems using multiple optical wavelengths, but they have to use optical filters or demultiplexers. The frequency selectivity of the optical filters is not as good as the electrical ones, so the density is lower in these systems.

A IM-DD system that has been implemented during the project based on a InGaAsP/InP DFB 1.55 µm wavelength laser diode (type FLD 150 F2RH from Fujitsu). This laser can be operated up to 6 GHz, covering in excess the range specified for the circuit. The laser has been mounted on a KAP-10 module (from Radians Innova). This module is equipped with a Peltier cooler and a thermistor to enable temperature stabilization, besides of a collimating lens to focus the light beam inside the fibre. The laser module is connected to a temperature controller to stabilize operating temperature at a fixed value, and the laser is biased by a precision current source with a DC level. The modulation index of the signal is about 80%. The receiver block diagram of this system is shown in figure 1.

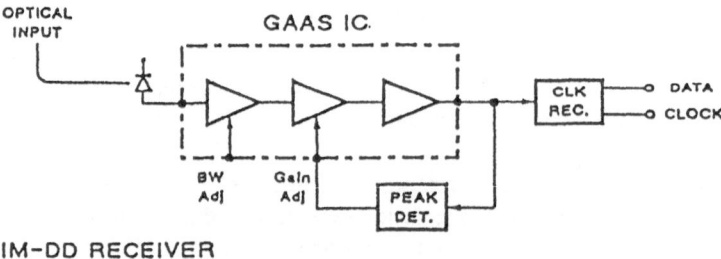

Figure 1 : Block diagram on an IM-DD receiver

2.2   Coherent System

For the coherent systems the information modulates one of the parameters of the optical wave, the amplitude (ASK), the phase (PSK) or the frequency (FSK). To recover the data, the optical input signal is mixed with another locally generated by a laser. Depending on the frequency of the local oscillator the system can be homodyne or heterodyne. It is homodyne when the local oscillator is of the same frequency than the transmitter, so the spectrum is translated directly to baseband. In the heterodyne case, the frequencies of the laser are different, and the spectrum is translated to an intermediate frequency. The basic principles for this kind of receivers are the same used in radio.

By extension all systems that include a local oscillator (LO) are normally called coherent, although there are systems that do not detect the phase of the signal and are not coherent in a strict way.

The homodyne systems have very strong requirement on the lasers used, so semiconductor lasers with the present technology are not well suited. In the case of the heterodyne systems the required bandwidth of the electronic circuits is significantly larger than for homodyne systems, while the theoretical sensitivity is 3 dB lower. The main advantage of the heterodyne is that the control of the phase and frequency of the laser is not so critical, although also has to be very well controlled.

The two reasons to use the coherent systems are the sensitivity, and the high density multichannel. Although the theoretical sensitivity of the direct detecton systems is better than most of the coherent systems, in real life the measured sensitivity of coherent systems is better by 10 to 20 dB. The reason for this increase is that the power of the signal produced in the mix is proportional to the square root of the product of the powers of the incoming signal and the local oscillator. Thus by increasing the local oscillator power, the power of the input signal is also increased.

The block diagram of a basic coherent receiver system is presented in figure 2. The implementation of coherent demonstrators has involved the development of specific electronic circuitry. Besides, a number of optical parts like couplers and polarization controllers are necessary to implement this type of systems. The increase in complexitiy is compensated by a better sensitivity, and the fact of being suitable for multichannel applications. Lasers used to build the demonstrators are LD 5171 type DFB tunable laser from GMMT. System performance is highly dependent on the linewidth and phase noise of the transmitter and local oscillator lasers. Another important factor is the relation between the LO output power and the i_{enc} of the receiver.

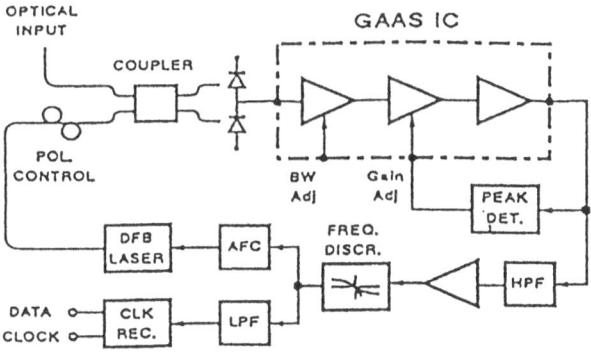

Figure 2 : Block diagram of a basic coherent receiver

3.   Transimpedance Amplifier MMICs

The circuit is composed of three sections: a transimpedance preamplifier, a controllable gain section, and 50 Ω output buffer. An additional buffer provides also off monitor output, as it is shown in figure 3 where dot outlined blocks are low-frequency external loops. DC coupling has been used along the circuit, and self-biasing techniques have been applied to each one of the sections. The power consumption of the circuit is about 2.4 W, thermal aspects have been taken into account to ensure an adequate reliability. Actual chip size is 3.9 mm by 3.1 mm. A microphotograph of the circuit is shown in figure 4.

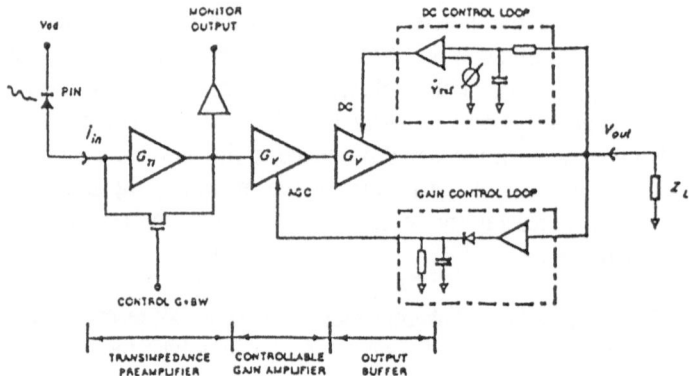

Figure 3 : Block schematic of the circuit

A cascode based transimpedance preamplifier topology has been developed achieving a good compromise between noise, bandwidth and gain characteristics, while obtaining at the same time the required bias point stability. The usual feedback resistor R_F has been substituted by a 25 μm gate width MESFET working in the linear region. This allows to perform a bandwidth adjustment while maintaining constant the $G \cdot BW$ product. The feedback loop impedance was designed to allow bandwidth settings in the range 120 MHz to 2.4 GHz (second run), with an equivalent resistance of 8.7 KΩ and 820 Ω, respectively. The bandwidth control margin is shown in figure 5, where transimpedance gain at the monitor ouput for $C_d = 0.6$ pF is plotted.

Feedback control in the controllable gain section has been achieved by mean of MESFET transistors used as variable resistors (3 x 100 μm). In this way, a control margin in excess of 40 dB has been simulated and measured, as can be seen in figure 6. The design was realized using GMMT foundry facilities.

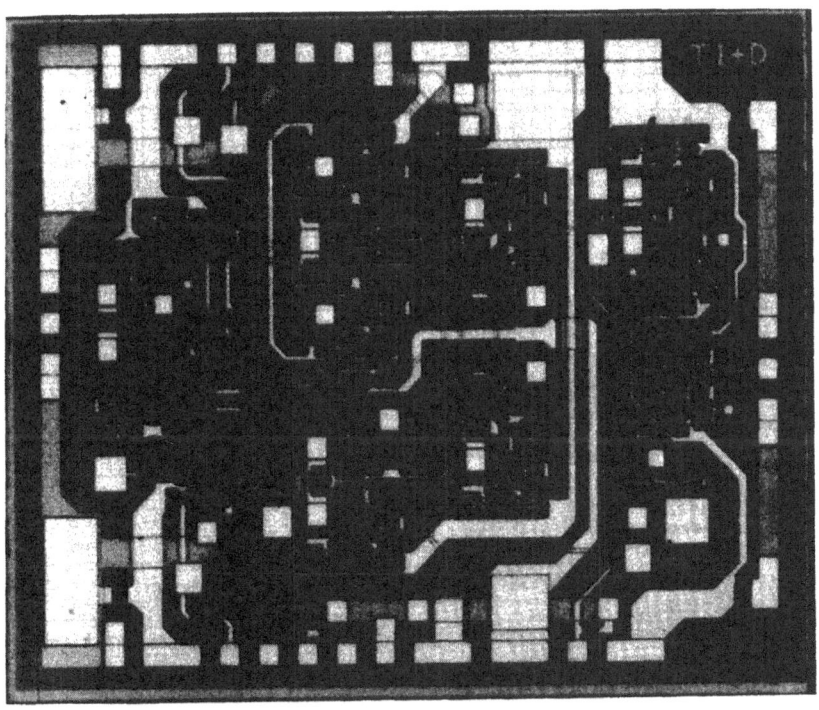

Figure 4 : Microphotograph of the circuit

Specifications of the circuit have evolved at the same time the project was being developed. The most relevant change from initial requirements has been the maximum adjustable bandwidth that has been increased from 1.9 GHz to 2.4 GHz. Less importance has been given to the available gain that has been slightly reduced. Control margins of both G · BW and overall gain have been achieved as it was initially requested, and state of the art performances in noise values have been obtained.

Significant experience has been extracted from the first run, and applied to the redesign, as it is shown in Table 1, where a close correlation between simulation and results in the second run can be observed. Noise data in this table represent the average value in the specified band of interet with $C_d = 0.6$ pF. The value of i_{enc} obtained without loading the input of the amplifier with the photodiode would be lower, as absolute noise currents at high frequencies considering input load capacitance are significantly higher.

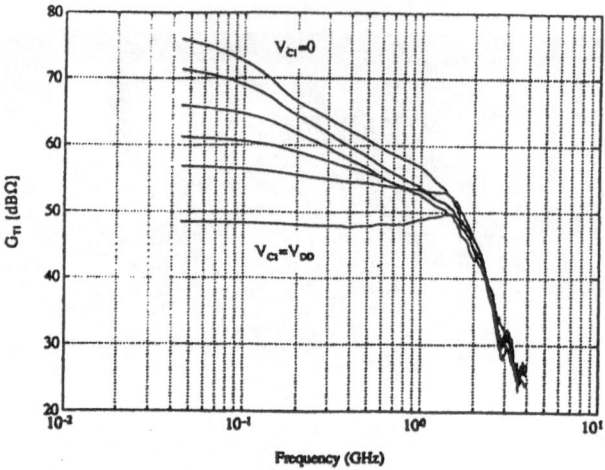

Figure 5 : Transimpedance gain at monitor output with $C_d = 0.6$ pF

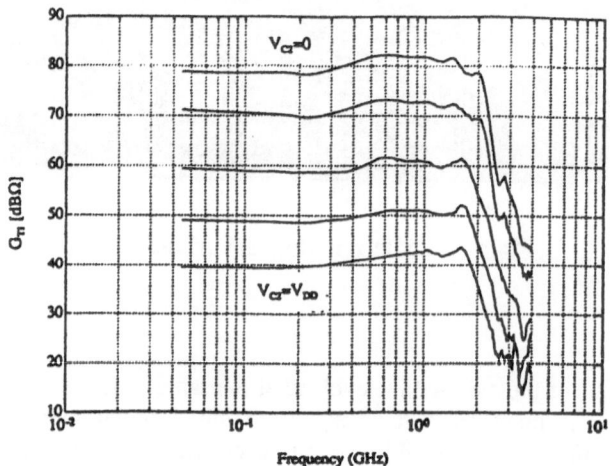

Figure 6 : G_{TI} at the output showing gain control margin

WAFER RUN	REF	BW [GHz]	G_{TI} [dBΩ]	i_{enc} @ BW = 1.9GHz [pA/√Hz]
I	SIMULATION	DC to .12-2.0	121-98	< 8.2
	TARGET	DC to .12-1.9	106-85	< 9
	ACHIEVED	DC to .12-1.75	108-87	6.9
II	SIMULATION	DC to .12-2.5	110-82	6.5
	TARGET	DC to .12-2.4	106-78	< 8
	ACHIEVED	DC to .12-2.4	107-80	6.6

Table 1 : Comparison of targets, simulation values and measured results

A number of other transimpedance amplifiers were investigated: a cascode plus differential stage structure, a three inverter chain with reactive feedback, a cascode with peaking inductors, a cascode with active feedback and AGC and, finally, a cascode with active feedback. All five designs were DC coupled.

The cascode plus differential stage structure was chosen for the final design. Some improvements (a redesign of the bias network including active loads with shunting capacitors) were added. Figure 7 shows the schematic of the chip.

Figure 8 shows the chip. A number of samples were DC and AC (S-parameters, transimpedance gain, equivalent noise current) characterized on-wafer, on-chip and on-carrier at different temperatures. Optical/Electrical measurements were also performed; figure 9 shows the corresponding test fixture. Table 2 gives a resume of the desired, simulated and measured result values. Figure 10 shows in detail some of the measured results. An excellent performance has been obtained for this final version of the transimpedance amplifier.

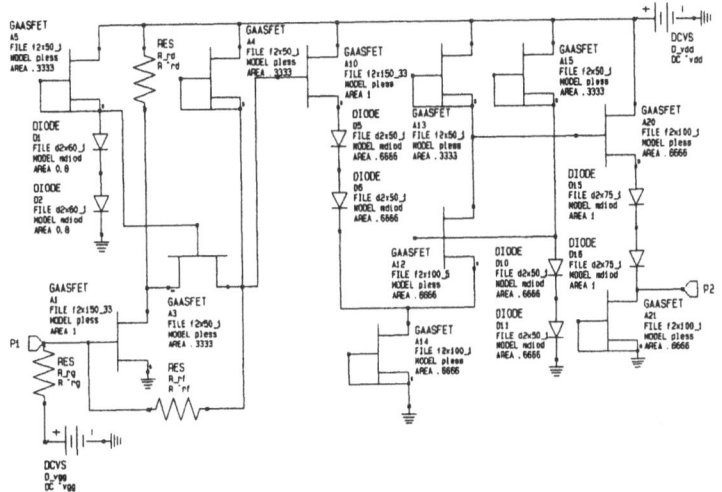

Fig. 7 : Schematic of the Transimpedance Amplifier Final Design

4. System Test

System measurements performed on the different demonstrators that have been described are summarzied in table 3. In order to verify results, sensitivity values presented are those required to achieve a BER = 10^{-9}, and have been obtained using a 2^7-1 pseudorandom bit-stream.

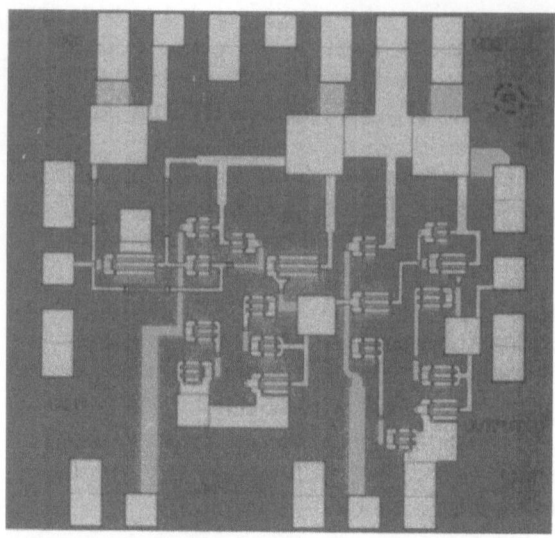

Fig. 8

Transimpedance Amplifier

Chip

TRANSIMPEDANCE AMPLIFIER	TARGET	SIMULATED VALUES	MEASURED RESULTS
Transimpedance Gain (dBΩ)	69,5	70,7	69,6
Bandwidth (DC to ...) (GHz)	> 2.0	2.6	2.4
Equivalent Input Noise Current (pA/√Hz)	≤ 6	5,2	5,5
Output VSWR	--	< 1,5	< 1,5
Maximum Output Voltage (v)	1	--	1,5
Electrical Dynamic Range (dB)	50	--	50,5
DC Power (mW)	≤ 1000	930	913
OPTICAL RECEIVER	TARGET	SIMULATED VALUES	MEASURED RESULTS
Global Responsibity (A/W, dB)	--	--	31,8
Bandwidth (DC to ...) (GHz)	1,9	2,1	2,01
Optical Dynamic Range (dB)	25	--	25,3

Table 2 : Comparison of targets, simulation and test results

SYSTEM TYPE	IM-DD			COHERENT		
BIT-RATE (MBit/s)	155	622	2488	155 NRZ	155 AMI	622 SF
SENSITIVITY (dBm)	- 36.1	-31.5	- 22.4	- 41.7	- 38.2	- 34.6

Table 3 : Measured sensitivity for BER = 10^{-9} in several system demonstrators

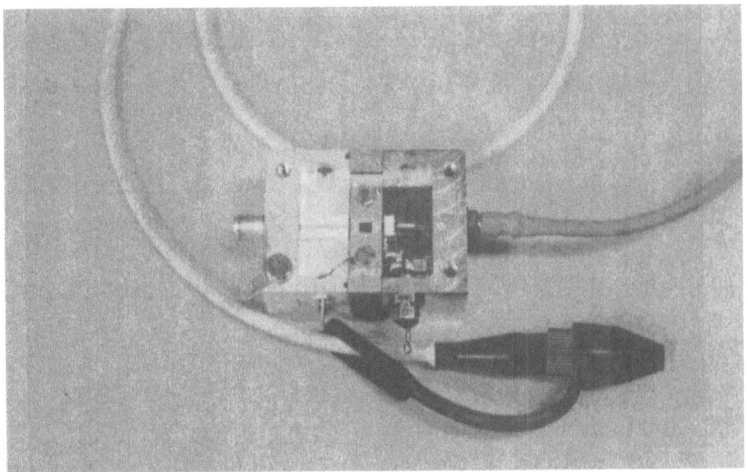

Fig. 9 : Optical Receiver Set Up

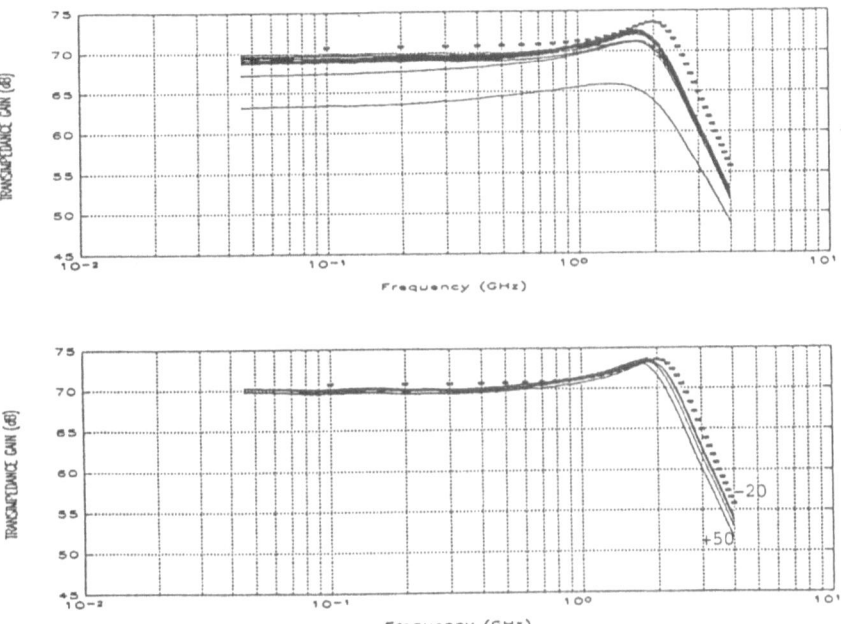

Fig. 10 : Measured results
a) Transimpedance Gain (18 samples measured on-wafer)
b) Temperature Dependence of the Transimpedance Gain

3.1   Advanced HFETs and Pseudomorphic HFETs

S. Gourrier

Philips Microwave Limeil, F-94453 Limeil Brevannes Cedex, France

1. **INTRODUCTION : the HFET concept and its situtation within ES-PRIT**

    HFETs (HEMTs, TEGFETs, MODFETs...) and pseudomorphic HFETs (PM HFETs) are now key components for specific applications in millimetre wave systems or in very low noise receivers.

    Although similar in many respects to MESFETs, their fabrication requires specific knowledge and expertise especially in the area of epitaxial growth and sub-half micron lithography.

    The concept (1) was pioneered in 1980 in Japan by Fujitsu (Mori and Ando), in Europe at Thomson (1), and in the US at Bell Laboratories and Rockwell.

    The principle is based on the modulation doping of heterostructures (1) (see figure 1).

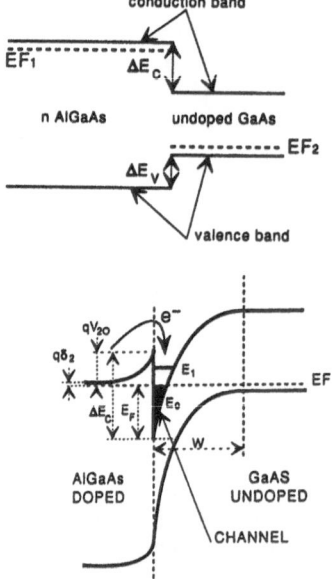

Figure 1 : Band structure of a (doped) GaAlAs (undoped)
GaAs heterostructure before (top) and after (bottom) equilibrium

Considering an heterostructure formed by a wide bandgap material (e.g. GaAlAs) on top of a smaller bandgap material (e.g. GaAs), if the wide bandgap material is doped with donors, electrons will be transferred to the smaller bandgap material (which is undoped) until an equilibrium is reached. Assuming the Fermi level in the wide bandgap material pinned on the donor level a potential well appears at the heterointerface. The electrons emitted from the donors of the wide bandgap material will be confined on one direction in this potential well. A 2-dimensional electron gas is formed. The main characteristics of the device can be deduced from this configuration :

▶ The electrons move in an undoped material and therefore have a high low field mobility. At room temperature, the mobility is limited by phonon scattering and is therefore comparable to that in high purity materials. At low temperature the mobility is limited by Coulomb scattering from ionized impurities and can reach extremely high values (above 10^7 cm^2/V x 1) if a "spacer" layer is used to separate the donor layer from the electron channel.

▶ The electrons are confined along the vertical dimension. This leads to a better control of the channel charge density by the applied gate voltage (see figure 2). The presence of the confinement even relatively close to pinch-off shows that the device will have gain at low current and is therefore suitable for low noise operation. Increasing the electron confinement will therefore be beneficial and this is at the origin of the pseudomorphic HFET concept.

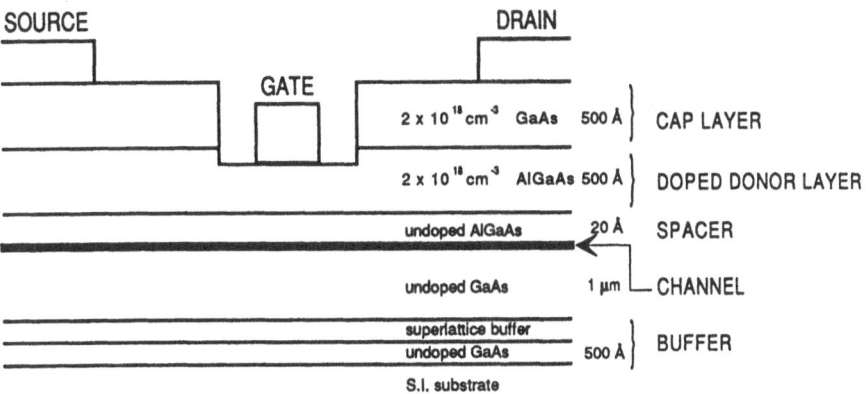

Figure 2 : Typical structure of a HFET device

Figure 2 shows a typical HFET structure which appears very similar to a MESFET.

Three main structures have been developed :

- "conventional" HFET GaAlAs/GaAs
- pseudomorphic HFET GaAlAs/GaInAs/GaAs where a thin $Ga_{1-x}In_xAs$ layer is inserted between the donor layer and the GaAs layer. This improves the confinement at the heterointerfaces and leads to higher electron velocities because of the more favourable properties of GaInAs.
- InP based HFET with a $Al_{52}In_{48}As$ donor layer on top of a $Ga_{0.47}In_{0.53}As$ undoped layer. The confinement and electron dynamics in this structure are superior to those of the pseudomorphic device.

At the end of the 80's the main reported achievements came from Japon and above all from the US because of the very strong support from the military Administration to HFET development in the US. The European industry recognized very soon the importance of HFET technology as one of the key enabling technologies. The need for collaborative schemes in this area led to the definition of sizeable workpackages within several ESPRIT programmes. A key aspect in these actions was the coherent approach between material suppliers, foundries, modellers, circuit designers and end users.

At the time of writing (october 93), it is clear that the competitive position of the European III-IV industry in this area has strongly improved compared with the situation at the end of the 80's. Several state of the art results have been obtained in terms of MMICs and discrete low noise HEMTs are now produced with competitive performance and price in Europe.

This chapter will review the main achievements obtained during several ESPRIT programmes. Some are now finished (GIANTS, COSMIC), some are nearing completion (AIMS) and others are underway (MANPOWER, CLASSIC). It is divided along the main activity areas : materials, discrete devices (technology and performance) and MMICs.

The presentation will also follow the main evolutions of the HFET development work within ESPRIT during the last few years. The dominant and most widely used device is now the pseudomorphic GaAlAs/GaInAs/GaAs (PM HFET) which clearly brings a definite advantage without any detrimental effect on material growth, technology and reliability. The InP based HFET has a greater performance potential but very few studies have been devoted to it within ESPRIT programmes.

In terms of applications and device type, as in the rest of the world, the developments resulted in a first phase in the definition of technologies for low noise high f_T HFETs (GIANTS, COSMIC) validated by a number of state of the art MMIC demonstrators (AIMS). The present phase is now devoted to the development of HFET for power and highly non linear applications (MANPOWER, CLASSIC).

The paper presents the most significant progress in epitaxial materials, devices and demonstrators. Further developments are also outlined.

2. PROGRESS IN EPITAXIAL MATERIALS AND DEVICES WITHIN ESPRIT

▶ **2.1 - Key parameters of a HFET**
The development efforts aim at the optimization of several key parameters of the device :

- intrinsic current gain cut-off frequency f_T :
$$f_T = \frac{gm}{2\pi Cgs}$$
- maximum oscillation frequency f_{max}
$$f_{max} = \frac{f_T}{2[g_d(Rg + Rs) + 2\pi f_T CdgRg]^{1/2}}$$
- noise figure NF :
$$NF = 10 \log_{10}[1 + kfCgs(\frac{Rs + Rg}{gm})^{1/2}]$$

- gm = intrinsic transconductance
- Cgs = gate source capacitance
- Gd = drain conductnce
- Rg = gate resistance
- Rs = source resistance
- k = fit factor

▶ **2.2 - Epitaxial materials**

A key to a successful HFET technology is obviously the quality of epitaxial material which controls the following parameters :
- Transconductance through the quality of the heterointerface
- Access resistance through the doping and thickness of the cap layer.

ESPRIT has significantly contributed to the advances in this field. The topic will also be reviewed in more details in chapters 3.4, 3.5 and 3.6.

MBE has been and is still presently the most widely used technique (from Picogiga and internal sources). MOCVD was also used especially in conjunction with the development of the multiwafer PLANET reactor which was succesfully transferred to industry during the ESPRIT programme PLANET (see chapter 3.4) (2).

▶ **2.3 - Device technology**
The fabrication technology is generally similar to the one for MESFET, except for the gate material which is in certain cases TiAl. A T-shaped (or mushroom) gate is often used in order to keep Rg at acceptable values for gate length below 0.5 μm. In this case electron beam lithography is generally used.

Siemens within COSMIC developed a completely different approach which is adapted for mass production (see figure 3).

KEY DATA @ 12 GHz

	NF_{min} (dB)	G_{ass} (dB)
● **AlGaAs/GaAs HEMT :**		
Chip	0.6	10
Ceramic package	0.7	11
Plastic package	0.8	9.5

← Passivation

← Epilayer
(AlGaAs/GaAs or
AlGaAs/InGaAs)

Gate cross section

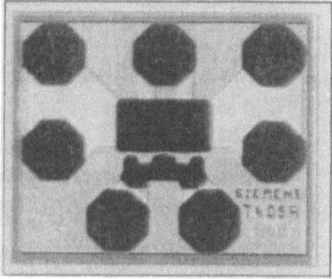

Chip photograph

Figure 3 : Gate cross section, chip photograph and performance of Siemens 0.25 μm HEMT (data obtained by Siemens within ESPRIT project COSMIC)

The sub-half micron dimension of the gate is not obtained with e-beam lithography but results from a first i line stepper 0.5 μm lithography followed by the formation of internal SiN spacers to give 0.2 - 0.25 μm. A key issue is of course the statistical control of the gate which, according to Siemens, is now mastered. A refractory gate self aligned structure is used. The latest version of this technology shows at 12 GHz a noise figure of 0.6 dB and an associated gain of 11 dB. Plastic and ceramic packaged devices are available (7) (8).

Another key aspect of HFET processing is the threshold voltage and saturation current control and uniformity.

Since the thickness of the active channel is very small, threshold voltage control through gate recess is extremely critical. Selective dry etching procedure with automatic etch stop (on GaAlAs for instance) are being developed to overcome the problem.

▶ **2.4 - Device performance**

Depending on the envisaged application several structures have been defined and optimised :

- low noise : a homogeneously doped GaAlAs layer is used (see figure 1) with a conventional structure (GaAs channel). An example has been shown in AIMS. Typical performance of a 0.3 μm x 120 μm FET is (figure 4) (5) :

  - gm = 350 mS/mm
  - $I_{DS\ MAX}$ = 320 mA/mm
  - F_T = 55 GHz
  - F_{max} = 110 GHz
  - NF = 0.55 dB, G = 11 dB (12 GHz)
  - NF = 0.64 dB, G = 9 dB (18 GHz)

- ultra low noise and very high frequency : InP based HEMTs. The first developments were carried out in GIANTs. In this programme Philips demonstrated InP based lattice matched HEMTs with gm values up to 600 mS/mm (0.6 μm gate length) and cut off frequencies up to 100 GHz for a 0.25 μm gate. This to our knowledge remains the only work in ESPRIT on InP HFETs (6).

- low noise and medium power : a pseudomorphic structure with a single heterojunction (SH) is now used for this purpose. Compared with conventional HEMTs, PMHEMT's have the following advantages :

  - higher carrier density (2 x 10^{12} cm^{-2} instead of 1.3 x 10^{12} cm^{-2})
  - wider range of biasing conditions for maximum transconductance and minimum noise
  - smaller drain conductance

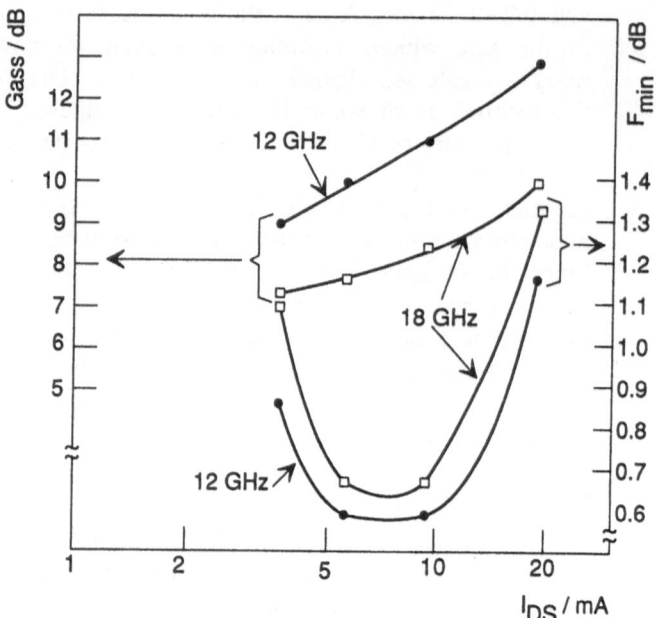

*Figure 4 : Minimum noise figure F_{in} and associated gain
Gass of a conventional 0.3 μm gate HEMT as a function of
drain curren
I_{DS} (data obtained by Daimler Benz,
Research Centre Ulm within ESPRIT project AIMS).*

A further improvement is to use a pulse doped donor layer in
the GaAlAs which brings an increase in carrier density and
breakdown voltage. The structure is suitable both for medium
power and low noise circuits (figure 5).
Typical performance is (0.25 μm gate length)   (5) :

- gm = 700 mS/mm
- $I_{ID\ MAX}$ = 650 mA/mm
- F_T = 100 GHz, F_{MAX} = 160 GHz
- NF = 0.6 dB, G = 14 dB (12 GHz)
- NF = 1 dB (26 GHz)

A similar structure was used to fabricate a power HFET of
12 x 75 μm gate width (see figure 6) which produced 380 mW
at 28 GHz (with 4.9 dB gain and 33 % power added
efficiency).

- high power : this is a development presently carried out in
MANPOWER and CLASSIC. A double pulse doped structure (double heterojunction DH) is studied in CLASSIC for
this purpose by Daimler Benz and Thomson.   An
optimisation of this structure is presently under way (8).

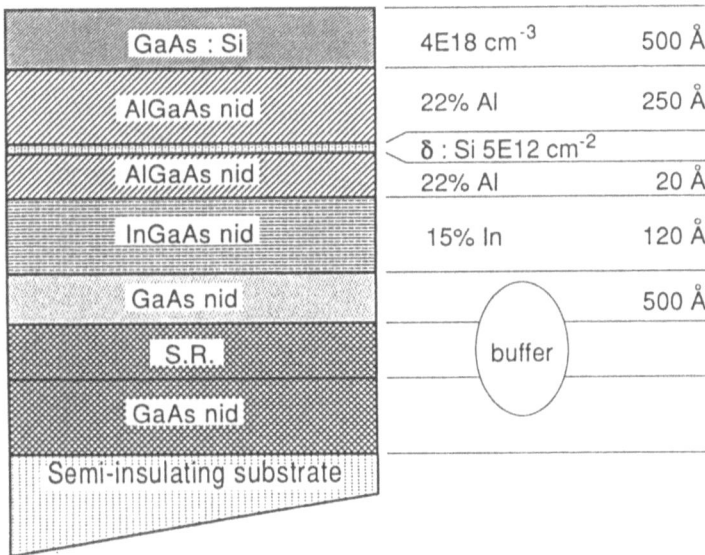

Figure 5 : Pulse doped layer sequence used for HEMT for low noise and medium power applications (data obtained by Daimler Benz Research Centre Ulm within ESPRIT project AIMS). This structure is used in the components shown in figures 6 and 8)

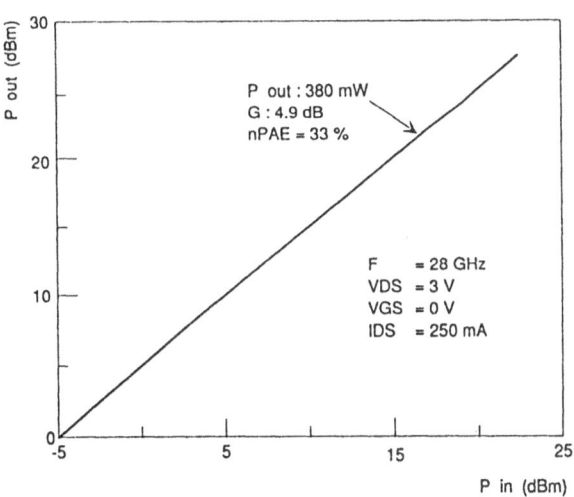

Figure 6 : Output power as a function of input power at 28 GHz of a 0.25 μm gate HEMT of gate periphery 12 x 75 μm (data obtained by Lille University within ESPRIT project AIMS)

3. PROGRESS IN HFET CIRCUIT DEMONSTRATORS WITHIN ESPRIT

▶ 3.1 - Small signal MMICs

The majority of the demonstrators are on microstrip structure. However a large effort is devoted to coplanar circuitry by Daimler Benz in AIMS and CLASSIC. Table 1 summarizes the most significant results for linear small signal MMICs :

ESPRIT programme (ref.)	Company (Foundry/Year)	Device	Type of circuit	Performance
GIANT (4)	PHILIPS 1991	InP Based HEMT	60 GHz LNA	NF = 4.5 dB G = 14 dB (figure 7)
GIANT (4)	GMMT 1991	Pseudomorphic HEMT	2 - 30 GHz amplifier	G = 11 dB
AIMS (9)	DB 1993	S.H. pseudomorphic HEMT	28-30 GHz Coplanar low noise amplifier	NF = 1.9 - 2.3 dB
AIMS (9)	DB 1993	S.H. pseudomorphic HEMT	35 GHz LNA	G = 17 dB
AIMS (9)	THOMSON 1993	S.H. pseudomorphic HEMT	20.2-21.2 GHz LNA	NF = 2.7 dB, G = 18 dB
COSMIC (7)	SIEMENS 1992	pseudomorphic HEMT	11-12.5 GHz LNA	NF = 1.3 dB G = 19 dB
COSMIC (7)	SIEMENS 1992	pseudomorphic HEMT	21.2-23 GHz LNA	NF = 2.5 dB G = 15dB
AIMS (9)	DB 1993	dual gate HEMT	35 GHz gain controllable amplifier	G = 7-17 dB at 35 GHz

Table 1. Representative results obtained within ESPRIT on HEMT based small signal MMICs (SH means single heterojunction, pulse doped)

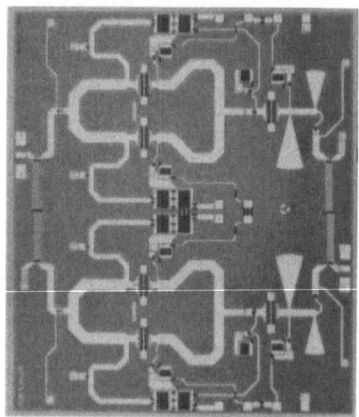

Type	Balance, 2 stages
Process	SH-PM-HFET
Frequency	27.5-29.5 GHz
Linear gain	10.0 dB
Output power (at 1 dB compression point)	27-28 dBm
Bias	4.5 V / 1100 mA
Chip size	3.5 x 4.0 mm²
Pulse	10 %, 3 µs

Figure 7 : Chip photograph and measured and simulated gain of a 60 GHz InP based HEMT amplifier (data obtained by Philips within SPRIT project GIANTS)

► **3.2 - Large signal MMICs**

*As for power and highly non linear applications the "unified" struc-
ture defined in collaboration in AIMS by Thomson and Daimler-
Benz is, as already mentioned, suitable for medium power
application. Several MMICs have been designed and manufactured
within AIMS by Thomson and Daimler. Table 2 summarizes the
most significant results. A similar development of power HFET is
now carried out by Siemens within MANPOWER (10).*

ESPRIT programme (ref.)	Company (Foundry/Year)	Device	Type of circuit/ device	Performance
AIMS (9)	THOMSON 1993	S.H. pseudomorphic HEMT	20.2-21.2 GHz SSPA	P_{out} = 28-29 dBm (pulsed) G (linear) = 13 dB
AIMS (9)	THOMSON (1993)	S.H. pseudomorphic HEMT	27.5-28.5 GHz SSPA	P_{out} = 24.5-25 dBm (pulsed) G (linear) = 11 dB (figure 8)
AIMS (9)	DAIMLER-BENZ (1993)	S.H. pseudomorphic HEMT	35 GHz SSPA	P_{out} = 21 dBm G = 6 dB

Table 2. Representative results obtained within ESPRIT on HEMT based
power MMICs

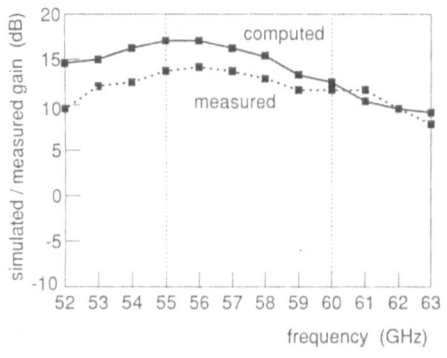 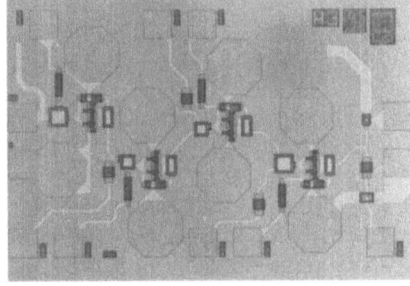

*Figure 8 : Chip photograph and performance of a 30 GHz power
amplifier (data obtained by Thomson Composants Spécifiques
within ESPRIT project AIMS)*

4. FURTHER DEVELOPMENTS IN HFETs

Since AIMS is now nearing completion (end of 1993), the future work carried out in ESPRIT on HFET will be essentially within MANPOWER and CLASSIC. These two programmes have 2 main objectives :

▶ development of new technologies (and in particular HFET technologies) for power applications, with the associated tasks on modelling and characterization.

▶ demonstration of the industrial relevance and manufacturability of these processes. This is of course, especially applicable to MESFET processes as studied in MANPOWER but also to a lesser extent to the HFET processes in CLASSIC and MANPOWER.

In terms of innovation, CLASSIC is probably the most challenging programme since it addresses "components for large signal 60 GHz integrated circuits". Key aspects of the future work will be the optimisation of the double pulse doped structure and the accurate characterization and modelling up to 60 GHz. A 0.15 μm technology is being developed in this project. The feasibility of low noise processes based on 0.1 μm HEMT has already been demonstrated in the Basic Research Action NANOFET (11).

Finally further actions on manufacturing science are planned within the EuroGaAs framework programme.

5. CONCLUSION

ESPRIT has contributed to the achievement of state of the art results by the whole European III-V semiconductor industry.

Low noise and medium power demonstrators up to 40 GHz have already been demonstrated. A high level of expertise in material growth, lithography, fabrication processes and design exist, and Europe is in a good position to face the international competition. The present activity aims now at increasing the frequency of operation up to 60 GHz and developing power processes.

Further extensive development efforts will be necessary in the future and Esprit and more generally Eurogaas will play a very important role in this respect.

REFERENCES

1) D. Delagebeaudeuf et al., Electron. Letters 16, 1980, p. 667
See also C. Weisbuch and B. Vinter "Quantuum Semiconductors Structures" (Academic Press, 1991)

2) P.M. Frijlink et al., Jl of Crystal Growth 107, 1991, p. 166

3) COSMIC : final technical report (1993)

4) F. Ponse and D. Berger : Proceedings of 2nd ESA Electronic Components Conference (ESTEC, 1993).

5) AIMS : 1st six-monthly report (1990)

6) GIANTS : final technical report (1991)

7) AIMS : 3rd technical report (1991)
see also H. Dambkes and LP. Schmidt, Proceedings of GaAs 92 (Estec, april 1992)

8) CLASSIC : 2nd six-monthly report (1993)

9) AIMS : 6th six-monthly report (1993)

10) MANPOWER : 2nd six-montly report (1993)

11) Basic Research Action NANOFET : final report (1992)

3.2 Advanced HBTs

D. Pons
Laboratoire Central de Recherches, Thomson-CSF

1. HBT technology development in Europe

With the exception of earlier works at Plessey Research* and at the French telecommunication research agency** (1), most of the III-V Heterojunction Bipolar Transistor (HBT) technology development effort in Europe is recent. HBT activities started around 1990 almost simultaneously at Siemens and Daimler-Benz in Germany, and at Thomson-CSF in France. This is somewhat later than in the USA and in Japan, where R&D started in the mid 80's in the research laboratories of Texas Instruments, Rockwell, TRW, and Fujitsu, NTT, NEC, etc. However the European HBT activity has received a decisive impetus thanks to the strong support of both civilian and military european cooperative programmes. The lag of the European microelectronic industry in this field could be rapidly caught up, if this effort were maintained.

Concerning HBT technology the most important civilian cooperative programme is probably the ESPRIT 5032 "AIMS" project. AIMS allocated a substantial part of its ressources to the development of HBT technology in the research laboratories of Thomson-CSF and of Daimler-Benz for millimeter-wave applications. Another on-going ESPRIT programme in which HBT technology is evaluated for mobile telecommunication circuits is the ESPRIT 6050 project "MANPOWER"; Siemens is in charge of the HBT technology development within MANPOWER. Table 1 give some important characteristics of these programmes.

There is also mentioned in this table, a European (IEPG) military cooperative programme, in which X-band high-power HBT amplifiers will be fabricated for future phased array radars. Please note that this table does not give the exhaustive list of the partners involved in these three programmes, but only those directly in charge of the HBT technolgy development. Additional and essential efforts on modelling, assessment, and circuit design are also performed by the other partners of these programmes. In AIMS the HBT technology development work is based on a sound cooperation between Thomson-CSF and Daimler-Benz, with very close

---

* Now GEC-Marconi Materials and Technology - GMMT

** CNET Laboratory of France Telecom

device processes, allowing common mask-sets. In MANPOWER, the HBT technology is developped by Siemens, the circuit design activity is shared between Siemens and Dassault Electronique, some reliability studies will be performed by Alcatel Telettra and Padova University and finally electrical and thermal modelling are performed by Dublin University College and Turino University.

The description of the HBT activity in Europe would not be complete without mentionning the work of the CNET laboratory of France Telecom for optical fiber telecommunication and mobile telecommunication circuits and the work of the Fraunhofer Institut in Freiburg (Germany) in MO-CVD epitaxy of GaInP/GaAs HBT structures.

2. HBT principles and comparison with other devices

The HBT technology should be considered as relatively young compared that of much more mature devices like : standard silicon bipolar junction transistors (BJTs), GaAs field effect transistors (MESFETs), or GaAlAs/Ga(In)As heterojunction transistors (HFETs or HEMTs, see section 3.1). So it could be useful to remind the reader of the basic characteristics of HBTs as compared to these other devices. III-V HBTs are the III-V device counterparts of the silicon BJTs. In HBTs, contrary to conventional BJTs, the emitter layer is a semiconductor with a larger band gap than of the base layer. The induced valence band discontinuity at the emitter-base heterojunction blocks hole injection (for a npn transistor) from the base to the emitter. This allows the use of an extremely high doping level for the base and a relatively low doping level for the emitter. As a result, the base resistance, which is one of the main microwave performance limitation in conventional BJTs, is much lower in HBTs ; this is also the case for the emitter-base capacitance.

The HBT structure is a vertical succession of layers of various compositions, thicknesses and dopant concentrations, grown by epitaxy. Typical HBT structures are given in Table 2. Structures A and B are grown on a GaAs substrate and the only difference between these two structures is the use of GaAlAs (A) or GaInP (B) as a wide band-gap emitter layer. We will come to this later on. Structures C and D are InP based structures. Structure C is potentially extremely fast because of the very high electronic velocities and of the strong velocity overshoot effects in the GaInAs collector layer; however it suffers from very low breakdown voltages, due to high ionization rates in GaInAs. To remedy this weakness, structure D has a wide band-gap collector layer and is called a "Double-Heterojunction Bipolar Transistor" (DHBT) ; the difficulty in the latter is to smooth the potential barrier for electron injection from the base to the collector layer.

Table 1. European cooperative programmes for HBT technology developments.

Programmes HBT makers	HBT	Circuits Applications
Civilian cooperative programmes		
ESPRIT 5032 "AIMS" '90-93		
Daimler-Benz (G.)	GaAlAs/GaAs	35GHz VCOs Airport surveillance radar
Thomson-CSF (F.)	GaInP/GaAs	14-28GHz VCOs Frequency synthesizer
ESPRIT 6050 "Manpower" '92-94		
Siemens (G.)	GaAlAs/GaAs	L-band power amplifiers Mobile telecommunications
Military cooperative programmes		
IEPG-TA1-CTP8 '92-94		
GMMT (UK)	GaAlAs/GaAs	X-Band power amplifiers
Defence Research A. (UK)		T/R modules for phased arrays
Daimler-Benz (G.)	GaInP/GaAs	
Thomson-CSF (F.)	GaInP/GaAs	

Table 2. Various III-V HBT structures on GaAs and on InP substrates. Also indicated, the possible nature of the base dopant and the epitaxial techniques which can be used.

layer		str. A	str. B	str. C	str. D
emit. contact	n+	GaAs	GaAs	GaInAs	GaInAs
emitter	n	GaAlAs	GaInP	AlInAs	AlInAs
base: dopant	p++	GaAs: C,Be	GaAs: C	GaInAs: Be,Zn	GaInAs: Be,Zn
collector	n-	GaAs	GaAs	GaInAs	InP
coll. contact	n+	GaAs	GaAs	GaInAs	GaInAs
substrate	S.I.	GaAs	GaAs	InP	InP
epitaxial technique		MBE MO-VPE GS-MBE	MO-VPE GS-MBE	MBE MO-VPE GS-MBE	MO-VPE GS-MBE

Most of the development effort in Europe, and especially in the on-going ESPRIT cooperative programmes, is devoted to the GaAs based structures A and B, so that structures C and D will not be considered in the following.

The following equations 1 - 2 give fairly realistic values for the current gain cut-off frequency F_T and for the maximum oscillation frequency (or power gain cut-off frequency) F_{max}:

$$F_T = 1/2\pi\tau, \text{ with } \tau = R_E C_{BE} + W_B^2/2D_B + W_C/2v \quad (1)$$
$$F_{max} = (F_T/8\pi R_B C_{BC})^{1/2}. \quad (2)$$

In Eq. 1 for the total transit time τ, the second right-hand side term is the diffusion time of electrons in a base of thickness W_B, with a minority carrier diffusion coefficient D_B ; the third term is the delay time due the transit of electrons with an average velocity v in the collector depletion region of thickness W_C.

Another important figure of merit is the DC current gain ß, which is equal to the ratio of the total minority carrier recombination time constant τ_R (combining bulk and surface, radiative and non-radiative recombination mechanisms) to the diffusion time of the electrons in the base:

$$ß = 2D_B \tau_R/W_B^2 \quad (3)$$

For the optimization of a HBT structure for a given application, there are two important trade-offs to consider:

- a thin base will give a short diffusion time constant in the base and therefore high F_T and ß ; but this will be at the expense of a high base resistance and this will reduce F_{max};

- a thin collector will give a short transit time in the collector layer and therefore a high F_T; but this will be at the expense of a low breakdown voltage.

Si BJTs may be optimized (with narrow emitter fingers, 0.5µm or narrower) for high F_T, with record values aroud 50 GHz (2), but F_{max} will be much lower (20 - 30 GHz) and the breakdown voltage will be lower than 3V. The base resistance limitation can be partly removed in advanced SiGe HBT, with demonstrated F_T / F_{max} values around 50 / 60 GHz, but with again very low breakdown voltages (2). For the most "aggressively" optimized SiGe HBTs, the record values for F_T exceed 100 GHz, but with considerably lower F_{max} (50 GHz) and weak breakdown voltages (3).

In comparison III-V HBTs, with relaxed geometries (typically 2μm wide emitter fingers, the finest lithographic feature) demonstrate routinely F_T / F_{max} values of the order 50 / 100 GHz, together with breakdown voltages around 20V. Record values for F_T / F_{max} exceeds 100 / 200 GHz (4).

Table 3 gives a brief principle comparison between HBTs and MESFETs or HFETs. Shortly, HBTs with 2 μm wide emitter fingers will have cut-off frequencies intermediate between those of 0.5 μm gate length MESFETs and 0.25 μm gate length HFETs or HEMTs, however with higher breakdown voltages. Although the microwave noise performances of HBTs are still very poorly documented, it seems that they exhibit higher microwave noise figures than MESFETs or HEMTs. To the contrary they give much weaker low-frequency noise power levels: the typical field of microwave applications for HBTs will be power amplification with expected high power-added efficiencies, or those circuits for which low frequency noise power levels are important, like oscillator circuits or analogue frequency dividers. This is certainly not an exhaustive list of interesting microwave applications and HBTs may also prove competitive for numerous analogue and digital applications (e.g. analogue-to-digital converters) or opto-electronic applications for which a material compatibility with laser diodes or photodiodes may be sought, favouring an InP-based technology.

Table 3. Comparison between HBTs and FETs (MESFETs and HEMTs)

	FETs	HBTs
Material	Ion-Implanted Epitaxy	Epitaxy only
Device structure Current flow	Horizontal Parallel to layers	Vertical Perpendicular to layers
Cut-off frequencies	defined by lithography (gate length)	defined by active layers thicknesses
Minimum feature size for similar Ft and Fmax	0.25 - 0.50 μm (Gate length)	2 - 3 μm (Emitter finger width)
Breakdown voltage	medium (lateral Schottky Gate-to-Drain junction)	high (vertical bipolar Base-Collector junction)
Microwave noise (main source)	low (channel resistance thermal noise)	high (base current shot noise)
Low-frequency noise (main source)	high (surface recomb. at gate recess/passivation layer interface)	low (bulk and surface recomb. in the emitter-base space-charge region)

3. Epitaxy of HBT structures

Three growth processes for HBT structures have been established in the MANPOWER and AIMS projects:

- MO-VPE epitaxy of GaAlAs/GaAs HBTs, with carbon doping of the base, by Siemens (MANPOWER);

- MBE epitaxy of GaAlAs/GaAs HBTs, with beryllium doping of the base, by Daimler-Benz (AIMS);

- MO-VPE epitaxy of GaInP/GaAs HBTs, with carbon doping of the base, by Thomson-CSF (AIMS); a study of the growth of GaInP/GaAs HBT using CBE has also been performed by Thomson-CSF, in the frame of the MORSE project, see section 3.5.

The base doping is one of the key issues of HBT structure growth. The acceptor dopant concentration has to be as high as possible in order to get a low base resistance R_B, while at the same time achieving two objectives. First, to keep a high quality material with long enough minority carrier recombination time constants τ_R in order to get sufficiently high DC current gains ß. Second, to avoid the diffusion of the acceptor into the emitter layer, which would have the effect to displace the electrical p/n junction away from the metallurgical heterojunction and ruin the device performances. The use of C doping appears clearly to-day the best choice with demonstrated negligible diffusion effects, even at very high dopant concentrations, of the order 10^{20} cm^{-3}. The choice of C doping imposes MO-VPE epitaxy, since only poor quality C-doped GaAs has been obtained so far by solid-sources MBE, from graphite evaporation. As a result of this obvious superiority of C doping over Be doping, Daimler-Benz decided to stop the development of Be doped, MBE grown GaAlAs/GaAs HBT and to concentrate on C doped GaInP/GaAs HBTs, relying on external supplies.

Another key issue is the use of a GaInP/GaAs structure (structure B in table 3.2.2) instead of GaAlAs/GaAs. GaInP is an alloy between InP and GaP; the lattice matching on GaAs is obtained with a 50% In or Ga composition. The use of GaInP instead of GaAlAs offers several advantages. Firstly, a much lower chemical reactivity to oxygen; one may expect that GaInP could be more efficiently passivated and that more reliable devices should result. Secondly, for npn HBT devices, the band-gap alignment is more favourable with a much lower conduction band offset ($\Delta E_C \leq 0.1$eV). Therefore, a composition grading is not necessary at the emitter-base interface to smooth the energy spike which degrades the injection of electrons in the GaAlAs/GaAs system. Thirdly, it appears that GaInP may be prepared with a negligible concentration of deep levels (e.g. similar to the DX centers in GaAlAs) which is important when weak low-frequency noises are needed (as in oscillator circuits, for instance which are the demonstrator circuits of the AIMS project).

However, the most interesting advantage is technological: each of the GaInP and GaAs layers can be easily etched selectively with respect to the other with natural etch-stops at the hetero-interfaces. This etching selectivity facilitates greatly HBT device processing. In particular, it allows for an easy and excellent control of the emitter mesa etching, this fabrication step being one of the most critical. As a result, higher fabrication yields and device reproducibilities should result.

One additional interesting feature of C-doped GaInP/GaAs HBT structures is that C does not dope GaInP either n or p type (C is probably amphoteric in GaInP), so that there is a perfect coincidence between the electrical p-n junction and the metallurgical GaInP/GaAs heterojunction. This is contrary to what may happen in a GaAlAs/GaAs structure if the base dopant diffuses into the emitter layer during the epitaxial growth, device processing or device operation, which will induce a catastrophic degradation of the devices.

It is perhaps worth stressing that, in the frame of the AIMS project, Thomson-CSF was among the very first to develop and promote the GaInP/GaAs HBT technology, with the first demonstration of state of the art HBT microwave performances (5). This is now followed by numerous other laboratories.

The MO-VPE growth process for C-doped GaInP/GaAs HBT established by Thomson-CSF uses arsine, phosphine, tri-ethyl-gallium and tri-ethyl-indium. The n-type dopant is Si from silane for the GaAs and GaInP layers; the C doping precursor is CCl_4 for the base. One critical aspect for the growth of the GaInP layer is that this alloy can exhibit marked ordering effects, visible for instance in electron diffraction patterns observed by TEM. For a fixed In composition, the band gap energy, the etching behaviour of GaInP, and the band-gap discontinuities at the GaInP/GaAs heterojunction will vary with the amount of ordering.

This ordering is dependent on the growth conditions: growth temperature and III/V ratio. The optimization for the growth of these HBT structures required precise X-ray diffraction measurements for perfect lattice matching on GaAs, photoluminescence spectroscopy for band-gap measurements and constant feedback from device processing.

GaInP/GaAs HBT structures from external suppliers are also used now by Daimler-Benz in AIMS. Up to now, the best results have been obtained with wafers grown at the Fraunhofer Institute in Freiburg (IAF). The MO-VPE growth process established by IAF uses arsine, phosphine, tri-methyl-gallium and tri-methyl-indium. Silane is used for n-type doping. The arsine flow is switched-off for the growth of the base layer, which is grown and C-doped from tri-methyl-arsine. One interesting feature of the HBT structures optimized by IAF is the extremely thin GaInP emitter layer (of the order 100 - 200 Å): this thickness is enough to block the injections of the holes from the base to the emitter and, at the same time so thin, that it is almost transparent for the electrons. This renders the

113

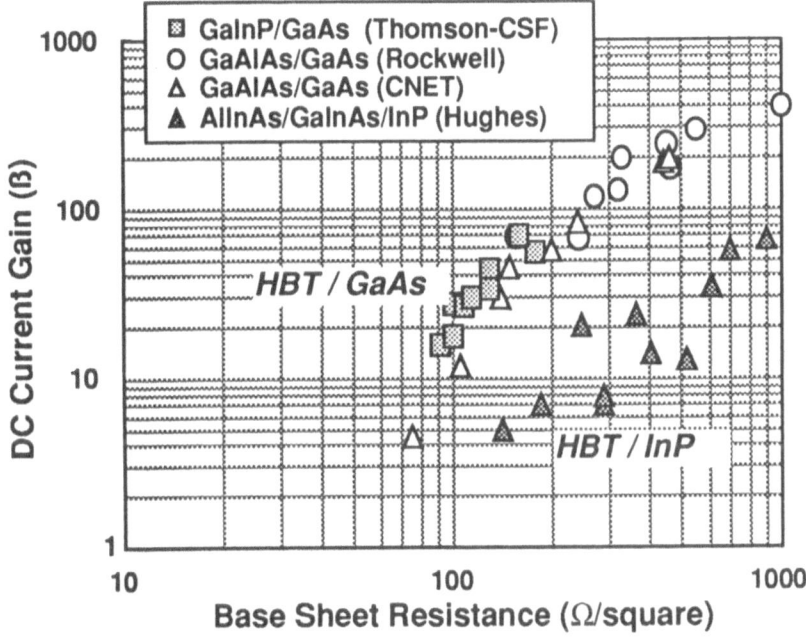

Fig. 1 Current gain vs base sheet resistance of large size GaInP/GaAs HBT as compared to the best published values for GaAlAs/GaAs HBT (C and Be doped) and for AlInAs/GaInAs/InP HBT.

ordering effects much less critical than for the HBT structures of Thomson-CSF with a much thicker GaInP layer (of the order 3000 Å). GaInP/GaAs HBT structures from IAF are also processed by Siemens, but in the frame of a German national programme.

Fig. 1 shows the DC current gain ß as a function of the base-sheet resistance of C-doped GaInP/GaAs HBTs grown and processed by Thomson-CSF. This data is compared to best published values for C- and Be- doped GaAlAs/GaAs HBTs (6-7) and for Be-doped AlInAs/GaInAs/InP HBTs (8). It can be observed that GaInP/GaAs HBTs have identical or even higher current gains than conventional GaAlAs/GaAs HBTs, especially for highly doped base layers with base sheet resistances approaching 100 Ω/square. It is also shown in this figure that with today's know-how, GaAs-based HBTs can have considerably lower base sheet resistances (certainly due to the use of carbon doping) than InP-based HBTs, and therefore, higher microwave performances.

114

4. HBT processing

The fabrication of the device requires to contact each of the three active layers, collector, base and emitter. Up to now, the best microwave performances have always been obtained with a device structure constituted by a stack of several mesas, one mesa per active layer. This multi-mesa structure allows minimum access resistances, minimum junction and parasitic capacitances and maximum breakdown voltages. This non-planar process has the drawback that air-bridges are necessary for connecting each layer, with anticipated difficulties for achieving high fabrication yields.

Daimler-Benz and Thomson-CSF have established very close processes for multi-mesa HBTs with self-aligned base contacts with respect to the emitter mesa, in order to minimize the base access resistance and the base-collector capacitance. The process exploits the full etching selectivity between GaAs and GaInP layers, as mentioned above. As a result, the emitter mesa etching and the self-alignment of the base contacts are easily controlled with extremely uniform and reproducible device performances: this would have been much more difficult to attain with conventional GaAlAs/GaAs structures. The process also includes multi-energy ion implantation for device isolation and reduction of the base-collector parasitic capacitance. The HBTs fabricated by Siemens have also a muli-mesa structure but are not self-aligned. A so-called "collector trench" technique has been established in order to reduce the base-collector parasitic capacitance by separating the collector contact from the inner intrinsic device. A planarization technique has also been developped, using a thick silicon nitride layer (9). Fig. 2 gives a sketch of a multi-mesa, self-aligned HBT of Thomson-CSF or Daimler-Benz.

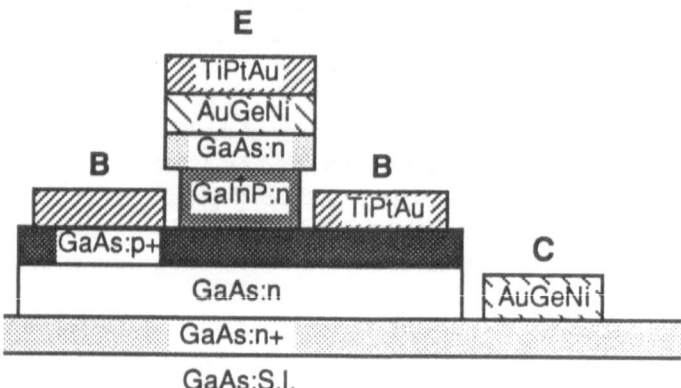

Fig.2 Sketch of a multi-mesa, self-aligned microwave GaInP/GaAs HBT from Thomson-CSF.

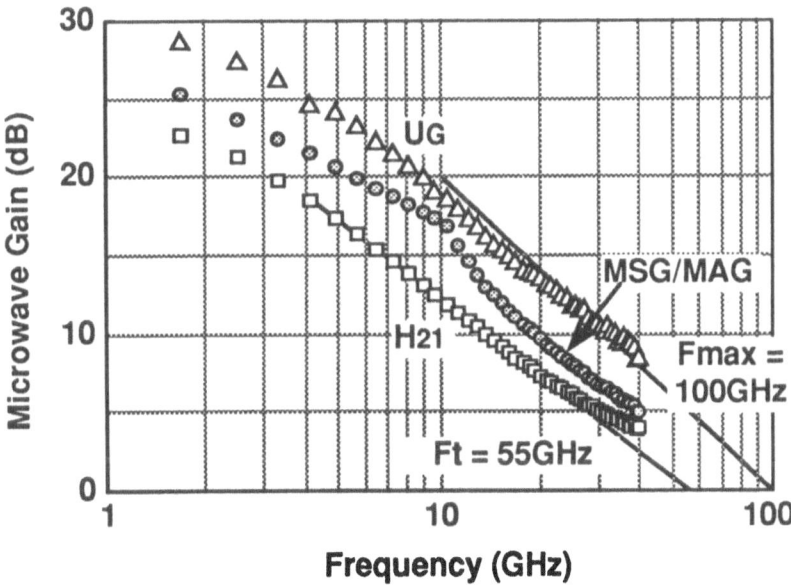

Fig. 3 On-wafer measurements of the current gain and power gains of a common-emitter self-aligned HBT ($2\times30\mu m^2$).

Comparable small-signal microwave performances have been obtained with one emitter-finger HBTs in the three laboratories of Siemens, Daimler-Benz and Thomson-CSF. Fig 3 gives the typical small-signal microwave current and power gains of a self-aligned, multi-mesa GaInP/GaAs HBT with one 2×30 μm^2 emitter-finger, from Thomson-CSF. Table 4 gives a summary of the microwave performances demonstrated by the three laboratories; these results may be considered as the world state of the art for HBT technology. The GaInP/GaAs HBT results of Daimler-Benz and Siemens have been obtained with layers grown at IAF. The results obtained by Siemens and Daimler-Benz allow to compare GaInP/GaAs and GaAlAs/GaAs HBTs. Being obtained with comparable device designs and technology maturities, GaInP/GaAs HBTs appear to allow better microwave performances. Daimler-Benz has compared the microwave performances of GaInP/GaAs with various collector thicknesses W_C: with W_C varying from 2500 to 7000 Å, f_T decreases from 80 to 40 GHz, but with similar f_{max}, between 100 and 110 GHz. The increase in the collector transit time is compensated by a reduction in the collector-base capacitance.

The low-frequency noise performances of GaInP/GaAs HBTs of Thomson-CSF have been measured at LAAS laboratory (CNRS, Toulouse). The input current noise spectral density a self-aligned HBT with one 2×30 μm^2 emitter finger is compared in Fig. 4 to those published for other HBT structures. It is shown to be

Table 4. Cut-off frequencies of GaAlAs/GaAs and GaInP/GaAs HBTs demonstrated by Siemens (MANPOWER), Daimler-Benz, and Thomson-CSF (AIMS). The GaInP/GaAs HBT results of Siemens (9) have been obtained in the frame of a German national programme.

Laboratory	Project	HBT structure	Ft (GHz)	Fmax (GHz)
Siemens	MANPOWER	GaAlAs/GaAs	50	50
(+ IAF)		GaInP/GaAs	40	90
Daimler-Benz	AIMS	GaAlAs/GaAs	50	75
(+IAF)		GaInP/GaAs	40-80	110
Thomson-CSF	AIMS	GaInP/GaAs	55	100

significantly lower than that of typical conventional GaAlAs/GaAs HBTs and comparable to those of AlInAs/GaInAs/InP HBTs and of specially designed GaAlAs/GaAs HBTs with a thin undoped GaAlAs guard-ring layer to decrease the surface recombination velocity at the extrinsic base surface (10).

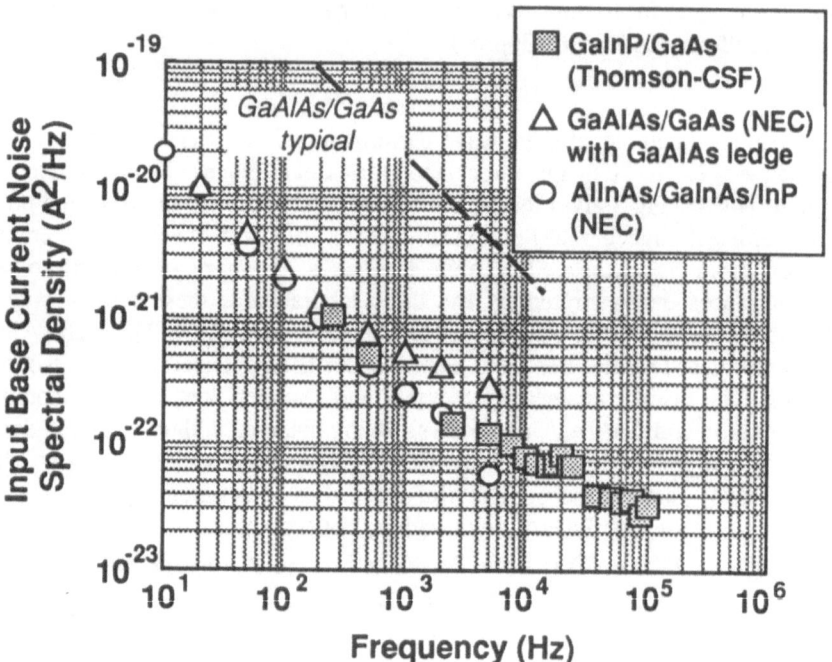

Fig.4 Comparison of the low-frequency noise performances of GaInP/GaAs HBT (Thomson-CSF), of conventional GaAlAs/GaAs HBT, GaAlAs/GaAs HBT with a GaAlAs passivation (ledge) layer (NEC) and of AlInAs/GaInAs/InP HBT (NEC).

5. HBT MMICs

The demonstrator HBT MMIC circuits of the AIMS and MANPOWER contracts are respectivelty millimetre-wave (14, 28 and 35 GHz) VCOs (voltage controlled oscillators), and L-band power amplifiers for mobile telecommunications. The VCOs are simple circuits integrating one or two one-emitter-finger HBTs, one or two Varactor diodes and passive elements (MIM capacitances, spiral inductances, metallic resistors, via-holes). The Varactor diodes are built using the p+/n base-collector junction and have high quality factors, due to the low access resistances. The passive elements are indentical to those used in standard MESFET MMIC processes. The first 14 and 28 GHz VCOs MMICs with GaInP/GaAs self-aligned HBTs have been fabricated by Thomson-CSF. Fig. 5 is a photograph of a 28 GHz VCO. The detailed performances of these VCOs and of the 35 GHz VCOs of Daimler-Benz will be published in the next 94 MTT-Symposium.

The fabrication of the L-band power amplifiers of Siemens is underway. Fig. 6 shows the lay-out of one of these amplifiers whose the target output power is 1.2W with more than 40% power-added efficiency at 3V bias.

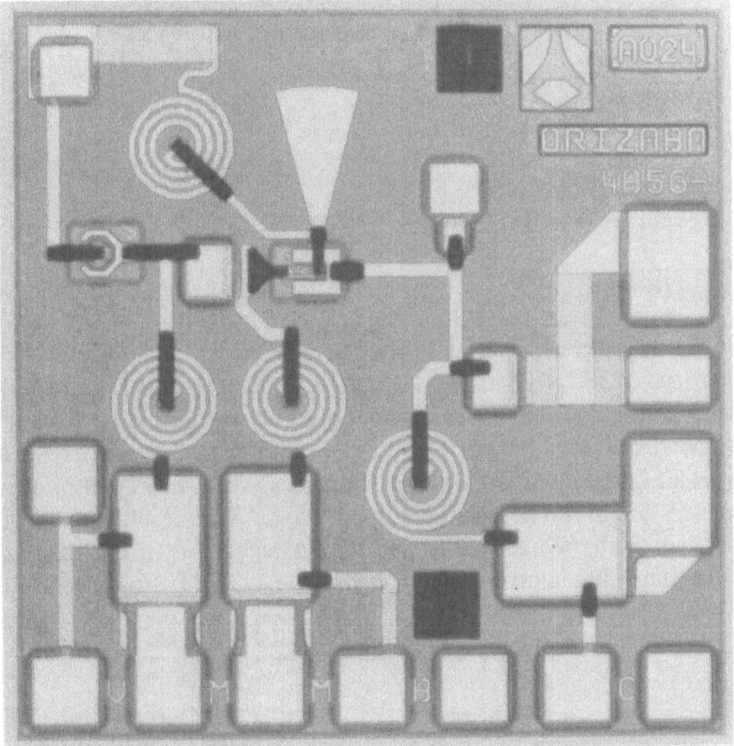

Fig.5 Photograph of a 28GHz GaInP/GaAs HBT VCO (AIMS).

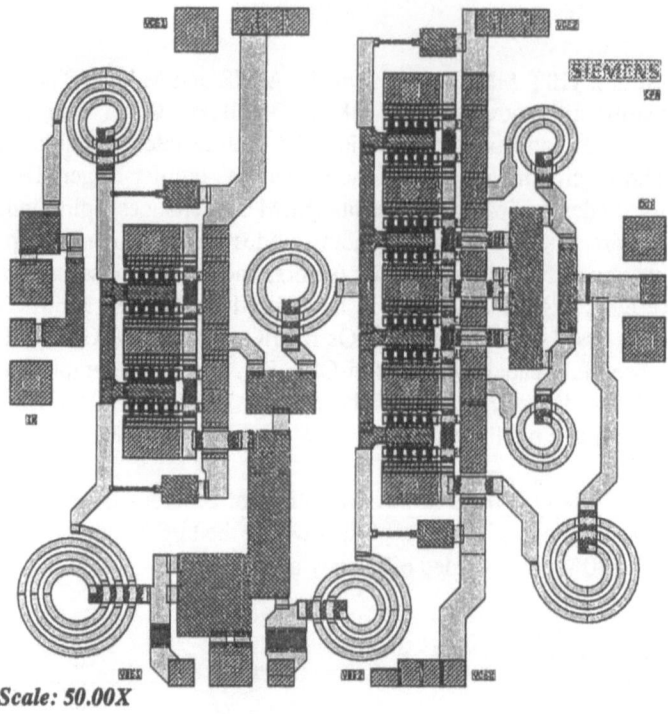

Scale: 50.00X

Fig. 6 Lay-out of a 1.2W HBT power amplifier for L-band mobile telecommunications (MANPOWER).

6. Conclusions

The two Esprit projects AIMS and MANPOWER have allowed the development of a microwave HBT technology in Europe. Epitaxial growth and fabrication processes for high performances carbon-doped GaAlAs/GaAs and GaInP/GaAs HBTs have been established. Similar microwave performances have been demonstrated at Siemens, Daimler-Benz and Thomson-CSF and these performances are at the best world state of the art. The results are particularly promising for GaInP/GaAs HBTs and the advantages of using this structure instead of the more conventional GaAlAs/GaAs HBT structure have been established.

References

(1) The HBT work of the CNET was also partly supported by an Esprit1 programme.

(2) see for instance: P.J. Zdebel, "Current status of high performance silicon bipolar technology", GaAs IC Symposium 92, technical digest p. 15.

(3) E. Crabbé et al, "113 GHz f_T graded-base SiGe HBTs" and A. Gruhle et al., "Base Thickness and High Frequency Performances of SiGe HBTs", both contributions at the 51st Annual Device Research Conference, 1993 (unpublished).

(4) see for instance: Hidenori Shimawaki et al, "High-f_{max} AlGaAs/InGaAs and AlGaAs/GaAs HBTs fabricated with MOMBE selective growth in extrinsic base regions", ibid.

(5) S.Delage et al, "First microwave characterization of LP-MOCVD grown GaInP/GaAs self-aligned HBT", Electronics Letters, 1991, vol. 27, p.253.

(6) J.L. Benchimol et al., Very high gain in carbon-doped base heterojunction bipolar transistor grown by chemical beam epitaxy", Electronics Letters, 1992, vol. 28, p. 1344.

(7) G.-W. Wang et al., "High-Performance MOCVD-grown AlGaAs/GaAs Heterojunction Bipolar Transistors with carbon-doped base", IEEE Electron Device Lett., 1993, vol. 12, p.347.

(8) M. Hafizi et al., "Dependence of DC current gain and f_{max} of AlInAs/GaInAs HBT's on base sheet resistance", IEEE Electron Device Lett., 1993, vol. 14, p. 323.

(9) P. Zwicknagl et al., "High speed non-selfaligned GaInP/GaAs-TEBT", Electronics Letters, 1992, vol. 28, p. 327.

(10) K. Honjo et al., "Hetero-guardring fully self-aligned HBT (HG-FST) for microwave and high-speed digital applications", NEC Res. & Develop., 1992, vol. 33, p. 324.

3.3 Advanced GaAs MMIC Technology

R W W Charlton and J A Turner
GEC-Marconi Materials Technology Ltd., Caswell, Northants NN12 8EQ, UK

1. Introduction

There are as many European GaAs MMIC process technologies as there are companies processing GaAs. Over the years each company has addressed its own needs by developing specific technologies; they all have the same aim however, of interconnecting active and passive circuit elements using transmission line structures. In this section we will describe some of the key elements of the various companies' technology and describe how some of these have been enhanced to achieve higher performance and above all, smaller chip sizes.

European MMIC Technologies All GaAs MMIC manufacturing processes require the fabrication of a number of key elements in order that the technology can produce the required microwave or millimetre wave circuit performance. These elements are

a) the active device (MESFET, HEMT or HBT)
b) the passive elements (inductors, capacitors and resistors)
c) the interconnections (air bridge or dielectric)
d) the transmission lines (microstrip or coplanar)

In general each company has tried to use where ever possible common circuit processing regimes for the range of GaAs based MMICs. Whether it be MESFET, HEMT or HBT, the policy is to have a common sequence of dielectric and metallisation layers above the active device thereby allowing a rapid introduction of new device technologies into the MMIC process.

The Active Devices History plays a major role in determining the technology used for a particular device. Once a technology is developed it takes a major crisis of confidence to get a company to change its fabrication philosophy, and so no two technologies are exactly the same. This of course makes true second sourcing extremely difficult.

At present the key active device is the GaAs MESFET and across Europe, this device, when used in MMICs, is produced in ion implanted material. Ion implantation is a very versatile technology and is able to produce the carrier density profiles necessary for a whole range of circuit functions. Material optimised for low noise, switches, low power consumption, power, and digital circuit operation can all be produced in this way - combinations of these implants can be produced selectively to allow optimised mixes of functions to be produced.

These profiles are however peculiar to each company as they depend to some extent on the preceding processing steps.

Power MESFETs are however moving away from the implantation technology towards Molecular Beam Epitaxy (MBE) for reasons of improved gain and efficiency. The shape of tail of the material carrier concentration profile is key to improved device performance and it is possible to produce much sharper/steeper profiles in MBE material which gives higher high frequency gain.

The device fabrication processes have polarised into two main technologies - one using an etched recess before placement of the gate, the other using a totally planar process where the drain current is determined by the original thickness of the ion implanted layer. Both techniques have demonstrated the capability of volume manufacture although, perhaps, the level of drain current control in the non etched technology is the greatest.

Passives and Interconnect Technology   The main difference in the MMIC process technology itself is in the method of interconnection of the active device and the passive matching and biasing components. Here either an air bridge or a metal-dielectric-metal technology is used. Air bridges contact the upper most layers of metal and allow low parasitic interconnections to be made between circuit elements. The metal-dielectric-metal process is more 'silicon-like' and makes uses of an interlayer dielectric to isolate upper and lower levels of metallisation. Each technique has its own peculiar advantages but perhaps the latter offers more freedom in circuit layout, has the ability to produce some novel multi level components, and is essentially a planar process.

The capacitors and inductors are in general produced in similar ways throughout the industry. Capacitors are usually the MIM variety making use of either silicon nitride or silicon dioxide as the insulator between two metal layers. Smaller value capacitors are sometimes produced using interdigitated fingers of metal and using edge coupling to define the capacitance. Inductors are single or multispiral metallisations either circular or rectangular in shape and are used for matching elements at low frequencies and for biasing networks. They are usually made in thick (> 1μm) metal layers to reduce loss and to allow considerable amounts of current to be passed through them when used to bias high output power devices.

The passive element and interconnect technologies described above apply to a distributed element/microstrip approach to circuit design. However, at the higher frequencies, such as those of the AIMS and CLASSIC Esprit projects, alternative circuit design concepts are being developed based on a coplanar wave guide structures; here the requirements of the MMIC technology are somewhat different. The needs for through substrate via connections are removed as grounding is effected on the top surface of the chip, substrate thickness does not determine the impedances of the transmission lines as it does in the microstrip approach, so thickness control is not important, and the use of air bridges is likely to be the preferred method for the interconnecting of the ground planes as any dielectric overlay process will add parasitic capacitance.

Advanced GaAs Based Technologies in ESPRIT Projects.   Many new technologies and circuit fabrication enhancements have been and are still being used in the four ESPRIT projects briefly described in Chapter 1. At the lower frequencies (12GHz and below) the aim of the technology has been to minimise chip size, to improve manufacturability and to introduce innovative fabrication steps to aid the achievement of the designed microwave performance.

One of the key tasks of GEC-Marconi Materials Technology in the COSMIC programme was to develop a high packing density technology so that chip size could be reduced. They achieved a four fold reduction in chip size for a DBS downconverter (Figure 3.3.1) by reducing line gate spacings to 6 microns and reducing through dielectric via sizes to 2 microns. This gave a 2.75 times size reduction in inductor spiral and a 3 times reduction in the size of a typical transistor.

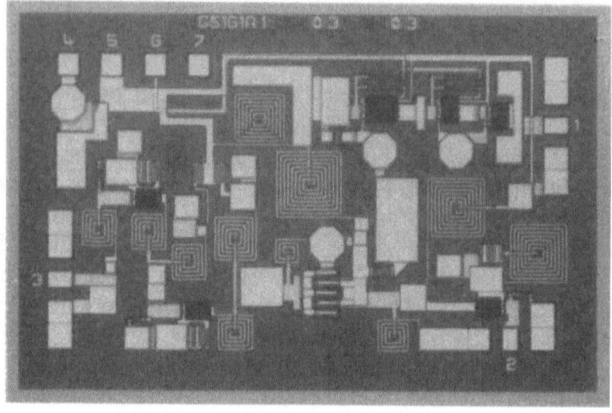

Figure 3.3.1   Down Converter Circuit for DBS with a $2mm^2$ chip size

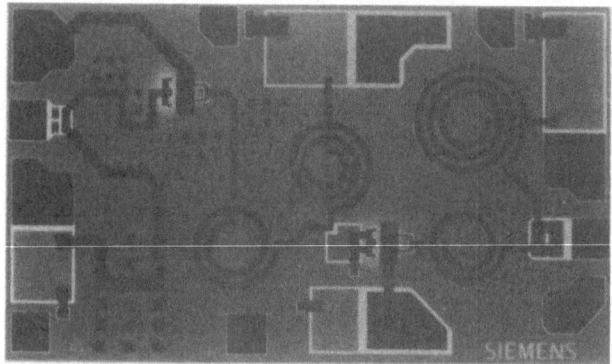

Figure 3.3.2   A HEMT Low Noise Amplifier Circuit Demonstrating the Structured Array Approach

On the same project Siemens demonstrated the versatility of their structured array approach to prototype circuit fabrication. This technique allows the tuning on wafer of critical parts of the matching circuitry by air-bridge connecting additional lengths of transmission line. It also facilitates the controlled movement of the centre frequency of a particular circuit, allowing the same basic circuit design to be used over a number of discrete frequency bands. This is particularly valuable for higher frequency circuits where perhaps the design data is a little less mature (figure 3.3.2).

The performance specifications for the demonstrator circuits in CLASSIC calls for the fabrication of 0,15 microns gate length pseudomorphic HEMTs (PMFET). The materials and device fabrication technology is being developed on this programme. Picogiga have the task of supplying the necessary molecular beam epitaxial grown material into the project. The very fine geometry of the gate is defined utilising an electron beam lithography process, that allows a low resistance (mushroom shaped) gate metallisation to be achieved (Figure 3.3.3). The same technology is being used to produce the high frequency Schottky diodes required on the project.

Figure 3.3.3   A Scanning Electron Microscope Picture of a Mushroom Gate as Used in the PMHEMT Circuits

124

The objectives of the MANPOWER project are concentrated on achieving reasonable levels of output power, circa 1 - 2W, from applications in L Band (1.8GHz) through to 27.5GHz. The full gambit of available devices are being utilised to achieve the demonstrator objectives. At the lower frequency the MESFET is the main work horse though an examination of the potential benefits of the heterojunction bipolar transistor (HBT) are being examined. At 14GHz the MESFET is being used to achieve the required 1 - 2 watts of output power. In this part of the programme an innovative 'bath-tub' heat sinking technology is being developed to improve the removal of heat from the active device (Figure 3.3.4). The manufacturing viability of this technique is being examined. At frequencies above 20GHz the pseudomorphic HEMT is the active device in the MMIC technology and where 100mw of output power is required.

Low noise performance at frequencies between 26 and 35GHz are the objective of the AIMS project and circuits are being produced using two device types. The pseudomorphic HEMT is the active device in the low noise MMIC amplifier where noise figures as low as 1.3dB at 26GHz were obtained (Figures 3.3.6 and 3.3.7). The heterojunction bipolar is being used to produce a low phase noise oscillator at 35GHz. The programme is comparing the microstrip and coplanar approaches to circuit design and IMST (Germany) are determining the models to be used in the coplanar circuit designs.

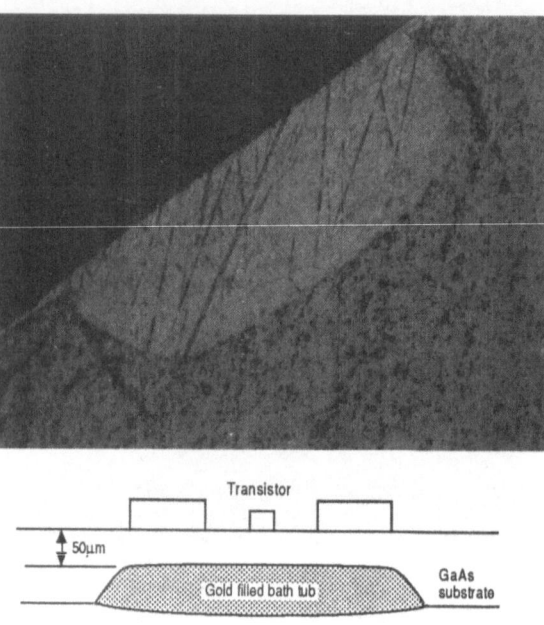

Figure 3.3.4   A Section Through a Plated 'Bath-tub' Which Gives Selective Heat Sinking to the Transistor Positioned Directly Above

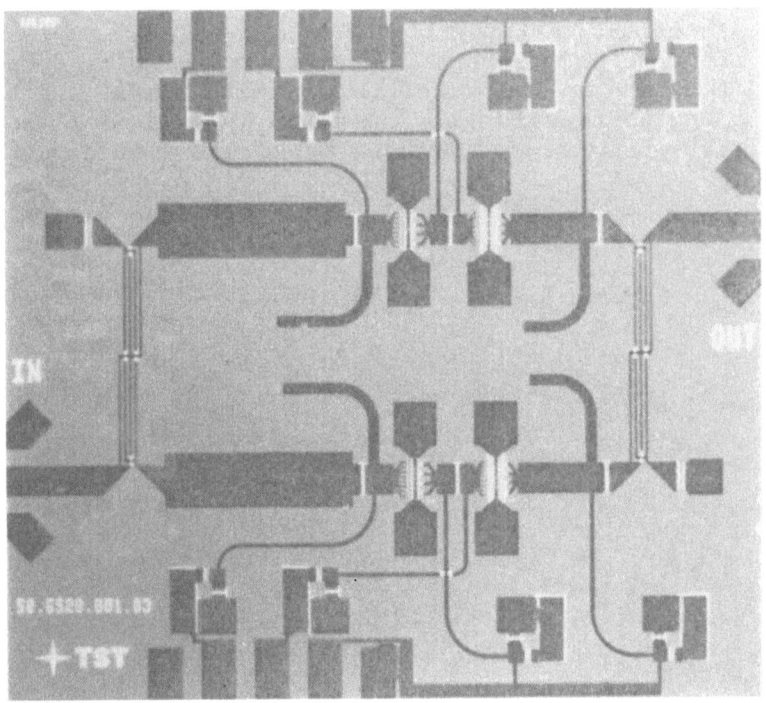

Figure 3.3.5   A Broadband Balanced PMHEMT MMIC Giving 8dB Gain
Over the Band 26-40GHz

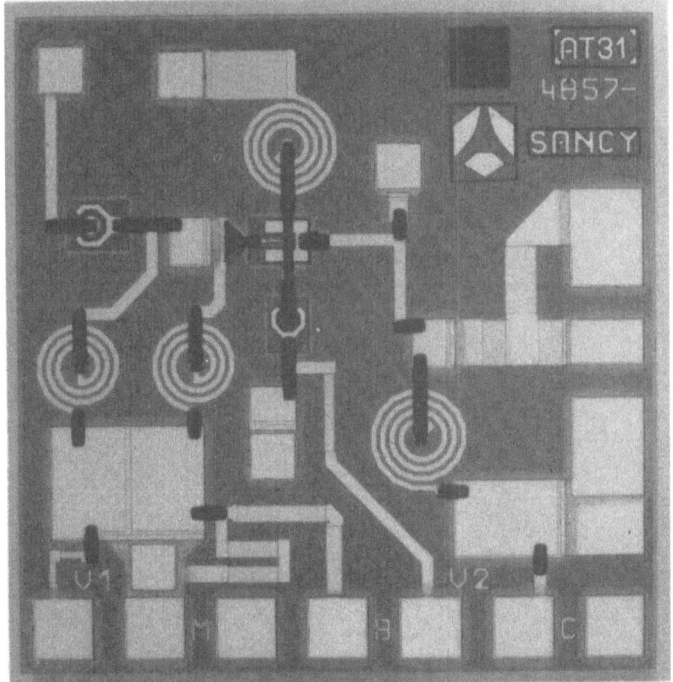

Figure 3.3.6   A Varactor Tuned HBT Oscillator Operating Over 26-32GHz

These brief resumes of the technologies being used and/or developed on ESPRIT programmes demonstrate the depth of the achievements of European chip makers in these highly sophisticated semiconductor areas. Enhancements to existing processes and the development of new ones ensures that European Industry maintains contact with the performances achieved elsewhere in the world.

3.4 High Throughput Multiwave MOVPE Reactor

H. Jürgensen, AIXTRON Semiconductor Technologies, Germany
P. Frijlink, Laboratoires d'Electronique Philips, France

1 Achievements

1992 was the year when multiple 4" wafers with uniformities in the 1% range could be already fabricated reproducibly, whereas five years ago, for the first time an MOVPE reactor has reached a level of development to produce multiple 2" wafers to be of use in HEMT fabrication. The time in between has brought many improvements on the original Philips design, resulting also in the AIX 2400 design, a reactor that can handle 5 x 4", 8 x 3", or 11 x 2" wafers at a time. A European Community ESPRIT program (called PLANET) generated many improvements required for multiple 3" or optoelectronic applications, as i.e. required in another ESPRIT program HIRED for high power GaInP visible lasers. Other programs were also directed toward InP based materials. All programs and cooperations had in common to improve the productivity and process versatility of the Planetary Reactor. These industrial requirements have been addressed in particular and are listed in the following:

- Production of batches of identical wafers
- Availability of monitor wafers with identical properties as in main batch
- Reduced defect densities compared to MBE of comparable batch size and epi-structure
- Multiple 2", 3", or 4" batches
- Superior process flexibility with Phosphorous compounds for laser and LED applications
- Frictionless rotation, eliminating feedthroughs. Rotation speed adjustable
- Adjustable temperature uniformity over full plates and symmetry in profile
- Flexible use of multiple 2", 3", or 4" wafers from run to run or within a run
- Abrupt interfaces in heterostructures realized by straight forward laminar flow design with no dead-volume
- All metal reactor for safety
- Warm wall inner deposition area avoiding particle formation

2 Introduction

High performance ICs and OEICs rely on complex epitaxial heterostructures with tight bandgap engineering. Related development and production requires homogeneous material, manufacturing of batches of comparable wafers, insitu multiwafer fabrication, reduced surface contamination, and defect density along with proven heterostructure processes for many combinations of III–V materials. The application of substrate rotation always seemed to cause technical problems, such as particle generation, mechanical feedthroughs etc., rather than to improve the process from a real production point of view.

MOVPE has evolved and has established itself as a unique epitaxial technique yielding 2D structures for extremely sophisticated device applications or thin films with more robust requirements as for solar cells, etc. For small structures, layer thickness of < 50 Å or a few atomic layers, MOVPE is ideally suited for R&D applications as well as a production tool for manufacturing of device material.

A leading edge system for MOVPE production of wafers is the Planetary Reactor using a Gas Foil Rotation Technique. This reactor type has been developed by Philips LEP, France, /1/, and is exclusively licensed to AIXTRON, Germany.

This paper is summarizing the breakthrough in III–V multiwafer MOVPE mass production applications using the Planetary Multiwafer Reactor with Gas Foil Rotation (5x3"/7x2" or 5x4"/8x3") which was originally developed and patented by LEP for growth of GaAs/AlGaAs heterostructures and has been used successfully since then for HEMT production, laser fabrication, and GaInP deposition.

In similar Planetary Reactors, GaAs and InP based materials for a wide range of optoelectronic applications have been produced. The use of low pressures is not only advantageous for the handling of P-containing compounds and reduction of overall gas consumption, but also allows to drastically reduce the amount of H_2 required for driving the wafer support. The variation of thickness in these multiwafer systems is reduced to the order of 1–2% for GaAs, AlGaAs, GaInP, InP, GaInAs, and GaInAsP (1.55, 1.3, 1.05 μm). Thus, one major advantage in comparison to MBE is that these reactors are capable to handle both GaAs and InP based processes with high concentration of phosphorus. For production of visible lasers (AlGaInP), GaInP HBTs, or complex solar cell structures, these processes have also been developed. This finally makes this MOVPE technology by far superior for production application than MBE.

The relatively simple realization possible for such a variety of different wafer sizes emphasizes the efforts to create even larger reaction chambers capable for large area epitaxial growth of different material systems. The use of silicon wafers as substrates for advanced device structures is no longer limited to 2" or 3" wafers, also 4", 6", and 8" wafers can be used, and rectangular arrangements for solar cell production are developed.

Due to the circular arrangement of the substrates and a rotational movement during the epitaxial process, a very high uniformity on the wafers and from wafer to wafer is obtained. At the 5th IC-MOVPE Conference held June 1990, in Aachen, P. M. Frijlink of Philips LEP, the inventor of this Gas Foil Rotation Technique, presented results for inter- and intra-layer uniformities of thickness, doping, and composition /2/.

With the exception of minor edge effects – he reported how to avoid even these small effects – all uniformity results were in the 1% range. In a compositional measurement of the In-content in lattice matched GaInAs, with 53% In on InP, Frijlink presented that the deviation from the mean value is < 0.05% at five different points on each four different wafers grown simultaneously.

Being asked by Mme Razeghi (Thomson CSF, now Northwestern University, Evanston IL) about the difference between the data from MOVPE and MBE grown, Frijlink's answers led the delegates to the point that today's MOVPE with Planetary Reactors and Gas Foil Rotation Techniques confers device quality comparable to that obtained with MBE. He explained that his group at LEP had demonstrated this by obtaining samples of both MOVPE and MBE wafers and processed them in the same device processing line at LEP.

The situation that MOVPE wafers are at least as good as MBE wafers for device production has become apparent principally due to the use of this Planetary Reactor with Gas Foil Rotation Technique as a multiwafer production system. In addition to the excellent results in wafer to wafer uniformity, Frijlink noted that this approach has outstanding cost-effectiveness and productivity advantage over MBE which is not that far developed for multiwafer production.

An overview on GaAlAs/GaAs HEMT data achieved in industrial laboratories will be given also in the course of this review paper /3/. 0.5 μm gate HEMT structures (35 GHz cut-off frequency) with mean g_m value of 285 mS/mm and deviations of ± 1.9 % at a yield of 93 % have been obtained. The IC relevant surfscan measurement is detecting about 10 defects (> 0.18 μm) on a 2" wafer which is comparable to the base substrate wafer.

Pseudomorphic $Ga_{0.82}In_{0.18}As$ (150 Å) HEMT structures could be produced with variation of sheet carrier concentrations of only ± 1 % . Two dimensional electron gas mobilities in a $GaAs/Al_{0.38}Ga_{0.62}As$ structure with 350 Å spacer layer exhibit low temperature mobilities exceeding 660.000 cm^2/Vs. These processes are already partly transferred into a GaAs foundry service and are commercially available.

The development of the reactor continues. AIXTRON will soon be announcing a larger system with twice the capacity of today's AIX 2000/2400 Planetary System and with the same or even better uniformity quality.

3 Multiwafer Reactors: The Concepts

3.1 The Market Driving Force

Industrial exploitation of multi quantum well and HFET structures for application in discrete devices such as laser diodes for CD or optical storage systems, or low noise FET in satellite communication have emphasized the development of epitaxial reactor systems with the capability to grow on several wafers at a time. The application of AlGaAs/GaAs in LSI/VLSI with high speed and low power dissipation requires new technological breakthroughs not only in the field of epitaxial growth. In 1986, a Japanese group reported new ideas of the geometrical configuration between the source and the substrate in a MBE system and tried to improve the growth conditions for highly uniform epitaxial layers on 3" diameter semi-insulating GaAs substrate which should make high throughput and high quality possible /4/. However, the surface defect problem of MBE-grown layers remains a serious one for circuit fabrication on LSI level complexity. The occurring surface irregularities are called oval defects. Depending on growth conditions, the density of the oval defects was 500 – 3.000 cm^{-2}. The oval defects have a lateral dimension in the micrometer range and they seriously affect the current-voltage characteristic of a HFET. The origin of these surface defects in MBE seems to be the gas sources. The challenges from a material viewpoint must be centered on the reduction of these oval defects on MBE wafers.

3.2 The Planetary Reactor Concept

The Multiwafer Planetary Reactor developed by P. M. Frijlink (Fig. 1) is based on the Gas Foil Rotation Technique. The newly developed Planetary Epitaxial Reactor System employs a novel concept of crystal rotation that is specially suited for growing ultrathin films for optoelectronic and micro-wave devices. Originally developed by Philips LEP (France) and under license to AIXTRON, the system is proven to have large-scale manufac-turing capabilities and to result in device quality comparable to MBE tech-niques in the case of HEMTs, but additionally, a variety of extremely thin compound films and structures have been synthesized, such as InP, GaInP, and GaInAsP /5,6/ which are not available with MBE.

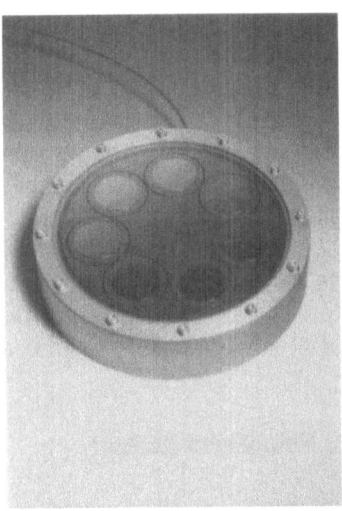

Fig. 1 Planetary Multiwafer Reactor Model

In the horizontal reactor, the wafers are arranged on a circular susceptor which rotates around its central axis through which the gas flow of group III organometallics with the dopants and a hydrogen carrier flow with the group V hydrides is provided. A graphite susceptor plate supporting the wafers is levitating on a layer of H_2. The same gas flow is also causing the rotation of the wafer support by means of shear forces. Each individual wafer is rotating by the same technique, thus causing planetary motion. The speed of rotation is directly controlled by computerized mass flow controllers. This technique has been demonstrated to result in unlimited symmetry between the wafers in each run. The 2", 3", or 4" capability of the reactors is only defined by the geometry of the graphite discs and thus can be changed in a simple procedure between the runs.

The ability to extract better performance or to design complex device structures depends on thin-film growth quality, which, in turn, critically depends on equipment functionality. Major strides in improving film thick-ness and compositional uniformity have been achieved using AIXTRON's advanced-design horizontal low-pressure MOVPE reactor. The reactor overcomes gas–phase depletion problems and geometry–related non-uniformities by using the unique gas–bearing technology.

3.3 The Low Pressure Advantages

Low pressure MOVPE in a horizontal reactor has proven to be the most effective method for large scale production of sophisticated III–V or II–VI devices. Using a Planetary Reactor Concept with Gas Foil Rotation Technique, AIXTRON incorporates the most advanced technology in its multi-wafer systems.

At first used for atmospheric pressure reactors, AIXTRON succeeded in producing low pressure reactors /10/ providing all the advantages of the above techniques combined with the low pressure capabilities to handle all phosphorus containing compounds, to have better control on In compounds, and to reduce overall gas consumption. The advantages have been verified in a joint ESPRIT program, presented at the Int. Conf. on VGE at Nagoya 1991 /11/ and confirmed at several customer laboratories /12/.

3.4 The Advantages in Manufacturability

One major advantage of the Planetary Reactor is the Gas Foil Rotation Technique. No mechanical drives are necessary to rotate the wafers as well as the susceptor as a whole. This makes it very easy to change the susceptors, to grow materials of different sizes, e.g. 2", 3", 4", or 6". The relatively simple realization possible for such a variety of different sizes emphasizes the effort to create even larger reaction cells. Even the use of silicon wafers as substrates for advanced device structures is no longer limited to the 2" or 3" sizes.

MBE is not yet a proven multiwafer technology. There has been no inde-pendent publication demonstrating reproducible proven multiwafer results from MBE. There appears to be a clear advantage in MOVPE as a proven multiwafer production technology which is already used in a GaAs foundry service, the Philips Microwave facility in Limeil /12,13/, which is cooperating with LEP; and merchant epiwafer suppliers have also invested in such equipment.

Today defect density for production type MBE is worse than that for MOVPE. For MOVPE, there are proven results available which demonstrate data of less than 1 per cm^2 /14/. In general, as worst case, it is always better than 10 per cm^2 which in this case is a clear and unbeaten advantage of MOVPE and is a process specific feature.

Run to run reproducibility in MOVPE has been published /1/, but there are as yet no data available from MBE. In MOVPE it is better than 1% in thickness, doping, and composition (depending on structure and composition).

The need for in-situ monitoring in MOVPE is not an absolute necessity as it is in MBE. In the users' experience, this is due to the excellent control in the MOVPE process. Additionally, the need for such extra equipment makes MBE even more expensive.

There is an additional advantage to MOVPE which needs to be mentioned, too: there is the capability of graded structure control to an extremely flexible extent, just by precise control and ramping of mass flow controllers with any kind of mathematical functions /15/. This is not possible with MBE, due to the lack of temperature control required for such structures.

3.5 The Reactor Features

The AIX 2000 (5x3"/7x2") and AIX 2400 (5x4"/8x3") systems consist of the reactor itself, a gas mixing system, and a computerized control and safety unit (Fig. 2).

Fig. 2 Front View of the AIX 2000/2400 MOVPE System

The electropolished tubes, automatic inert gas welding, and leak check certification before and after installation combined with AIXTRON's outstanding safety standards, define the high quality of the system. High safety demands and high quality of wafers are strongly correlated.

The manually driven or computerized process control unit runs specific processes with user-specified data. In case of any failure to the system, a

fixed-wired electronic program runs the machine with a safe set of data so that no further damage can be caused.

In the computerized version, the user interface is a proprietary editor, specially developed by AIXTRON to provide the most convenient data handling. The interface operating system is MS-DOS, the system itself is running under OS9, the most efficient real-time operating system for complex time critical applications.

The main advantages are:
- Very low operating costs
- Excellent homogeneity in film thickness and composition
- Excellent electrical and optical quality of the films
- Compact, modular system
- High capacity
- High up-time
- High versatility.

A 3-dimensional schematic of the reactor cell is given in Fig. 3. The cross section shows the main graphite susceptor, the satellite discs, the IR lamp reflector underneath, the double O-ring leak monitoring system, and water cooling of the entire metal contained reactor.

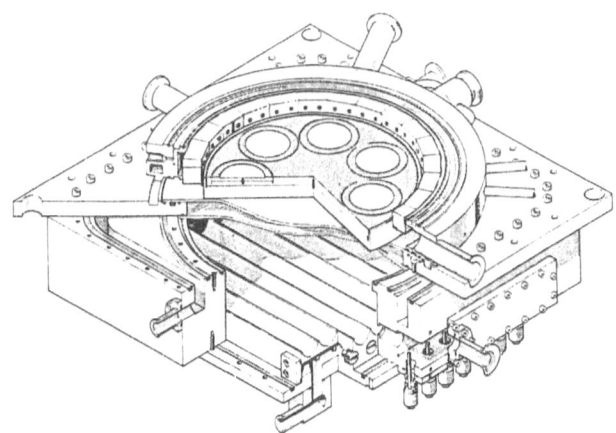

Fig. 3 Schematic of the AIX 2000/2400 Planetary Multiwafer Reactor Cell

A schematic of the top view of the different susceptor configuration of the AIX 2400 is given in Fig. 4.

As can be seen, the main disc is available in 3 versions that can be replaced: 5x4", 8x3", or 12x2" wafers.

8 x 3" 11 x 2"

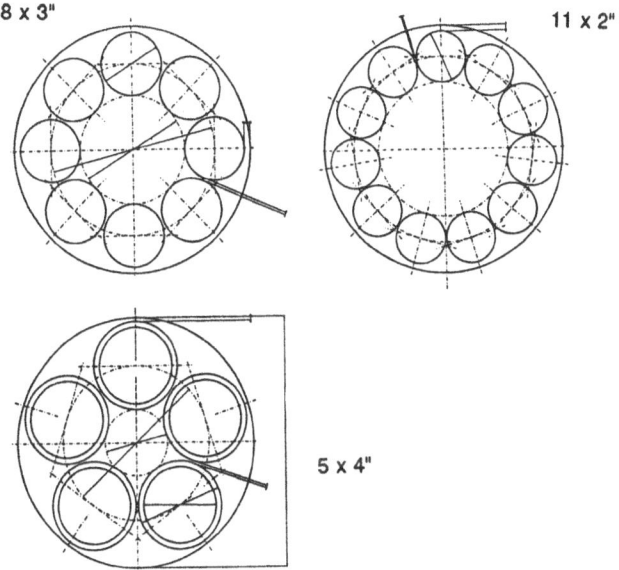

5 x 4"

Fig. 4 Schematic of the different AIX 2400 Susceptor Configurations

The capabilities and features of the AIX 2000/2400 systems are listed below:

- Stainless steel reactor with run/vent inlet manifold /6,7/
- Horizontal flow conditions with planetary motion of substrates /16/
- Susceptor for seven x 2" wafers or five x 3" wafers (AIX 2000) and five x 4" or eight x 3" wafers (AIX 2400), without changes in flow characteristics
- Uniform substrate temperature
- Short heating time: 5 to 7 min.
- Short process times for high output: 2–4 runs per 8 hour day possible (depending on layer thickness)
- Ultra fast switching times (0.1 sec) are available by using AIXTRON's unique, flexible, purged manifold with zero-dead space (pat. pending)
- Very accurate process control: step programmable, 0.1 sec increments /6/ (optional: function programming)
- High speed 32 bit computer with floppy disks and hard disks (color graphics optional)
- Total flow rate: standard 25 l/min to 50 l/min
- Fast mass flow controllers and electronic pressure regulators: response time 1.5 sec
- High purity gas blending system: tubings and components electropolished; only VCR connectors or orbital welding is used

- Accurate temperature control for MO sources: better than 0.05°C
- Low pressure option including rotary pump and optional roots blower with Fomblin oil and chemical oil filter
- Safety interlock system operating even at computer down times
- Glovebox, double load lock system.

4 General Material Quality

4.1 Uniformities

Several multiwafer geometries have been applied to MOVPE growth of III–V compounds. In 1988, the new type of Planetary MOVPE Reactor was introduced for simultaneous growth of very uniform epitaxial layers on 7 wafers in Planetary motion, using a circular growth chamber with horizontal geometry and radial gas flow /1/. In this concept, the radial flow implies a decrease of the gas velocity along the flow direction, making it possible to realize an almost linear decrease of the growth rate with increasing radius, by depletion of the gas phase. Each wafer is turning around its own axis, the almost linear decrease yields a slightly concave or convex epilayer thickness profile over the wafers. The use of separate introduction of III and V compounds directly into the hot zone of the growth chamber, thus avoiding the parasitic side reactions so deleterious for uniformity, is also exploited to fine-tune the concavity/convexity, thus permitting to obtain very uniform thickness profiles on the wafers, with a variation of typically ±1% thickness and ±1% in doping over a 2" wafer and between the wafers (Fig. 5). This uniformity is obtained with a high degree of depletion of the gas phase: about 30% of the available group III material deposits from the gas as it flows along its path of 2" or 3" over the wafer, thus permitting a high efficiency in the use of the available group III materials: 0.16 Å/s of GaAs per cc/min of saturated TMG vapor at 0°C.

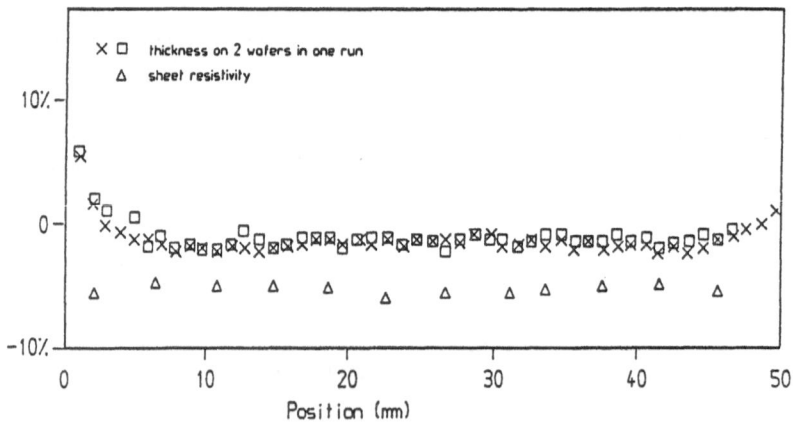

Fig. 5 Uniformity of GaAs Thickness and Sheet Resistivity

This high efficiency is particularly important for the growth of In alloys in multiwafer reactors, as partial pressures of In precursors at room temperature are small and the price of those precursors is high, thus limiting their practical quantity and the practically available stable massflow from In precursor sources. This reactor concept is particularly attractive for the growth of In alloys because of the high efficiency and the avoidance of parasitic side reactions by separate introduction of III and V precursors directly in the hot growth chamber. The suppression of side reactions is also improved by the geometrical consequences of radial gas flow: that is the high gas velocity and low gas volume per wafer upstream of the wafers reducing the transit time of the mixture to the wafer.

Throughout the data below, the same or similar reactor was used with the same vector flow rate values to obtain identical uniformity in thickness and doping. This uniformity proved to be reliable over the 670 growth runs actually carried out in this equipment. Although the form of the growth chamber itself was not essentially modified, a molybdenum tunnel in the form of a ring surrounding the susceptor, collecting the gas from the growth chamber through regularly spaced holes in its inner circumference, was included to reduce the frequency of cleanings of the equipment. The actual frequency of cleaning of any part of the reactor chamber is since then once every 150 growth runs, representing about 120 μm of cumulated layer thickness and a maximum of 1050 wafers, showing the effectiveness of ceiling temperature control.

The rotation speeds are 0.5 turns/s for the main rotation and 2 turns/s for the satellite rotation. The Gas Foil Rotation proved totally reliable over 670 growth runs. No influence of the actual value of the rotation speed on the layer properties was detected.

For the evaluation of layer thickness uniformity, 0.2 μm thick $Al_{0.5}GaAs$ and 0.8–1.5 μm thick GaAs layers were grown on a (100) oriented GaAs substrate under different growth conditions at PRLE Eindhoven /12/. After growth, half of the wafer was covered by wax and the wafer was exposed to an etchant which blocks totally on the AlGaAs layer. After removing of the wax, the thickness was measured using a moving stylus apparatus. Fig. 6 shows the layer thickness variation across a 2" wafer grown at 200 mbar. Growth performed at 100 mbar and 43 l/min gave similar results. The results at 100 and 200 mbar give excellent results in layer thickness control. Moreover, these results show that the total flow can be reduced by a factor of 2 without having any influence on the excellent homogeneity of thickness.

Results of the computer simulations are given in Fig. 7. It shows the calculated growth rate profiles on the susceptor from the center to the edge of the susceptor as function of the flow rate. The nominal flow is 43 l/min (curve 10, curve 9 is for a flow of .9x43 l/min etc.).

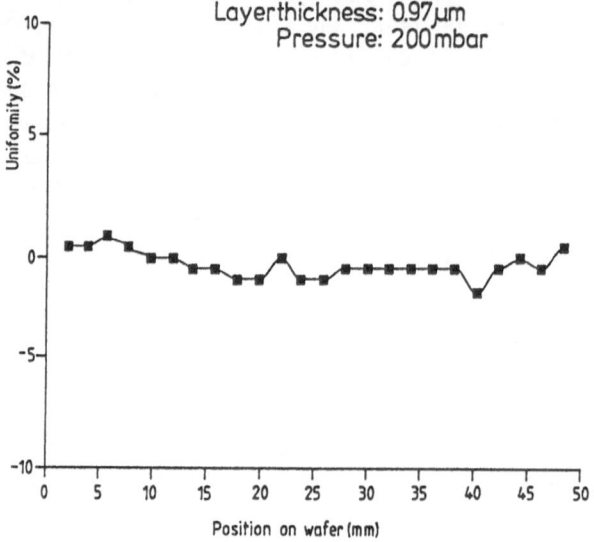

Fig. 6 Layer Thickness Variation over 2" Wafers Grown at Low Pressure

Proving the uniformity of epi-structures is usually the task of sophisticated testing and analysis equipment. Today's III-V technology is moving towards multiwafer epitaxial systems in mass production which calls for quick and large-area techniques to evaluate the thickness uniformity, most beneficially using non-destructive techniques. Specially designed GaAs/AlAs multilayer structures with proper choosen optical thickness show quarterwave stack like behaviour in the visible region of the light spectrum with a surprisingly sharp and colour-selective reflection behaviour.

These structures stem from the application as mirror planes in vertical cavity surface emitting laser diodes for visible light and other applications. Due to the wavelength sensitivity of the human eye, these structures can be simply evaluated by visual inspection immediately after the epitaxial growth run, prior to unloading from the reactor with a thickness uniformity resolution in the range of ± 0.5% (see Fig. 8)

Wafers fulfilling this thickness uniformity requirement show a uniform colour in reflection across the entire wafer surface. Larger variations would result in the formation of concentric rings of different colour or streamline images from depletion effects, dependent on the geometry of the reactor. A 10 period alternating structure of 33nm GaAs/41nm AlAs (Bragg reflector) is resulting in a yellow-green colour on the wafers, used for demonstration

is resulting in a yellow–green colour on the wafers, used for demonstration of the excellent uniformity of epitaxial layers in the AIXTRON Planetary reactors AIX 2000/2400 (see Fig. 9).

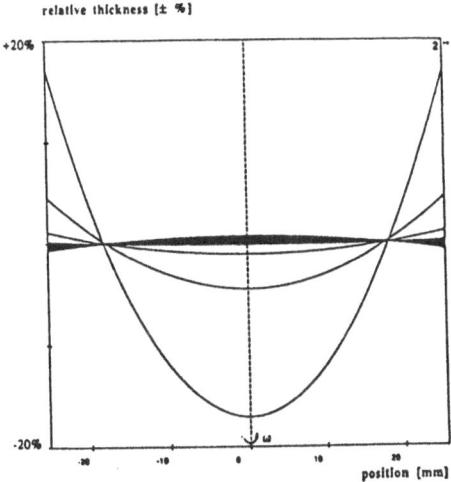

Curve	Total flow rate[*]	Standard deviation (%)
1	$0.1\ Q_{nom}$	±17.1531
2	$0.2\ Q_{nom}$	± 4.5550
3	$0.3\ Q_{nom}$	± 1.1335
4	$0.4\ Q_{nom}$	± 0.1867
5	$0.5\ Q_{nom}$	± 0.0583
6	$0.6\ Q_{nom}$	± 0.1294
7	$0.7\ Q_{nom}$	± 0.2050
8	$0.8\ Q_{nom}$	± 0.3308
9	$0.9\ Q_{nom}$	± 0.4982
10	$1.0\ Q_{nom}$	± 0.6795

[*] Nominal flow rate $Q_{nom} = 43\ l/min.$

Fig. 7 Calculated Normalized Thickness Profile on Rotating Wafer

To show how this homogeneity is for seven 2" wafers, in one run seven Bragg–reflectors have been grown consisting of 20 layers of alternating GaAs (300 Å) and AlAs (340 Å), giving for all seven wafers exactly the same green color indicating a homogeneity in thickness within 1% for all seven wafers (Fig. 10). These Bragg–reflectors were grown at 200mbar.

AIXTRON as leading equipment manufacturer of MOVPE systems is involved in several cooperations with customers in the investigation of complex optoelectronic structures. Resulting from this experience, AIXTRON is today using these Bragg reflector structures routinely as a non-destructive technique to proof the capabilities of its Planetary reactor systems AIX 2000 (for 7 x 2", 5 x 3" wafers) and AIX 2400 (for 5 x 4", 8 x 3", 15 x 2" wafers) with respect to uniformity over the single wafers, from wafer to wafer, or from run to run.

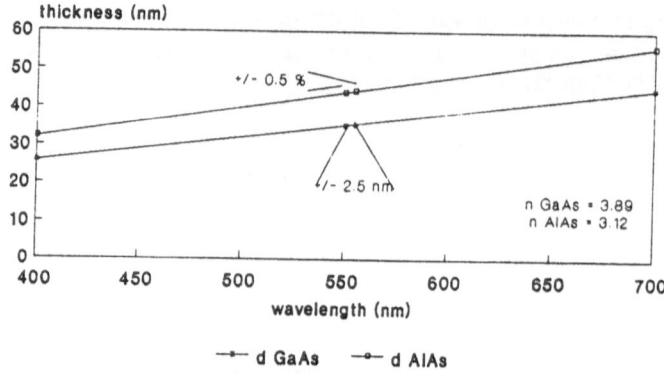

Fig. 8 Bragg Reflector Structures GaAs/AlAs Quarterwave Stack

Using these Bragg reflectors as simple tool for uniformity optimization, the outstanding material quality from the AIX 2400 multiwafer systems have been independently confirmed today by several U.S. and U.K. industrial laboratories demonstrating thickness uniformities in the range of ± 1%, sheet resistivity evaluations below 2%, with the unrivalled surface morphology advantage of the Planetary reactors giving residual particle defect densities below 10 per cm^2.

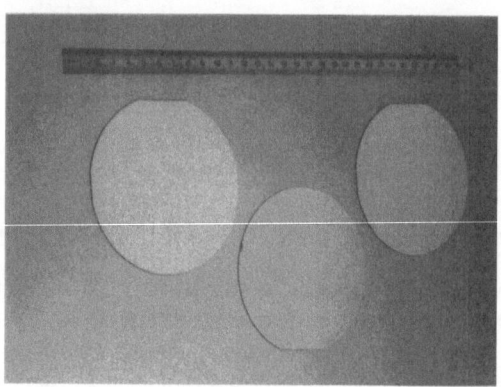

Fig. 9 Thickness Uniformity ±1% on 4" AlAs/GaAs Bragg Reflector Wafers

Meanwhile, the technique has been refined and high resolution characterization methods have been performed on 2", 3", and 4" wafers that show a single uniform colour visible to the human eye. These uniformly coloured wafers have a reflectivity characteristic measured by an automatic mapping reflectometer as in Fig. 11 (equipment FTP 500 by Sentech Instruments, measurement courtesy of Sentech Instruments, Berlin, Germany). This is confirming the uniformity in the order of ± 1% on 4" wafers.

Alcatel SEL, Stuttgart/Germany, has supported these data with high-resolution 4–crystal X–ray spectrometer measurements. These measure–

ments clearly confirm from evaluation of multiple satellite peaks a uniformity of 1–1.5% on the 4" wafers.

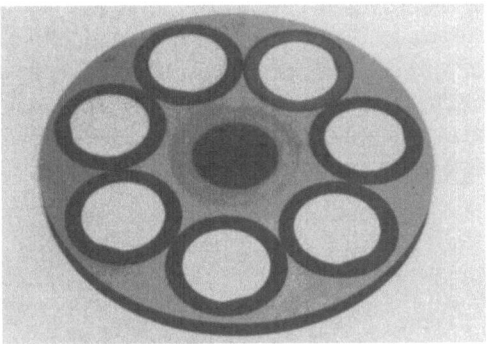

Fig. 10 Bragg Reflectors Grown with the AIX 2000 Showing Perfect Homogeneity

In order to evaluate the homogeneity in composition, $Al_{0.13}GaAs$ layers between $Al_{0.35}GaAs$ cladding layers have been grown, and the PL–wavelength at room temperature at 5 indicated places on the wafer was measured. Again the homogeneity is excellent (<0.5 nm) (Fig. 12).

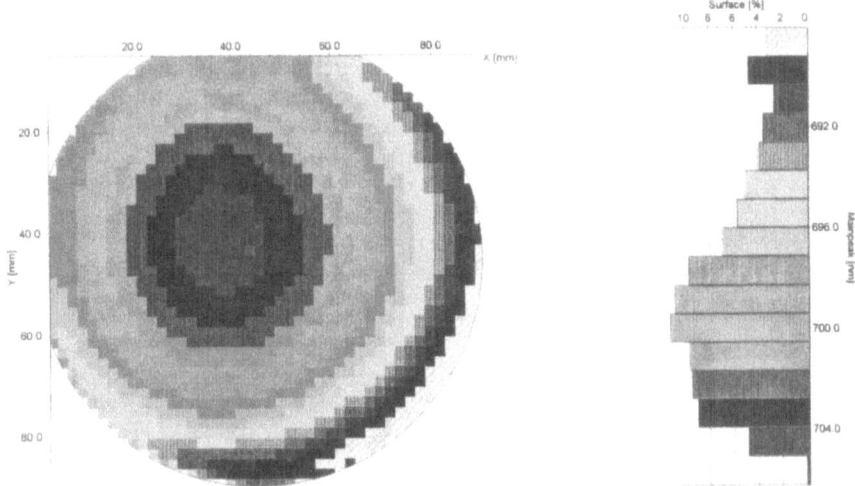

Fig. 11 Thickness Uniformity ± 1% on 4" AlAs/GaAs Bragg Reflector Wafers: Reflectivity Mapping (equipment by Sentech Instruments, measurement courtesy of Sentech Instruments, Berlin/Germany)

A layer of 6000 Å of $Ga_{0.47}In_{0.53}As$ was grown simultaneously on four 2" InP substrates. The substrate orientation was (100) oriented 3° off to the nearest (111) direction. The growth temperature was 620 °C, and the total

142

reactor pressure was 1 atm. The obtained growth rate was 1.4 Å/s. Double crystal X-ray diffraction rocking curves were determined at 5 points on each of the 4 wafers. Fig. 13 shows for each point the deviation of the In percentage from the mean value over these 20 points: All of the absolute values are within a range of 0.1%. Within each wafer, the absolute values are within a range of 0.05%. This very high uniformity compares favorably with the best reported values for single wafer reactors, and it settles the question whether the difference in diffusion coefficients of In and Ga species could be very deleterious to uniformity in composition.

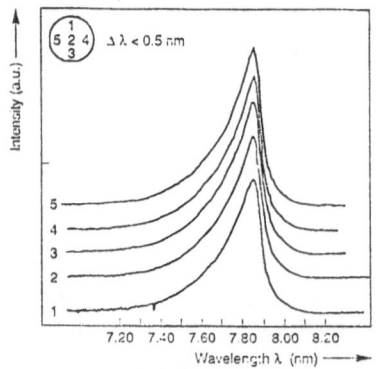

Fig. 12 Measurement of the PL-Wavelength at Room Temperature

Because material composition determines the wavelength of photon emission, measurements of the emissive wavelength serve as an indicator of material composition. In the GaInAsP system, emission-wavelength variations of less than 2 nm reflect excellent uniformity in composition across a single wafer.

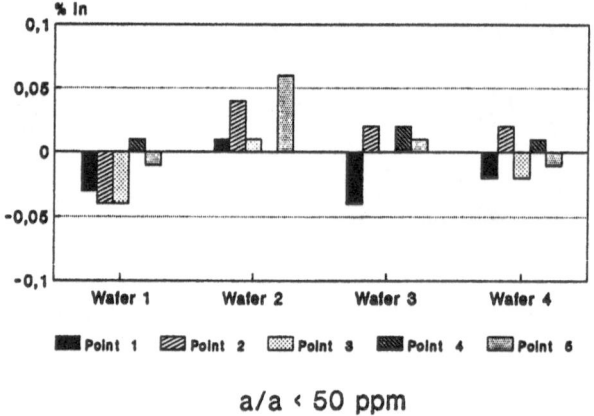

a/a ‹ 50 ppm

Fig. 13 Compositional Uniformity of GaInAs/InP

We may also deduct from this uniformity that parasitic elimination reactions between In containing species and As containing species are negligible. In view of the unoptimized and intentionally rather high V–III ratio of 216, the use of atmospheric pressure and the comparatively low total flow rate of 6 l/min per 2" wafer, we may conclude that parasitic elimination reactions do not easily occur in this reactor concept.

The determination of the film thickness was performed by using a selective etching technique and DEKTAK measurement. Alloy composition results were obtained by double crystal X–ray diffraction (DCXRD) or photoluminescence (Fig. 14). As shown in the histogram, wavelength variations of less than ± 2.8 nm over the entire load and between different wafers were obtained.

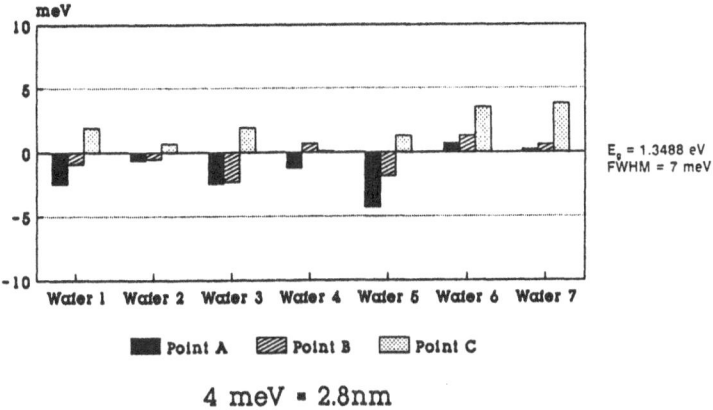

4 meV = 2.8nm

Fig. 14 4K PL Energy Shift of GaInAs/GaAs

As expected, the results on uniformity of film thickness, carrier concentration, and composition have been further improved. The variations in InP, GaInAs, and GaInAsP film thickness could be limited in the order of 1%. Wide bandgap materials in the Ga–In–As–P alloy system (λ = 1.05 μm) showed variations of less than 2 nm over the entire area of a 2"–wafer [7]. Electron mobilities of over 100.000 cm^2/Vs (77K) and over 10.000 cm^2/Vs (300K) for the $Ga_{0.47}In_{0.53}As$/InP structures are achievable.

The application of the technique to InP based compounds including ternaries and quaternaries is briefly described below. GaInAs/InP double heterostructures have been grown on 2" wafers and were evaluated by surface profiling after selectively etching grid structures in the InP. The variation of InP film thickness was less than ± 1,5% over 80% of the area.

144

InP layers have been doped intentionally with Si by injecting SiH_4 into the reactor during growth. The carrier concentration was determined by the Van der Pauw method taking into account the film thickness evaluation. The variation is less than 1.5×10^{15} cm^{-3} over the entire area (average absolute concentration 3.36×10^{16} cm^{-3}). The carrier mobility around 3.500 cm^2/Vs at 300 K compares well with literature data.

AIXTRON is using a Lehighton Sheet Resistance Mapping System to characterize all material grown in its application laboratories, in the framework of the confidential customers cooperation programs and EC/government funded projects. This equipment gives a non–destructive sheet resistance mapping with resolutions to better than 0.5%, dependent on the structures. This system is routinely used to characterize the uniformity in electrical properties and confirmed data of 0.5–2%.

The possibility of growing GaInAs was one of the requirements for preparing the structure for InP film thickness evaluation by selective etching. The variation of GaInAs film thickness over a 2" wafer is shown in Fig. 15. Deviations less than $\pm$ 1% appear over 80% of the area (2.5 mm edge excluded). The distribution of lattice mismatch measured along two perpendicular axes of the wafer is within 2×10^{-4}.

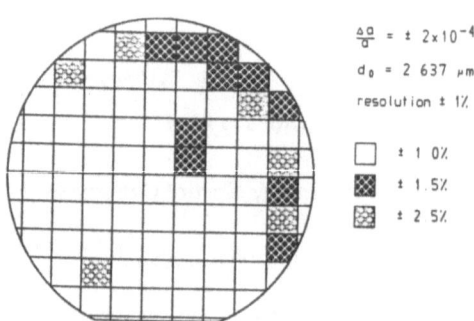

Fig. 15 Variation of GaInAs Film Thickness over a 2" Wafer

Finally, the quaternary material has been investigated. For this example, the 1.3 μm quaternary composition was chosen. This composition, used for waveguide and laser production, is more sensitive to variations of the growth process parameters than the longer wavelength material.

The thickness homogeneity shows a pattern similar to that achieved for InP. The variation of PL wavelength is plotted in Fig. 16 and indicates a rotational symmetry which can be further improved by slightly modifying the gas flow parameters. Excluding a 2.5 mm rim, the variation of the wavelength is within 3 nm.

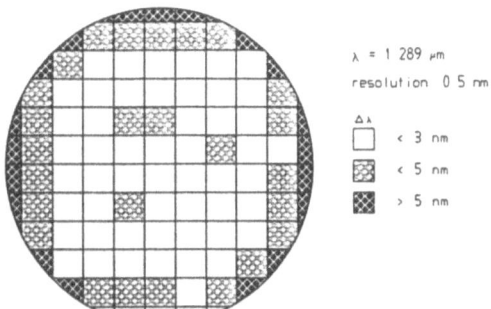

Fig. 16 Variation of PL Wavelength of InGaAsP over a 2" Wafer

GaInP/GaAs structures for optoelectronic applications have been developed with compositional uniformity characterized by PL of 660 nm ± 1 nm across full 2" wafers /17/.

4.2 Quantum Well Properties

In order to characterize the capability of the reactor to grow well defined quantum wells, a 100 Å GaAs layer was grown between two $Al_{0.5}GaAs$ layers. The 4K photoluminescence spectrum obtained on this structure shows a FWHM of 3.0 meV /14/.

Several similar quantum well structures of different thicknesses were grown separated by 330 Å thick $Al_{0.45}GaAs$ layers. The structures were grown on an exactly oriented (100) s.i. GaAs substrate. QW down to about 7 Å can be detected in the spectrum. The FWHM can be explained by an interface-sharpness of 1 mono-layer /12/.

For measurements of the non-radiative recombination time in undoped GaAs and $Al_{0.13}GaAs$ by studying the PL-decay, wafers were grown with 2 μm thick GaAs between $Al_{0.25}GaAs$ cladding layers and 2 μm thick $Al_{0.13}GaAs$ between $Al_{0.35}GaAs$ cladding layers. The GaAs showed a non-radiative recombination time of 700 nsec and for $Al_{0.13}GaAs$ 190 nsec, both measured at room temperature /12/. This indicated that this material is an excellent basis for growing GaAs/AlGaAs laser structures, which is one aim for this production scale MOVPE reactor.

4.3 Defect Densities

The preparation of the substrates prior to growth has been improved to reduce the amount of defects after epitaxy. The amount of surface defects were measured with a surfscan 4500 over the full range from 0.18 μm to infinity. The wafers in Fig. 17 show the record low value, and Fig. 18 shows the values on seven 2" wafers of one typical run. The record low value – one defect per 2 cm^2 – does not yet characterize the normal surfscan cleanliness, but it is nevertheless a very important result because it indicates that the local MOVPE growth process in this reactor does itself not add more than one defect per 2 cm^2, and probably even less. This is contrary to MBE growth process of HEMTs which adds 50 to 100 oval defects per cm^2 during growth itself.

The MOVPE PLANET Reactor

Standard GaAs/GaAlAs HEMTs

Dust contamination

```
PARTICLES TOT:     10
PARTICLES/cm²: 0.67
AREA

HISTOGRAM:
0.18- 0.41:             1
0.41- 0.64:             1
0.64- 0.87:             0
0.87- 1.10:             0
1.10- 1.33:             0
1.33- 1.56:             0
1.56- 1.79:             2
1.79- 2.02:             0
2.02- 2.25:             1
2.25- 2.48:             0
2.48- UP:               5
MEAN:           1.9100
STD.DEV:        45.15%

HAZE AVG.TOTAL:   6ppm
HAZE REGION:        23%

EXCLUSION:              2
MAX.SIZE:            2.56
THRESHOLD:    d   0.12
PARTCL:     0.18-   2.56
BIN:               0.01
HAZE:          9-   256
```

Total number of particles = 10 over 2"

Fig. 17 Defect Measurements on 2" Wafers

The actual average value in the Planetary Reactor is depicted in Fig. 19. It demonstrates the defect density measured on wafers of 25 subsequent runs.

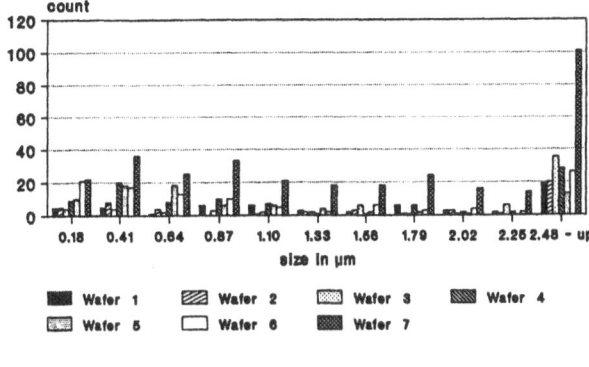

7.4 /sqcm

Fig. 18 Statistical Evaluation of Defect Counts on Full 7x2" Wafer Runs

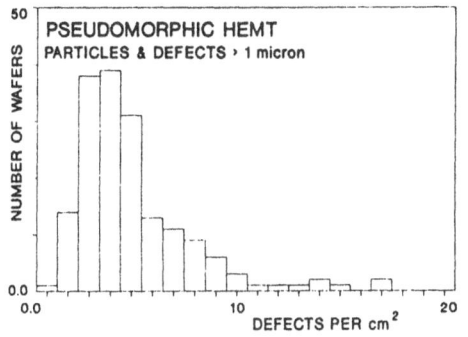

Fig. 19 Pseudomorphic HEMT: Count of Particles and Defects over 25 Runs

5 Device Production: Characteristics

5.1 HEMT

The use of a multiwafer reactor can only be justified by the necessity to feed device processing. The Planetary Reactor has been used for epitaxy of HEMT structures. Si_2H_6 as a dopant source, trimethylgallium, trimethyl-aluminium, and arsine was used. The capability of this reactor to grow good AlGaAs interfaces with a very highly mobile 2D electron gas was already shown in /1/, by a Shubnikov-de Haas measurement on the epilayer of growth run number 26. The same structure was grown at growth run number 402. The mobility of the sample was 8900 cm^2/Vs at 273°K, 170.000 cm^2/Vs at 77°K, and 655.000 cm^2/Vs at 4°K with cooling and measurement in the dark, with a sheet electron concentration of 3.9 x 10^{11}/cm^2 /14/. The 2-dimensional electron gas contained in GaAs/$Al_{0.28}Ga_{0.72}$As heterostructures for HEMT transistors typically has a room temperature mobility of 6.000 cm^2/Vs at a sheet electron concentration of 1.3 x 10^{12}/cm^2. Fig. 20 plots the value of a two dimensional electron gas mobility as a function of temperature, in the case of a 350 Å spacer layer between the GaAs region and the doping layer in Al$_{0.38}$Ga$_{0.62}$As. The values are the best reported so far for MOVPE materials.

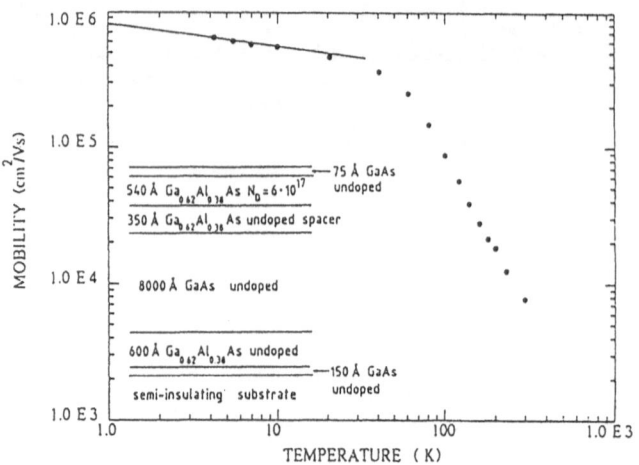

Fig. 20 Hall Mobility vs. Temperature with a Cooling Rate of 10K/Min.

HEMT transistors with 0.5 μm recessed gate process, AuGeNi ohmic contacts, and TiPtAu gate had a maximum transconductance of 320 mS/mm. The epitaxial layer structure of Fig. 21 has been used as conventional HEMT structure. It is about optimum in the sense that it combines good performance on F$_t$, noise figure, and inverse voltage breakdown, even in the case of small subsequent process deviations. Devices (L$_G$ = 0.5 μm) fabricated from layer structures with room temperature mobilities of 6.000 cm^2/Vs at a sheet carrier concentration of 1.3 x 10^{12}cm^{-2} show transconductances of 320 mS/mm and a current gain cut-off frequency of 35 GHz. In HFET structures grown in this reactor, electron mobilities of 720.000 cm^2/Vs and a carrier concentration of 3.9x10^{11}cm^{-2} were measured at 1.5 K by Shubnikov–de Haas measurements.

STANDARD HEMT EPILAYER:	PSEUDOMORPHIC HEMT EPILAYER:
500 Å GaAs 1.5E18	500 Å GaAs N$_D$=3E18
230 Å GaAlAs 28% undoped	230 Å GaAlAs 22% undoped
250 Å GaAlAs 28% N$_D$=1.8E18	250 Å GaAlAs 22% N$_D$=2E18
20 Å GaAlAs 28% undoped	30 Å GaAlAs 22% undoped
	100 Å GaInAs 20% undoped
5000 Å GaAs undoped	3500 Å GaAs undoped
SUBSTRATE	SUBSTRATE

Fig. 21 Conventional and Pseudomorphic (PM) HEMT
Epitaxial Layer Structure

Fig. 22 shows the obtained sheet resistances over eight runs, measured on five points per wafer by a Tencor apparatus for contactless measurements. A statistical evaluation of g_m-transconductances of HEMTs on a 2" wafer shows a standard deviation of 1.9% of the mean value. The g_m standard deviation is only 11 mS/mm. This value accounts for the material inhomogeneity itself, but also for any variation coming from the technology.

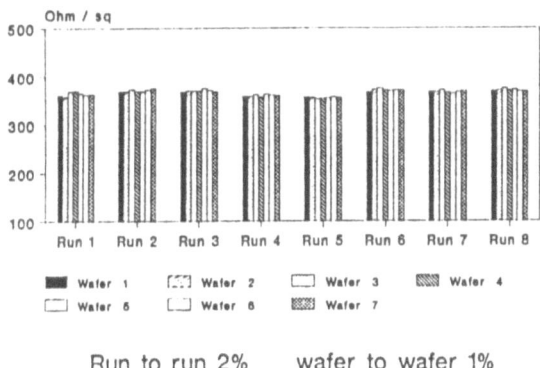

Run to run 2%   wafer to wafer 1%

Fig. 22 Sheet Resistance Statistic over 8 Runs

The current gain cut-off frequency is shown in Fig. 23. The RF performance statistics is also reported, in view of the unique design of the reactor. High frequency measurements were done on 9 points of some wafers by Cascade microprobe and HP S-parameter bridge from 1 to 26 GHz.

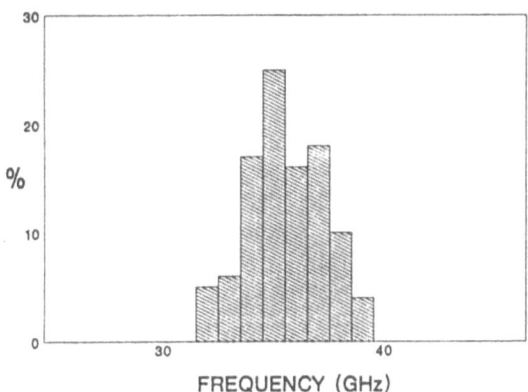

Fig. 23 Current-Gain Cut-Off Frequency in Conventional HEMT

Fig. 24 shows the obtained transconductance and extrinsic unity current gain cut-off frequencies as a function of the bias current. The noise figure at 12 GHz was measured on 5 transistors mounted in package of three different runs. The obtained values are typically 1.1 - 1.2 dB /14/.

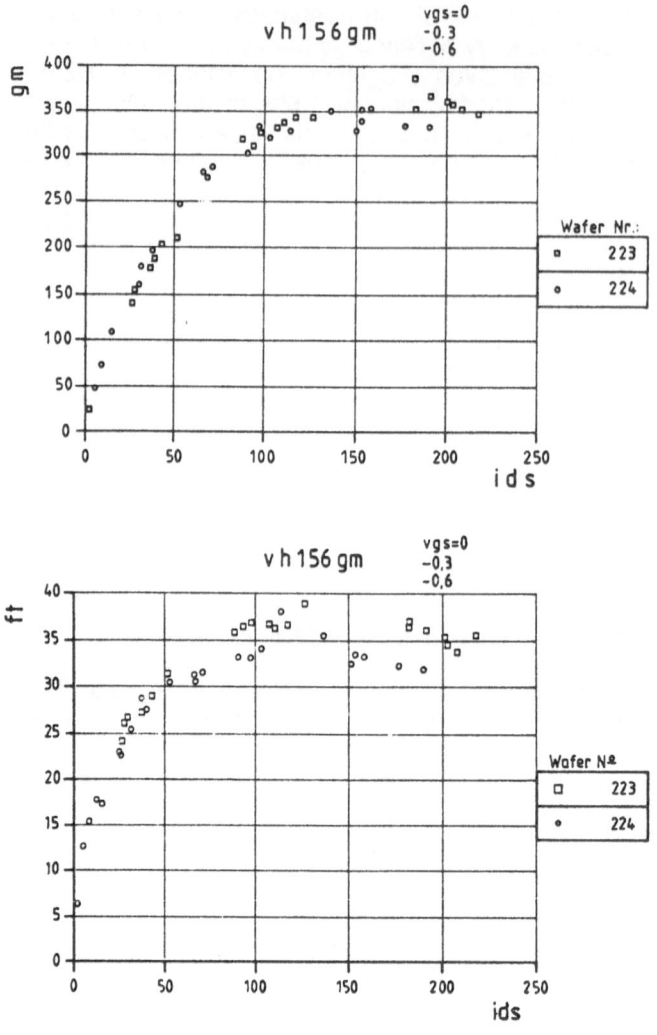

Fig. 24 Transconductance and Extrinsic Unity Current-Gain Cut-Off Frequencies

5.2 Pseudomorphic HEMT

Studies on the manufacturing of pseudomorphic HEMT material was carried out /14,18,11/. After a large number of trials, the structure of Fig. 21 was chosen and realized. The Hall mobility on such structures is typically 6.000 cm²/Vs with an N_s of 2.15 x 10^{12} electrons/cm² at 300 K, and 17.000 cm²/Vs at 77K. The N_s at 77K is typically only 0.03 x 10^{12} electrons/cm² less than at room temperature, indicating absence of parallel conduction through the AlGaAs. It is interesting to note that the same structure without GaInAs layer gives only a N_S of 1.3×10^{12} electrons/cm², indicating that the pseudomorphic structure makes much more efficient use of its donors. In order to assess the realized In content, trial runs were done in which the structure on top of the GaInAs layer was replaced by an undoped GaAs layer of 500 Å thickness. The photoluminescence spectrum of a pseudomorphic $Ga_{82}In_{18}As$ 150 Å quantum well is given in Fig. 25, while the variation of the peak wavelength over a given wafer is reported in Fig. 14. All the other six 2" wafers present a similar variation, with a slightly lower emission wavelength at the center. A relevant test of homogeneity has been made on such a pseudomorphic HEMT structure on GaAs, including a 150 Å $Ga_{0.82}In_{0.18}As$ active layer with a $Al_{0.28}Ga_{0.71}As$ layer. The sheet carrier concentration N_S has been measured using an automatic TENCOR system on 5 locations of the seven 2" wafers. The variations are within ± 1%.

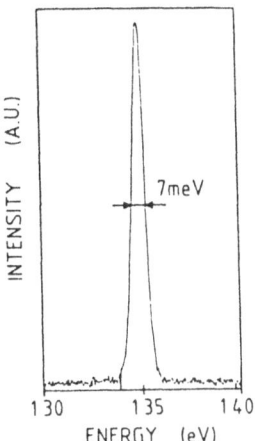

Fig. 25 Photoluminescence Spectrum of Pseudomorphic HEMT

A wafer with the complete structure was processed into FETs. Transconductances in excess of 400 mS/mm are obtained and gate leakage current is excellent. High frequency S–parameter measurements from 1 to 26 GHz have been done on 30 transistors scattered over this wafer, using a Cascade microprobe /11/. Fig. 26 shows a histogram of the obtained extrinsic unity current gain cut–off frequencies F_t.

The RF performance shows a mean value of the cut–off frequency equal to 35 GHz /14/. A homogeneity test has been made on a pseudomorphic HEMT structure on GaAs including a 150 Å $Ga_{0.82}In_{0.18}As$ active layer with a $Ga_{0.72}Al_{0.28}As$ layer. The measurement of the sheet carrier concentration on 5 locations of seven 2" wafers shows a variation within ± 1%.

The uniformity performance of the reactor on the pseudomorphic HEMT structure without caplayer is illustrated by figures 27 und 28. The sheet resistance of the layer is practically equal to the sheet resistance of the 2 dimensional electron gas in the strained $Ga_{0.78}In_{0.22}As$ layer at the hetero-junction interface, because the 300K and 77 K sheet electron concentrations only differ by less than 3%. The uniformity of the sheet resistance is typically ± 0.5%, measured by Tencor eddy current equipment on 27 points per wafer. This performance compares favourably to the previously reported performance on 2" wafers. The difference between wafers of the same run is typically ± 1%, identical to the previously reported performance on 2" wafers.

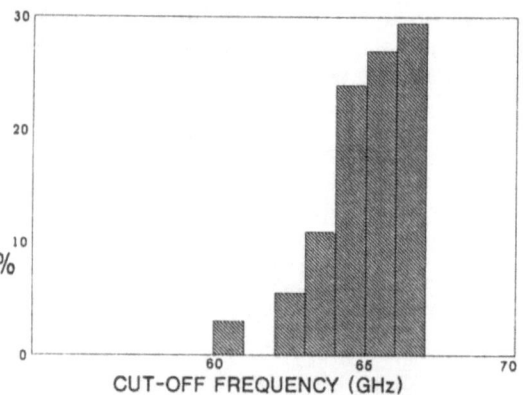

Fig. 26 Histogram of Cut–Off Frequencies of 0.25 μm Pseudomorphic HEMT

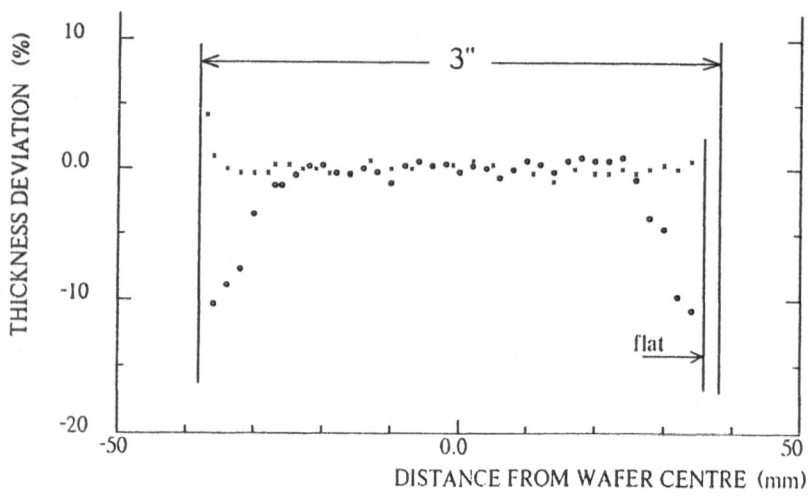

Fig. 27 Thickness Uniformity on 3" Wafers: Initial and Final Result

The obtained threshold voltage uniformity on the pseudomorphic HEMT test structure show that the edge uniformity problem of the doping has practically been solved. The repeatability of the measurement method is ± 50 mV. Statistical treatment on several wafers indicates a threshold voltage uniformity of ± 50 mV up to 4 mm from the edge, which translates to a maximum contribution to the threshold voltage of the recessed gate FET of ± 33 mV.

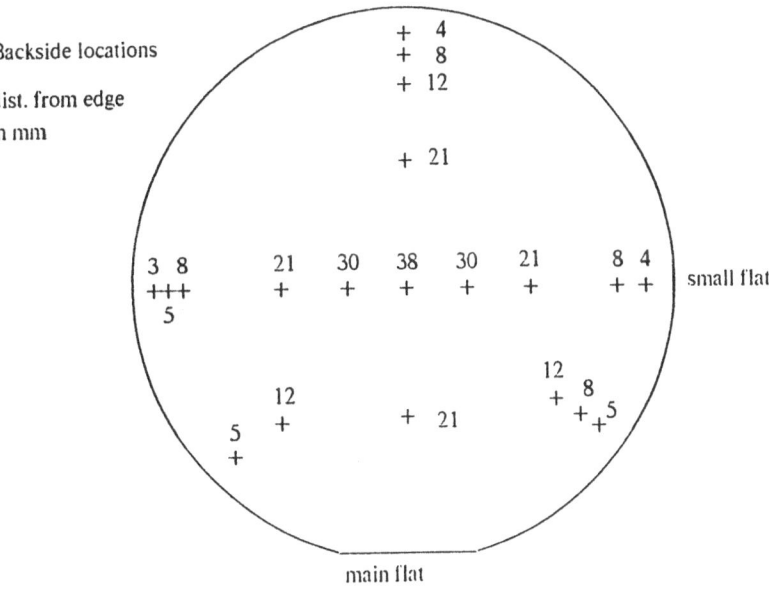

Fig. 28 Threshold Voltage Uniformity on Pseudomorphic HEMT Test Structure without Recess Etch, by Mercury Ball C–V Measurement
The Repeatability of the Measurement Method is ±50mV

5.3 Lasers

Using MOVPE techniques, scientists report that MQW (GaInP/AlGaInP) devices exhibit, for the first time, low CW threshold current (86 mA at 633 nm and room-temperature).

Improved doping control, bandgap engineering, and device modelling have enabled the realization of visible-emitting AlGaInP/GaInP lasers in the 600 nm range, with high beam quality required for optical storage applications /8,9/.

An accurate determination of the excess component to the threshold current density and its dependence on temperature and cavity length was studied in MOVPE-fabricated GaInP/AlGaInP diode lasers.

Philips is a supplier using advanced, fully-industrialized processes, the most advanced being MOVPE. GaAs/AlGaAs lasers grown with this reactor at 200 mbar exhibit threshold current density of 400 A/cm^2 /11/. The visible-wavelength laser diode, made by MOVPE, is the latest development in their range of opto-electronic products /8,9,19/.

5.3.1 Red visible lasers

Compared with IR lasers, the shorter-wavelength light from a visible laser makes higher density optical recording possible. And the development of new materials sensitive to visible light means visible lasers could be used in high-speed laser printers.

The Philips visible laser research program is one of continuous advancement. Present R&D efforts concentrate on:

*   Shorter wavelength lasers (in January 1990, Philips announced the world's first CW operated 633 nm laser diode, in October 1990 the first AlGaInP 555 nm laser /20/)
*   More symmetrical far fields (less beam-shaping)
*   Single longitudinal-mode types
*   Increased optical output power
*   Reduced threshold current.

Philips developed a completely new structure for their visible laser diodes: the "ridge-waveguide double heterostructure". The double heterostructure consists of an undoped GaInP active layer sandwiched between p- and n-AlGaInP cladding layers, grown by MOVPE on a n-GaAs substrate. On top, a p-GaInP inserted layer and a p-GaAs contacting layer are grown, and the ridge structure is formed by selective etching of the contacting layer /19, Fig. 29/.

Metallization is the final stage in the process. The different contact properties of the etched and non-etched areas ensure lateral current-confinement. The ridge-confined current flows to the center of the active layer, providing efficient pumping of the laser which results in a relatively low threshold current.

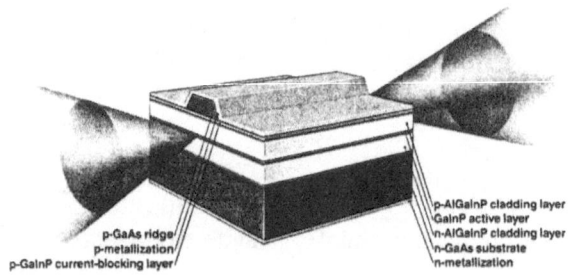

Fig. 29 Ridge Waveguide Double Heterostructure AlGaInP/GaAs Visible Laser (Philips)

Unlike other laser diode structures, this double heterostructure is manufactured by a one-step MOVPE process and a single etching stage. This produces diodes with unrivalled performance and reliability (more than 200,000 hours).

Furthermore, Philips' expertise in mass production with MOVPE enables fabrication with a high wafer-to-wafer quality, yield, and uniformity.

Emitting red light of 675 nm or 630 nm wavelength, this semiconductor laser can replace the Helium-Neon gas laser in many applications, having several distinct advantages over the latter:

- Low power consumption
- Direct modulation
- Rugged construction
- Lightweight and small-size
- High reliability
- High efficiency.

Where tight control of beam parameters is needed, an accurate collimation or focusing system is required. Consequently, Philips developed the visible-wavelength laser collimator pen. The pen, with integrated diode and optics, is made from thermally-matched materials to ensure optical precision over a wide temperature range. For application such as barcode readers, only the focusing distance, working depth, and spot size have to be specified.

The laser diode is fabricated using AIXTRON MOVPE reactors because it offers excellent control over device structures /12,19/. Compared to other crystal growth techniques, MOVPE allows for finer control of the thickness and uniformity of the active layer and also makes possible the high-volume production of visible-wavelength lasers (and of multiple quantum-well lasers). This guarantees that the laser performs reliably and exactly to specification. With over five year's experience in using MOVPE for IR lasers, Philips has unrivalled expertise in this advanced technology.

Current applications for visible lasers are:
- Barcode readers
- Non-contact measurement instruments
- Target marking/pointers
- Laser printers
- Data storage systems.

5.3.2 Green lasers (555 nm)

In October 1990, the Philips Research Laboratories Eindhoven made another unrivaled announcement in "Lasers and Optronics": Green light of 555 nm emitted by a semiconductor laser: this was the latest record value for short wavelength semiconductor lasers. Members of the Philips Research Laboratories Eindhoven, A. Valster, M. N. Finke, M. J. B. Boermans, J. M. M. van der Heijden, C. J. G. R. Spreuwenberg, and C. T. H. F. Liedenbaum succeeded in this aim /20/.

The active layer of the MQW laser with DH structure consists of 16 flat layers, each approx. 14 nm thick. These $Ga_xIn_{1-x}P$ layers are separated by Al containing layers of 4 nm thickness. This type of MQW structure is very similar to that of the 633 nm semiconductor lasers, but the layers themselves are much thinner. By MOVPE it was possible to grow these extremely thin layers well constructed with a special uniformity for thickness and composition. Only this made the fabrication of this laser possible.

The green laser provides a continuous power of 3 mW and is characterized by a differential cross section of 0.4 mW/mA. In contrast to the fair red emitting laser, the green one is emitting light up to now exclusively at temperatures far below 300K. At liquid helium temperature, this laser emits at 555 nm, at slightly higher temperature 113K the color is shifted to 559 nm. Due to this temperature dependency and the necessity for cooling, the "green laser" is used up to now only for scientific application.

6 Conclusion

The past years have been very fruitful in the move toward concretisation of the possibility of really useful and highly competitive application of MOCVD and the PLANET Reactor to high speed electronic devices and availability of a production technology for optoelectronics and for (Al)GaInP structures for LEDs or lasers.

The highlights are:

- Successful improvements of several features of the reactors, for easy handling and increased reproducibility, verified on HEMT, pseudomorphic HEMT, and laser structures.
- Tests of the epitaxial layer structure for conventional HEMT which were exploited by two different processing groups, showing reproducibility and in-line reliability of the epitaxy with this reactor.
- The first achievement of excellent pseudomorphic HEMT material by MOVPE with the PLANET Reactor. Processing of wafers resulted in 0.25 μm FETs with F_t ranging up to 64.5 GHz with an average transconduction of 383 mS/mm.
- Achievement of a very low surface defect density on the HEMT wafers. Now, generally less than 10 defects/cm^2 on at least 5 of the 7 wafers are obtained, with several wafers in the range of 0.18 μm to infinity. Disregarding this record, the average defect numbers are already 10 to 30 times better than those obtained on conventional HEMT wafers made by MBE. The establishment of this fact contributes a lot to the competitiveness of MOVPE in general.

- GaAs/AlGaAs lasers with threshold current density of 400 A/cm^2.
- Demonstration of epitaxy of extremely uniform GaInAs lattice matched to InP.
- Successful fabrication of extremely uniform GaInP material ($\lambda < \pm$ 1nm) for optoelectronic applications.

The results obtained show that this reactor leads to very real and concretely applicable technology, state of the art on the international level which can be used as a solid building block for real production in III–V technology.

Acknowledgement

The authors wish to thank the group around H. Ambrosius at Philips Research Laboratories Eindhoven, and M. Peanasky of Hewlett Packard, San José, for their outstanding effort on material preparation, characterization, and discussion. They are also indebted to the AIXTRON System Development Group, especially G. Strauch and D. Schmitz, for the many important improvements of details on the reactor itself.

The work reviewed in this paper is partly supported by the Commission of the European Community under ESPRIT contract 5003 "PLANET" and 6134 "HIRED".

References

/1/ P. M. Frijlink, "A New Versatile, Large Size MOVPE Reactor", J. Crystal Growth 93 (1988) 207

/2/ P. M. Frijlink, J. L. Nicolas, P. Suchet, "Layer Uniformity in a Multi-wafer MOVPE Reactor for III–V Compounds", J. Crystal Growth 107 (1991) 166

/3/ G. M. Martin, P. M. Frijlink, "New Materials for High Performance III–V IC's and OEIC's. An Industrial Approach", Proc. of the 'SPIE's Int'l Conf. on Physical Concepts of Materials for Novel Optoelectronic Device Applications', Aachen (1990) 67

/4/ T. Ohori, K. Makiyama, M. Takikawa, N. Tomesakai, H. Tanaka, H. Itoh, K. Kasai, M. Suzuki, J. Komeno, Inst. Phys. Conf. Ser. No. 112 (1990) 495

/5/ G. Strauch, D. Schmitz, H. Jürgensen, M. Heyen, "Improvement of Large Area Homogeneity on InP Based III–V Layers by Using the Gas Foil Rotation Concept", IEEE Conf. Proc. of the '2nd Int'l Conf. on InP and Rel. Mats.', Denver CO (1990) 112

/6/ D. Grützmacher, J. Hergeth, F. Reinhardt, K. Wolter, P. Balk, "Mode of Growth in LP–MOVPE Deposition of GaInAs/InP Quantum Wells", J. Electron. Mat. 19/5 (1990) 471

/7/   M. Heyen, "New Concept for Multiwafer Production of Highly Uniform III-V Layers for Optoelectronic Applications by MOVPE", Proc. of the 'SPIE's Int'l Conf. on Physical Concepts of Materials for Novel Optoelectronic Device Applications', Aachen (1990) 146

/8/   A. Valster, J. v. d. Heijden, M. Boermans, M. Finke, "GaInP/AlGaInP Visible-Light Emitting Laser Diodes Grown by Metal Organic Vapour Phase Epitaxy", Philips J. Res. 45 (1990) 267

/9/   A. Valster. C. T. H. F. Liedenbaum, M. N. Finke, A. L. G. Severens, M. J. B. Boermans, D. W. E. Vandenhoudt, D. W. T. Bulle-Lie-uwma, "High Quality AlGaInP Alloys Grown by MOVPE on (311) B GaAs Substrates", J. Crystal Growth 107 (1991) 496

/10/   G. Strauch, D. Schmitz, H. Jürgensen, M. Heyen, "Improvement of Large Area Homogeneity on InP Based III-V Layers by Using the Gas Foil Rotation Concept", IEEE Conf. Proc. of the '2nd Int'l Conf. on InP and Rel. Mats.', Denver CO (1990) 112

/11/   P. M. Frijlink, J. L. Nicolas, H. P. M. M. Ambrosius, R. W. M. Linders, C. Waucquez, J. M. Marchal "The Radial Flow Planetary Reactor: Low Pressure versus Atmospheric Pressure MOVPE", J. Cryst. Growth 115 (1991) 203

/12/   H. P. M. M. Ambrosius, R. W. M. Linders, C. Waucquez, J. M. Marchal, "Low Pressure OMVPE of GaAs/AlGaAs in a Multi-Wafer Planetary OMVPE Reactor", Proc. 'Conf. on Advanced Processing and Characterization Technologies' (APCT '91), Clearwater Beach, FL/USA (1991)

/13/   P. Suchet, M. Iost, P. Gamand, A. Belache, J. Bellaiche, A. Collet, "Towards a Manufacturable Millimetre Wave HEMT Process: Device Optimisation, Statistical Results, and Circuit Fabrication", presented at WOCSDICE '91

/14/   ESPRIT PLANET 5003 "Multiwafer PLANET MOVPE Reactor", 1990 Annual Report

/15/   G. Strauch, H. Jürgensen, D. Schmitz, M. Heyen, "LP-MOCVD of GaAs/AlGaAs heterostructures for solar cells and photocathodes", presented at the 3rd European Workshop on MOVPE, Montpellier, France (1989)

/16/   D. Schmitz, G. Strauch, H. Jürgensen, M. Heyen, "Growth of Extremely Uniform III/V Compound Semiconductor Layers by LP-MOVPE by Application of the Gas Foil Technique for Substrate Rotation", J. of Crystal Growth 107 (1991) 188

/17/   M. Peanasky, Hewlett Packard San José, Private Communication

/18/   L. Hollan, P. M. Frijlink, H. Jürgensen, G. Strauch, G. A. Acket, H. P. M. Ambrosius, J. M. Marchal, Ch. Waucquez, J. F. Hernandez-Gil Gomez, and E. Calleja Pardo, "PLANET MOVPE Equipment: A Breakthrough in Industrial Production of Sophisticated Epitaxial Layers", Proc. 'ESPRIT Conference', Brussels (1991) 135

/19/   Philips Optoelectronic Center, Product Literature "Visible Wave-
         length Laser Diode and Collimator Pens"
/20/   Lasers and Optronics Announcement October 1990

3.5 Epitaxy with "Safer-to-Handle" Precursors

J P Hirtz, Thomson-CSF, 91404 Orsay Cedex, France
T Martin, Defence Research Agency, Malvern, Worces. WR14 3PS, UK

1 Introduction

Both arsine and phosphine are extremely hazardous not only because of their toxicity, but also because of their storage in high pressure gas cylinders. Danger associated with these group V hydrides has been the major driving force for the development of improved precursors for both epitaxial techniques Metalorganic Vapor Phase Epitaxy (MOVPE) and Chemical Beam Epitaxy (CBE). Two different approaches have been investigated in the last few years: in-situ arsine generators and group V metalorganic sources. The first solution offers the possiblity to provide a safer alternative to compressed gas cylinders whilst retaining the benefits of arsine, but arsine is still present. Metalorganic sources are usually liquid, with lower vapor pressure, resulting in lower concentration in case of dispersion. To be useful for producing high purity epitaxial materials, these alternative group V metalorganic sources must fulfill a number of requirements in terms of toxicity, vapour pressure, stability, purity, pyrolysis temperature and reduced V/III ratio.

One of the objectives of the ESPRIT project 5031 MORSE* [1] was to study the possibility to use metalorganic sources as both arsenic and phosphorus precursors in either MOVPE or CBE growth technologies. These studies on new precursors are aimed at the identification of optimum source materials which provide both the improved operational characteristics (toxicity, vapour pressure, stability and pyrolysis temperature) and also reduced unintentional impurity uptake in the resulting epitaxial layers. Another goal of the MORSE project was to look for better sources, when needed, for group III elements. While the research of new, safer group V sources is common to both MOCVD and CBE, this field of group III sources seems to be more critical to the CBE technique. For example, the use of conventional MOVPE precursors, trimethyl gallium (TMGa) and trimethyl aluminium (TMAl) for GaAs/GaAlAs structures, results in CBE, in unacceptably high unnitentional carbon incorporation levels ($\sim 10^{20}$ cm^3).

2 Research within ESPRIT on precursors for epitaxy with improved hazard ratings

Within the ESPRIT project 5031 MORSE*, 25 new precursors have been either synthesized, purified or evaluated in MOVPE and CBE growth of III-V devices structures. From this variety of As, P, Ga, Al and In sources, only 7 precursors were selected, in order to replace toxic or not optimum (in terms of vapor pressure, purity, chemical stability) standard precursors for both MOVPE and CBE growth of III-V devices structures. These "best" precursors are:

Group III elements:

*   for Al :   Diethyl aluminiumhydride-trimethylamine
                                                    (DEAlH-NMe$_3$)   [2]
*   for In :   Dimethylaminopropyldimethyl indium   (DADI)   [3]
*   for Ga :   Triisopropyl gallium   (TiPGa)   [4]   *(only for CBE)*

Group V elements:

*   for P :   Tertiarybutyl phosphine   (TBP)   [5], [6]
                    Biphosphinoethane   (BPE)   [7]
*   for As :   Tertiarybutyl arsine   (TBAs)   [5], [6]
                    Tridimethylamino arsenic   (DMAAs) [8]   *(only for CBE)*

C-doping source in CBE GaAs:

                    Trimethylgallium   (TMG)   *(only for CBE)*

For these metalorganic sources, operational characteristics and commercial informations are listed in Table 1. While most of these precursors have been used in MOVPE essentially for InP-based laser structures, they have been evaluated extensively for CBE growth of GaAs-based devices, both for optoelectronic applications (0.98 µm GaInP/GaInAs/GaAs lasers) and microwave devices applications (GaAlAs/GaAs HBT, GaInP/GaAs HBT, GaAlAs/GaAs HEMT). The major conclusions concerning the best liquid precursors to be used in either MOVPE or CBE can be summarized as follows:

2.1   Best Phosphorus precursors for the MOVPE growth of GaInAsP/InP devices structures

The vapour pressure of bisphosphinoethane (BPE) is about 10 times smaller than for TBP. However, since there are two P atoms in the BPE molecule, the penalty expected in "effective" phosphorous vapour pressure as compared to TBP is only a factor of five.

BPE appears superior to TBP for the light emitting properties of both single quaternary layers and complex heterostructures [9]. The lasers grown with BPE exhibit threshold current densities (fig. 1), with the best values reported world-wide for structures with the same number of quantum wells. They also had the highest internal efficiency and the lowest optical losses. The lasers grown with

Table 1

OPERATIONAL CHARACTERISTICS AND COMMERCIAL INFORMATIONS ON BEST SELECTED PRECURSORS

	TBP	BPE	TBAs	DMAAs	DADI	TIPGa	DEAIH-NMe3
Vapour pressure at 20°C (mbar)	290	25	140	1.8	0.3	0.15	0.15
Toxicity (LC$_{50}$* in ppm)	1100	?	70	No As-H bond	?	Not acutely toxic	Not acutely toxic
% of metal available (by weight)	35	65	56	36	50	35	19
Batch to batch evaluation, impurities	Good	Cl, Si	Good	Ether, Si	?	Good	Good
Source Purity (Metallic)	1 ppm	?	1ppm	5ppm	?	1ppm	1ppm
Precursors combinations	No known problem in MOCVD and CBE						
Stability under vacuum	Good						
Precracking	Yes	Yes	Yes	No	No	No	No
Pyrolysis temperature	> 500°C	> 500°C	> 500°C	< 400°C	< 400°C	< 400°C	< 400°C
Price (ECU / gram)	26	40	39	44	63	22	46
European production	Epichem SMI		Epichem	Epichem	Eprova Merck (GB)	Epichem SMI	Epichem SMI
Production in USA	Yes	Yes	Yes	Yes	No	No	No
Production in Japan	Yes	No	Yes	?	?	No	No
Patent related to synthesis or use	Japan	France	Japan	None known	Germany	None known	None known

* LC$_{50}$ is the lethal concentration -50 percent, i.e the median concentration of a chemical in air which, when inhaled for a specific time (typically 4 hours), will result in death for 50% of exposed animals. LC$_{50}$ is respectively 5 ppm and 10 ppm for arsine and phosphine.

ESPRIT 5031 MORSE

TBP came a close third (after PH$_3$) in threshold current densities but had higher internal efficiencies than the PH$_3$-grown lasers. V/III ratio lower than 10 can be used, as contrasted to more than 200 for PH$_3$. Therefore, the use of BPE instead of PH$_3$ gives a better result with no increase in the price of the precursors. Moreover, the cracking and scrubbing of large quantities of phosphine at the reactor output is also cost-intensive so that, as a whole, a BPE-based process may be cheaper than the standard PH$_3$-based one.

2.2 Best Arsenic precursor for the MOVPE growth of GaInAs, GaInAsP and GaInAsP/InP laser structures

The growth of InGaAs with TBAs was studied as a function of TBAs molar flow and temperature, growths with AsH$_3$ being also made under similar conditions for reference. Excellent morphology as well as very good layer uniformity were obtained. The necessary molar flows for TBAs and AsH$_3$ are quite similar (same conclusion as previously reached for TBP and PH$_3$). Quaternary layers with compositions in the 1.25 μm and 1.55 μm ranges were grown under similar operating conditions with the following combinations: TBAs and TBP, AsH$_3$ and TBP, AsH$_3$ and PH$_3$. The results were compared in terms of uniformity, luminescence intensity and linewidth. The best uniformity was obtained with the combination TBAs/TBP, but the best luminescence intensity and linewidth was obtained with AsH$_3$/PH$_3$. Broad area, multi-quantum-well (MQW) 1.55 μm GaInAsP/InP lasers have been grown for the first time in AP-MOVPE, with TBP and TBAs for the quaternary material in the wells, TBP and AsH$_3$ for the quaternary material in the barriers [5]. Laser characteristics (threshold current density, internal efficiency and optical losses) of those structures grown with TBAs are almost as good as characteristics of lasers grown with AsH$_3$.

2.3 Best Precursors for the CBE growth of GaAlAs/GaAs 2DEG structures

CBE-growth studies under the MORSE project concluded that the optimum group III precursors to be used for the CBE growth of GaAs/GaAlAs device structures were tri-isopropyl gallium (TIPGa) [4] and Diethyl Aluminiumhydride trimethylamine (DEAlH-NMe$_3$) [2]. Figure 2 is a plot of bulk GaAs properties comparing TiPGa/AsH$_3$ and TBAs/AsH$_3$ with solid souce MBE. The best bulk data obtained to date with TiPGa/AsH$_3$ and using H$_2$S as the intentional n-type dopant is n_d-n_a = 7 x 10^{14} cm^{-3}, μ_{77K} = 60,000 cm2 V^{-1}s^{-1} (implies n_a<3x10^{14} cm^{-3}). Best bulk data obtained using TIPGa/TBAs, n_d-n_a = 3 x 10^{15} cm^{-3}, μ_{77K} = 29,000 cm2 V^{-1}s^{-1} (implies n_a < 10^{15} cm^{-3}).

GaAs/GaAlAs two-dimensional electron gas (2-DEG) structures grown using this optimum precursors combination TIPGa/DEAlH-NMe$_3$/H$_2$S and arsine have exhibited an important marked improvement in 77K 2DEG mobility (μ_{77K}= 65,000 cm^2/V· s) compared to identical TEGa-grown devices (μ_{77} =37,000 cm^2/V· s). This is the highest mobility value yet reported for CBE-grown Al$_{0.3}$Ga$_{0.7}$As/GaAs 2DEG structures. The significant improvement in both GaAs and GaAlAs material quality performed using TiPGa and the safer alternative arsenic precursor TBAs, have generated the highest electron mobility values yet reported for TBAs-grown CBE GaAs material. GaAs growth data are very encouraging with the lowest n-type (H$_2$S) intentionally doped layer grown to date (n_d-n_a = 3x10^{15} cm^{-3}) already exhibiting a 77K mobility of 29,000 cm^2 V^{-1}s^{-1}, directly comparable with that recorded for corresponding AsH$_3$-grown material (fig. 2). This GaAs material is directly comparable in purity to that achieved previously when using high-toxicity arsine. With this TiPGa/DEAlH-NMe$_3$ precursors combination, the CBE growth of GaAs/GaAlAs HEMT device structures had led to an important improvement in device performances. For 0.7 μm gate HEMT, transconductances values of 250 mS mm^{-1} were measured at drain currents I_D = 200 mA/mm. For 0.5 μm devices, RF characterization reveals cutoff frequencies f_t = 22 GHz for the current gain and f_{max} = 40 GHz for the unilateral power gain.

Recently, tridimethylamino arsenic (DMAAs) has been reported, as a new possible As precursor for CBE [8]. This precursor is extremely attractive, in terms of epilayer purity, cracking efficiency. Furthermre, although no data has been reported yet, DMAAs looks very promising in terms of toxicity, thanks to the absence of direct Arsenic to Hydrogen bond. When compared to TBAs, another important advantage of DMAAs, is that no precracking is needed. Even for substrates temperature as low as 380°C, epitaxial GaAs can be grown without precracking the DMAAs. This is a real breakthrough for CBE technology, where none of the existing high temperature cracker cells are reliable enough in the cracking of both TBP and TBAs. Another interesting feature of DMAAs, is the ability for that molecule to remove carbon contamination at the substrate/epilayer interface during the wafer deoxidization. In a direct comparison of DMAAs, with other possible As precursors TBAs and AsH$_3$,for the CBE growth of GaAs (TEGa being the Gallium source) DMAAs leads to a very significant reduction in carbon incorporation. SIMS gives a carbon level in the detection limit of [C] ~ 4.10^{15} cm^{-3} for GaAs grown with DMAAs, [C]~ 1.10^{16} cm^{-3} for GaAs grown with AsH$_3$ and [C] ~5.10^{16} cm^{-3} for GaAs grown with TBAs. For some specific purposes, such as C-doped GaAs layers for the base region of HBTs, TBAs will be preferred to DMAAs, which will not allow high C doping. These first results look promising: DMAAs seems to be a very useful alternative source to arsine for CBE growth of high purity GaAs-based structures.

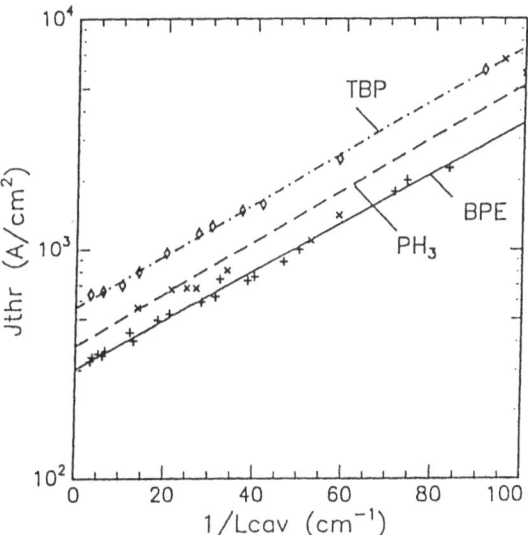

Fig. 1.   Threshold current densities as a function of the inverse cavity length for broad area lasers made from MQW wafers grown with PH$_3$, TBP, BPE.

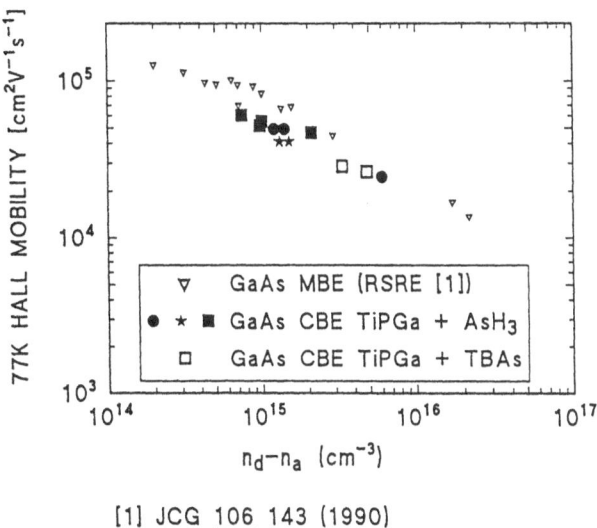

[1] JCG 106 143 (1990)

Fig. 2.   77K mobilities for GaAs grown by MBE,
CBE with the TiPGa + AsH$_3$ combination,
CBE with the TiPGa + TBAs combination.

2.4 CBE-growth of GaInP/GaAs HBT with TBP and TBAs

DC and microwave characteristics of GaInP/GaAs Heterojunction Bipolar Transistors (HBT) have been investigated [10] for structures grown by CBE using tertiarybutyl phosphine (TBP) and tertiarybutyl arsine (TBAs) precursors, trimethyl indium (TMI) and triethyl gallium (TEG). GaInP/GaAs HBT epitaxial structures have been grown on a 2" diameter semi-insulating GaAs substrate. The device structure is shown in Fig.1a. High carbon doping capability (up to $2. 10^{20}$ cm^{-3}) in the base is also possible with the combination of trimethyl gallium (TMG) and TBASs which should provide the base sheet resistance and thus the microwave gain. The n- type doping was achieved using H_2S. A self-aligned technology was used for fabrication. It employed a combination of wet and reactive-ion etching (RIE) to reduce base-emitter separation and trench mesa isolation to reduce device size and parasitics. The dc I_{CE}-V_{CE} characteristics of a 3 µm x 50 µm device shows a small $V_{CE, off}$ = 0.2 V due to the smalll ΔE_C,, fig. 3b . The maximum dc current gain for a base sheet resistance of 400 Ω/sq was $\approx$ 30, at a collector current density, $J_c \approx 22$ kA/cm^2 .The Gummel plots revealed base n(I_b) and collector n(I_c) current ideality factors of 1.8 and 1.2 respectively. The high n(I_b) seems to be related to RIE damage during base etching. In a separate run, non-self aligned HBT's fabricated using all wet etching on the same wafer gave typical n(I_b) and n(I_c) values of 1.2 and 1.1 respectively. Microwave measurements were also made using a Cascade on-wafer probe station from 0.5 - 25 GHz. The V_{CE} bias -dependent f_T and f_{max} characteristics obtained for a 3 µm x 50 µm self-aligned device showed higher f_T and f_{max} with V_{CE} over the measured J_c range of 2-40 kA/cm^2. The peak extrinsic f_T's at V_{CE} = 1,2 and 3V were 25, 32, 34 GHz at $J_c \approx 11$, 25 and 30 kA/cm^2 respectively. The corresponding peak f_{max} 's were 11, 14 and 15 GHz at $J_c \approx 11$, 25 and 30 kA/cm^2 .These preliminary results give some evidence of the suitability of the TBP and TBAs precursors with improved hazard ratings for the CBE growth of GaInP/GaAs HBT.

2.5 Best precursors for CBE-growth of GaAlAs/GaAs HBT

GaAlAs/GaAs HBT structures have been grown by CBE with ultra-high p-type doped GaAs base thicknesses controlled down to 0.08 µm. These structures, had already yielded values for f_T and f_{max} values as high as 34 GHz and 23 GHz, respectively for the 3x30 µm emitter-stripe devices, with corresponding output powers of 19.9 Wmm^{-1} at 8 GHz and 18.2 W mm^{-1} at 10 GHz. The output power data obtained are therefore already directly comparable with current state-of-the-art values reported for MBE-grown GaAs/GaAlAs HBT structures .

167

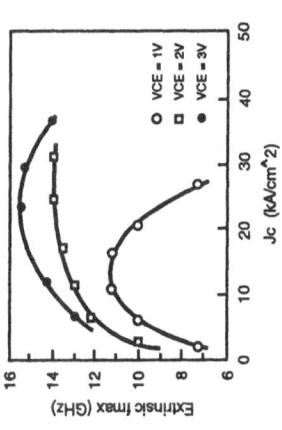

n GaAs	3×10^{18} cm^{-3}	2500Å
n GaInP	3×10^{18} cm^{-3}	1500Å
n- GaInP	3×10^{18} cm^{-3}	1500Å
p+ GaAs	5×10^{19} cm^{-3}	800Å
n- GaAs	8×10^{16} cm^{-3}	5000Å
n+ GaAs	7×10^{18} cm^{-3}	1000Å
n+ GaAs/GaInP Superlattice 3x(40Å/40Å)		
n+ GaAs	7×10^{18} cm^{-3}	4000Å
Semi-Insulating GaAs Substrate		

a) Device cross-section of GaInP/GaAs HBT

b) I_{CE}-V_{CE} characteristics of a 3mmx50mm self-aligned GaInP/GaAs HBT (Top curve: I_B=2.0mA, Step = 0.2mA)

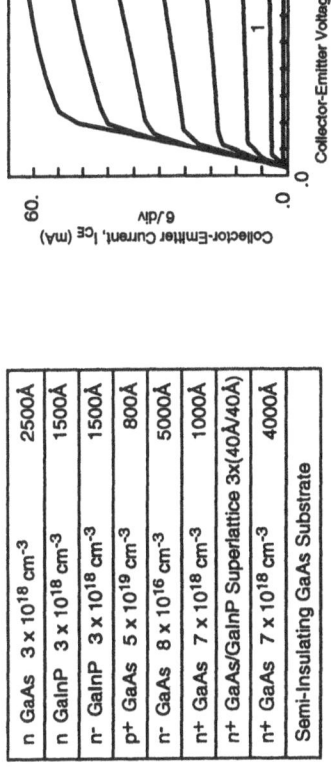

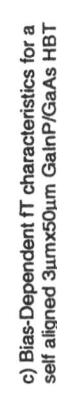

c) Bias-Dependent fT characteristics for a self aligned 3µmx50µm GaInP/GaAs HBT

d) Bias-Dependent fmax characteristics for a self aligned 3µmx50µm GaInP/GaAs HBT

Fig. 3. CBE GaInP/GaAs HBT grown with TBAs + TBP. Device structure and DC and RF performances

HBT wafers have also been grown to check the influence of the use of the new improved gallium precursor, TiPGa. These wafers were grown to a target specification (base width 0.08 μm, collector width 0.5μm) identical to that which previously yielded the state-of-art 10 GHz RF power output of 18.2 W/mm. Breakdown voltage and f_{max} values of 13V and 24 GHz, approaching the corresponding previous batch values of 18V and 23 GHz.

3   Conclusions and perspectives

For the replacement of the highly toxic arsine and phosphine currently used in both MOVPE and CBE, alternative liquid precursors for As and P have been studied within the ESPRIT project MORSE* . In combination with the most suitable group III precursors, it has been demonstrated that TBAs, TBP or BPE are capable of producing at least the same device quality as with AsH_3 and PH_3 , with much reduced safety requirements. In addition to the hazard issue, some of these combinations of optimized precursors had led to improved wafer uniformity (TBP, BPE, TBAs) and impurity uptake (TEAlH-NMe$_3$, TBP). As can be seen in Table 1, most of these safer to handle precursors are now commercially available in Europe.

In the future, a selection of optimum safer to handle precursors will certainly become be a essential part of each epitaxial process [11]. For example in MOVPE, GaAs-based HBTs are presently grown using CCl_4 as the carbon doping source for the GaAs base. The substitution of this Chloro-Fluoro-Carbon (CFC) source will certainly occur through the use of a metalorganic precursor of arsenic such as trimethyl arsine (TMAs) leading to a well-controlled carbon uptake. This research of metalorganic precursors will continue also in order to ideally match the requirements of each type of new specific growth technologies, such as selective-area or localized epitaxy, photon-assisted epitaxy.

* ESPRIT 5031 MORSE Consortium:

CNET (F), DRA (GB), FORTH (Gr), THOMSON-CSF (F) Prime, RIBER (F), University of Aachen (D), University of Padova (I), University of Stuttgart (D), SMI (F), ENSSPICAM (F)

Acknowledgements

This review paper has been prepared thanks to a strong interaction between all people who have successfully collaborated within the ESPRIT project MORSE. The financial support by the EEC is gratefully acknowledged.

169

References

[1]   J.P. Hirtz , Materials Science and Engineering; B17 (1993) 9-14

[2]   P.A. Lane, T. Martin, R.W. Freer, P.D.J.Calcott, C.R. Whitehouse, A.C. Jones and S. Rushworth; Appl Phys Lett 61 (1992), 285

[3]   A. Molassioti, M. Moser, A. Stapor, F. Scholz, M. Hostalek, L. Pohl; Appl. Phys. Lett. 61 (1992), 285

[4]   P.A. Lane, C.R. Whitehouse, T. Martin, M. Holton, G.M. Williams, A.G. Cullis, S.S. Gill, J.R. Dawsey, G. Ball, B.T. Hughes, M.A. Crouch, M.B. Allenson; J. Cryst. Growth 120 (1992), 245

[5]   A. Ougazzaden, R. Mellet, Y. Gao, E.V.K. Rao A. Mircea; Proc. of 4th Int. Conf. on InP & related compounds, Newport, R.I., USA, April 92, 36

[6]   J.C. Garcia, P. Maurel, J.P. Hirtz, Electron. Lett. 29 (1993), 432

[7]   A. Ougazzaden, A. Mircea, R. Mellet, G. Primot, C. Kazmierski, Y. Gao; Electron. Lett. 28, (1992), 1078

[8]   C.R. Abernathy, P.W. Wisk, D.A. Bohling and G.T. Muhr; Appl. Phys. Lett. 60 (1992) 2421

[9]   A. Ougazzaden, R. Mellet, Y. Gao, C. Kazmierski, D. Robein, A. Mircea; Electron. Lett. 27, (1991), 1005

[10]   G.I. Ng, D. Pavlidis, A. Samelis, D. Pehlke, J.C. Garcia and J.P. Hirtz; Accepted for presentation at DRC, June 93, Santa Barbara, USA

[11]   E. Yablonovitch, G.B. Strinfellow and G.E. Greene, Journal of Electronic Materials, 22, N°1, 1993, 49

4.1   Reliability Evaluation

F Magistrali and D Sala
Alcatel-Telettra, Via Trento 30, 20059 Vimercate, Italy

1. INTRODUCTION

This work summarizes the reliability activities performed during ESPRIT 5018 "COSMIC" project, that had the purpose to design, manufacture and test low noise, low level GaAs MMICs from L to K band, based on FET and HEMT technology; as part of the project, a quite large reliability testing plan was performed, based on testing facilities and resources from Alcatel-Telettra and on test structures, discrete devices and MMICs from the foundries.

The work is organized as follows: first the basic ideas for the reliability plan are presented, and the testing facilities are briefly described (for electrical and thermal characterization and for lifetesting); then the testing results are examined in details, for passive circuits, FETs, HEMTs and MMICs, including the failure analysis activities; finally, before coming to the conclusions, the most relevant failure mechanisms are discussed.

2. GENERAL PROGRAM

The problem to define a methodology for MMICs reliability determination is still open, even if some tentatives exist towards a standard approach [1]. Anyway, every MMICs reliability investigation is generally divided into two phases: first, reliability of active and passive components is investigated, by means of accelerated tests designed to stress specific aspects of each element [2]; afterwards, lifetesting of the final MMICs (or of structures that are functionally equivalent to them) is carried out, either in DC or RF conditions, with the aim to get reliability figures for the overall circuits.

This second point is the most critical and controversial, as a lot of discussion exist about suitability of DC or RF lifetesting [3], determination of the best temperature range, the failure criteria, etc.; in our case we followed our general philosophy to perform DC lifetesting, monitoring DC and RF parameters.

Our program was designed as follows:
- first, thermal and electrical step stresses are performed on active and passive elements, with the purposes to understand the limiting conditions for subsequent lifetests and to get a first rough idea of the main failure mechanisms;

- then, operating lifetests in DC conditions take place, with special care for FETs wich are subjected to three temperature tests (the lowest being Tch 175°C); as the tested devices have generally low power dissipation, the problem of channel temperature determination is not so critical and it can simply be solved by means of the so called △Vgs method;
- finally, MMICs are lifetested at high temperatures.

3. DEVELOPMENT AND REALIZATION OF TESTING FACILITIES

The importance of adeguate testing facilities, including test fixtures and electrical/thermal characterization methods, is well known for GaAs devices, featuring high sensitivity to ESDs and oscillations, and non-uniform thermal distribution [4].
We use hermetic metal or metal ceramic packages, sealed in inert atmosphere, with Au-Sn die attach and wedge-wedge bonding; this configuration does not introduce external effects during high temperature test, even if it minimize thermal resistance, that is a problem for low power devices, as we will see later. Metal DIL14 packages were used for test structures, while standard metal-ceramic FET-package, and Flat-packs were used for active devices.

3.1 Lifetesting hardware

A part from the very few problems posed by lifetesting of passive elements, which need only appropriate electrical connections to power supply and simple alarm circuits, it is known that lifetesting FETs is critical, due to their sensitivity to oscillations, ESDs and EOSs.
It is then needed to provide a complete lifetesting set up consisting in ovens equipped with several power supplies, individual control circuits (providing the proper bias point for each devices together with alarms and power shut down in case of failure), and individual jigs, which allow to lifetest and characterize FETs without direct access to the devices [5].
A jig suitable for low/medium power devices is reported in Fig. 1: 50 Ohm Au lines are available for drain and gate, with high temperature resistors and capacitors network to avoid high frequency oscillations; two ECCOSORB structures are placed very closed to the device for low frequency oscillations quelling and the three pins are on the rear side; the maximum working temperature is 200°C, but in our equipment it is limited to 160°C by the soldering method of the hardware around it. This jig was proved to be suitable for devices up to few Watts dissipated power, but the combination with metal packages gives origin to low case-to-ambient thermal resistance, so that the total channel to ambient overheating is not high for low power devices; if we consider that the channel-to-case thermal resistance is kept low by Au-Sn die-attach, we can

understand that the maximum available channel temperature is around 200°C for low power FETs and 175°C for HEMTs.

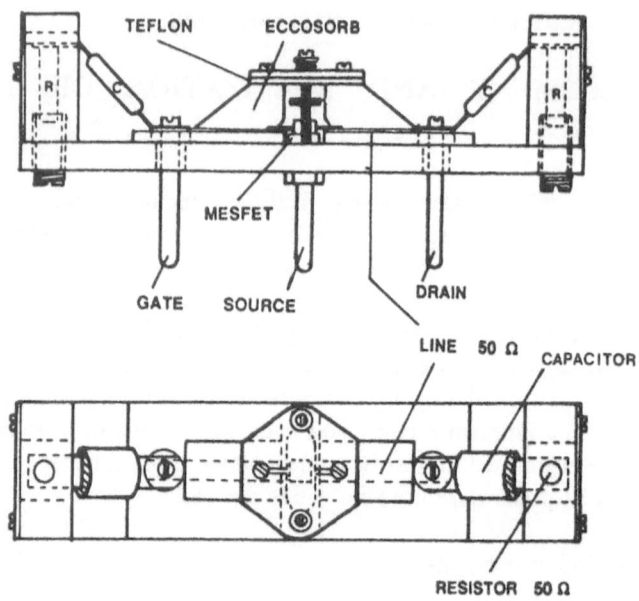

A different structure was then developed, allowing to reach 250°C case temperature. The basic idea is not to use ovens but local heaters, with the heat sink of the packages in direct mechanical contact to them; the structure works at room temperature and this means that everything, in particular soldered items, is kept at much lower temperature than the package itself; we had to avoid the ECCOSORB, that cannot withstand more than 200°C, but we solved the problem using AC terminations very close to the devices; by infrared analysis with the heater set to keep the case temperature at 250°C, we found that the soldering point closer to the package is at 145°C, much less than its maximum limit.
All the electrical tests with FETs and HEMTs proved that this solution was suitable for lifetesting, when very high temperatures are needed.

3.2 Electrical measurement facilities

Once the jigs are available, the problem of electrical measurement for FETs can easily be solved by a test system composed by a digital parameter analyser, a voltmeter, additional power supplies and mechanical arrangements for the jigs. Everything is controlled by a HP 300 computer which allows for automatic test, data storage and elaboration; a software was realized for DC characterization giving a complete set of parameters such as: transistor parameters (Idss, gm, Vp) at different operating points, forward and reverse gate diode characteristics and resistances (Rg, Rs, Rd and Rch); it is easy to change the software to set the appropriate measurement limit for each type of devices.

More detailed information are reported for example in [4] but here it is important to note that the measured parameters are expected to allow the identification of most important failure mechanisms.

3.3 Thermal characterization methods

As a general rule, for channel temperature determination of FETs, we rely on the measurement made by means of the so called △Vgs method, that, to our experience, is sufficient for low power devices.

For the evaluation of temperature distribution on large chips, or on packages, we can make use of our IR Thermography, based on a Hughes 4100P machine: this system has a lateral resolution of 35 μm, that is not sufficient for precise measurement of hot spots temperature in FETs active area, but it can give a good idea of temperature distribution, expecially on large chips.

4. PASSIVE ELEMENTS TESTING

Most of the passive structures are on a standard PCM, shown in fig.2; in this figure caption, and in the following of this work, M2 refers to the gate metal level (Ti/Pt/Au) realized both on the active channel and Semi-insulating material and M3 to the upper level interconnection (Ti/Pt/Au too); M1 is the ohmic contact level (Au/Ge/In).

When necessary, that is the case for interdigitated capacitors and long chain of narrow interconnecting vias (2x2 μm^2), discrete structures were used.

4.1 Thermal Step Stress

This test began at 175°C, with the temperature being increased by 25°C every 50 hrs, up to 350°C; at each step the elements were electrically characterized; the most important results are:

- dielectric integrity is maintained over a wide temperature range, up to 350°C for SiN and 325°C for polymide (whose curing temperature is 330°C);
- etched resistors showed no degradations at all;
- resistance of metal lines and interconnections slighty increased at very high temperature, with particular concern for the so called M2 level, which is the gate structure level (fig.3); some comments about that will be given in the following chapters, where an extensive description of gate degradation will be presented;

- surprisingly, ohmic contact structure, made by AuGeIn, showed little degradation.

As a general conclusion, it can be stated that temperature is not an important accelerating factor for passive elements.

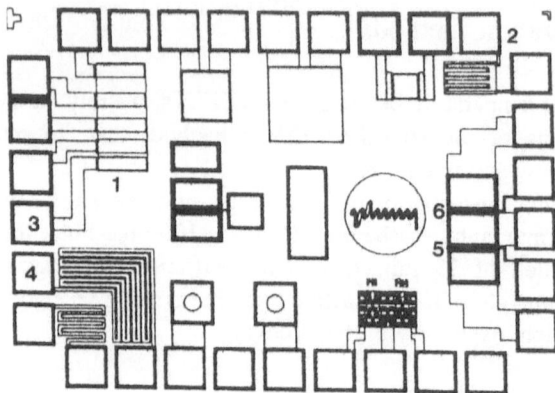

Fig. 2: PCM structure.
1) TLM
2) Metal line M2 (0.7 μm)
3) Metal line M3 (10 μm)
4) Metal line M2 (10 μm)
5) FET (5 μm)
6) FET (0.7 μm)

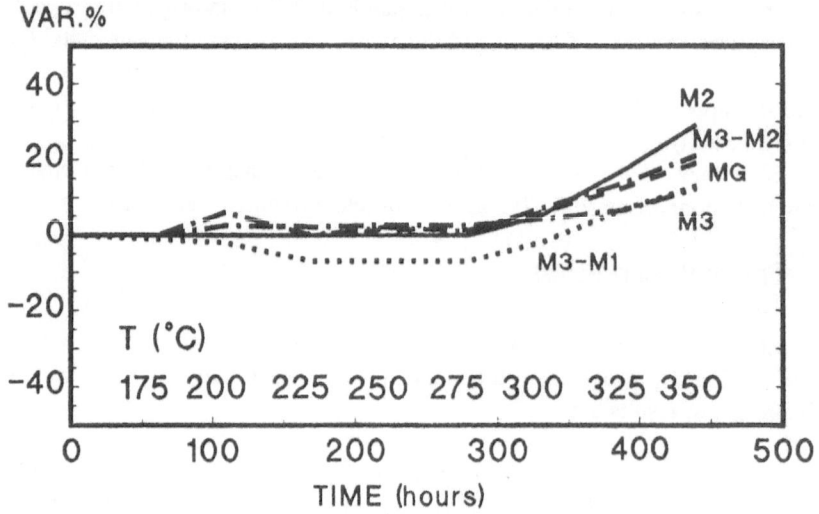

Fig. 3: Mean variation of metal lines and interconnection resistance values during thermal step stress.

4.2 Electrical Step Stress

To ascertain the robustness of passive devices against electrical stress, a step stress was realized, at Tamb = 160°C in the following electrical conditions.

DEVICE	STEP	FINAL VALUE
Capacitor	5V	40V
Interconnection	25mA	200mA
M3 line	25mA	200mA
M2 line	1.25mA	10mA

The duration of each step was again 50 hours.

Regarding the electrical conditions, it would be better to consider current density rather than absolute current: current density step were 1.5×10^5 A/cm² for interconnections (calculated into the interconnection area), 0.4×10^5 A/cm² for M3 and 5×10^5 A/cm² for M2.

No capacitor degradation was detected up to 40V.

For the other elements, the degradation behaviour is shown in Fig.4; it has to be noted that in this case, when the last final determined value was reached, it was decided to go further with the test, keeping the electrical stress constant; as shown, M3 and M2 lines has no degradation, up to current density of about 4×10^6 A/cm² (for M2).

A different behaviour was shown by interconnections (Fig.5): during the final lifetest at I = 200mA, the resistance of the interconnection chains between M3 and M1 progressively increased up to open circuit, while nothing happened at the interconnections between M3 and M2; the fact that the failures were found only at ohmic contacts suggested that the weak points were not the interconnections themselves but that something could have happened at ohmic metals.

Failure analysis confirmed that the failure mechanism was AuGeIn electromigration (Fig. 6): it is clear that material depletion/accumulation took place according to the electrons flow, as indicated in the figure, progressively increasing the contact resistance up to catastrophic failure.

Some considerations are due to the test temperature: by infrared analysis of a PCM with all the elements working in the same conditions of the lifetest, we found that the average overheating of the interconnections is about 55°C, while the hottest point of the PCM is determined by M3 lines which are at about 68°C over the ambient; M2 line overheating is about 45°C.

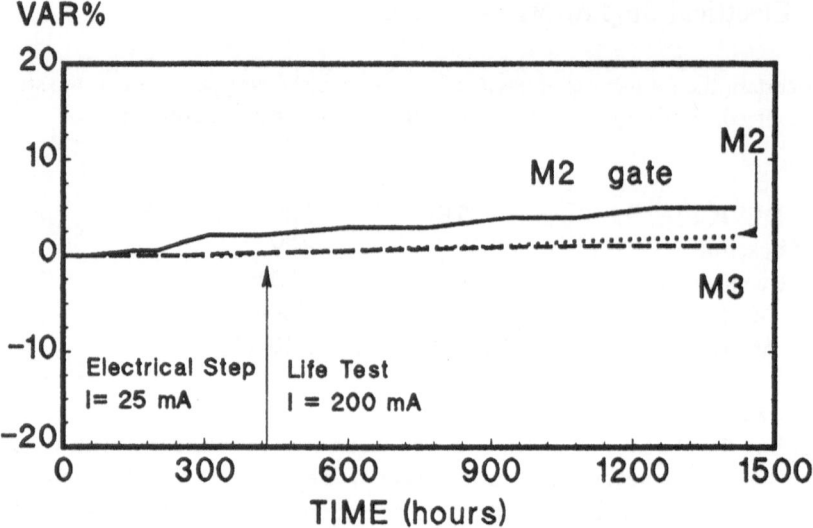

Fig. 4: Variation of metal lines resistance values during electrical step stress.

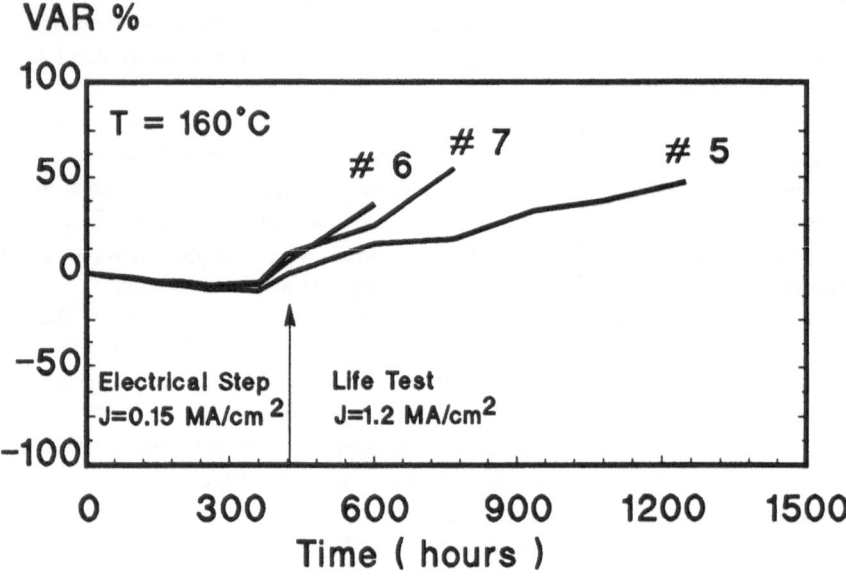

Fig. 5: Degradation of interconnects during electrical step stress and lifetest.

4.3 Operating Life Tests

At the end of step stress, life tests were run, at high temperature but at alectrical conditions inside (very close to) the maximum limits recommanded by the manufacturer; at the beginning these tests involved capacitors (MIM and interdigitated) and etched resistors with no noticeable degradation up to 4000 hours, Tamb=160°C.

After that, metal lines and interconnections were extensively tested, in the temperature/current density conditions summarized in fig. 7; again lines and M3/M2 interconnections (4x4 μm^2 and 2x2 μm^2) showed no degradation up to 4000 hours, confiming the stability of these structures; M3/M1 vias had a degradation behaviour similar to the one observed during electrical steps stress, summarized in fig. 8, and the failure mechanism was again M1 electromigration; these results will be widely discussed afterwards.

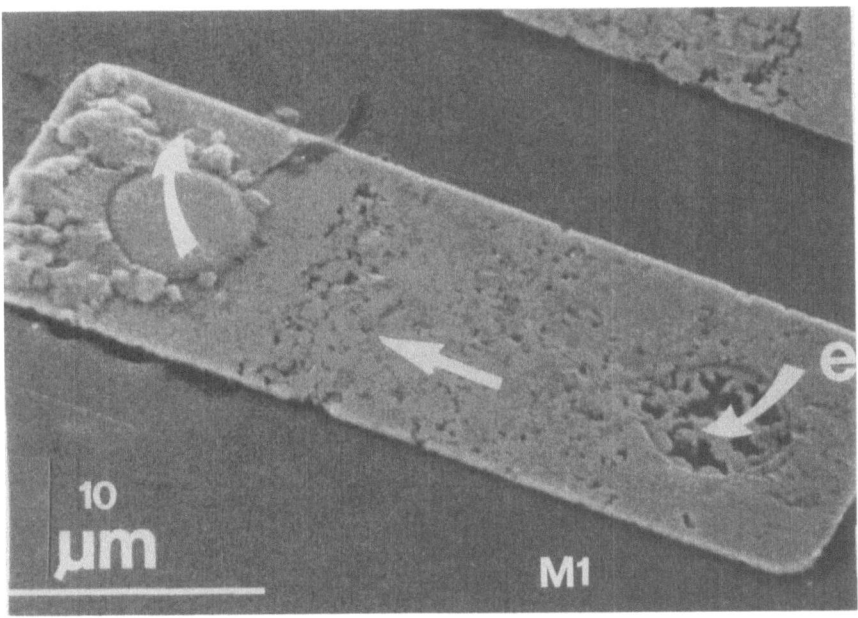

Fig. 6: SEM image of interconnection degraded during electrical step stress after removing metal M3. The arrows show the versus of the electron flow.
The electromigration phonomenon in M1 metal is according to the electron flow.

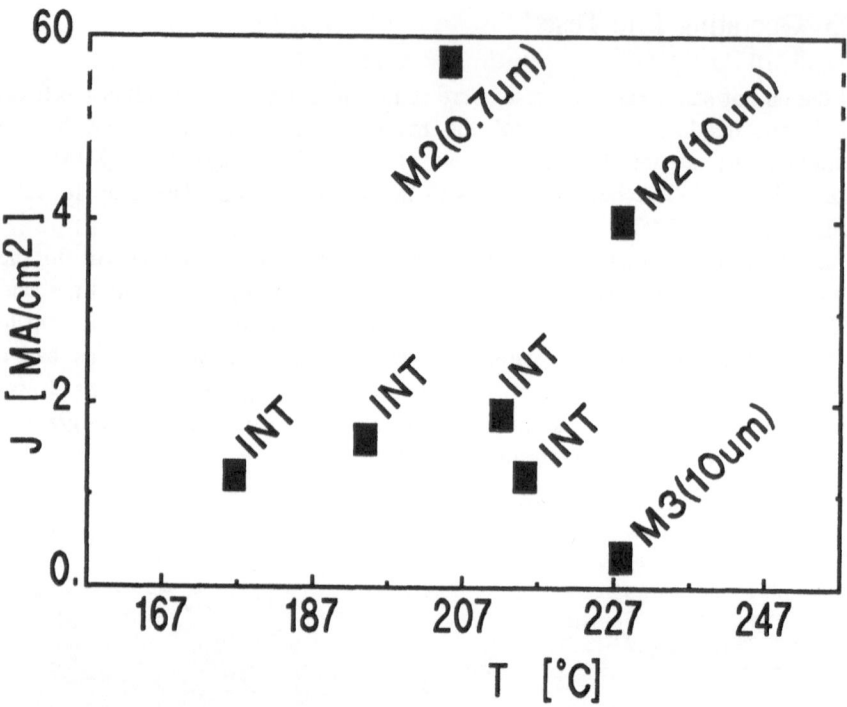

Fig. 7: Situation of the performed tests on test structures in term of current density and local temperature.

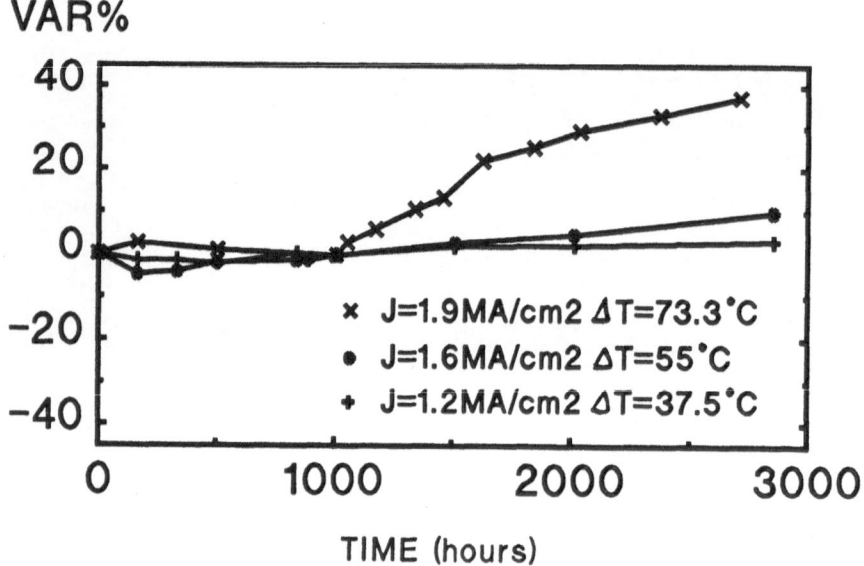

Fig. 8: Test results on M3/M1 vias.

5. ACTIVE ELEMENTS TESTING

FETs with 2 different technologies (based on Ti/Pt/Au and Cr/W/Au gate respectively) and HEMTs were subjected to a wide range of tests, to assess their reliability concerning temperature, electric field, forward current and power dissipation.

5.1 FET results (Ti/Pt/Au gate)

Thermal step stress, high forward gate current, high temperature reverse bias and operating life tests were performed in this case.

5.1.1 Thermal step stress

While performing this test on the PCMs described is the previous paragraph, we followed the degradation behaviour of the 2 FETs with 0.7 μm and 5 μm gate length; not only the classical FET parameters were measured but backgating characteristics were taken too, as they could be a problem for MMICs performance.

FET parametric degradation are shown in Fig. 9: such a behaviour has been reported several times to indicate gate-semiconductor interaction [6], and in our case, by measuring the other elements available on the PCMs, we could exclude degradation of the surface and of the ohmic contacts.

To get a more clear picture of the degradation behaviour, we performed 3 storage tests (at 300, 325 and 350 °C respectively) on similar samples, with the results summarized in fig. 10, in term of mean Idss degradation versus time; a fitting low is also reported, with no other meaning than a simple mathematical interpolation. By standard failure analysis methods (based on SEM, EBIC and the so called "back etching" [7]) we could not get any significant result, so that we decided to use cross-sectional Transmission Electron Microscope (TEM), which is normally used to analyze large area devices (such as memories) or wafers; sample preparation had to be accurately tailored to our devices, but at the end a complete hypothesis could be made about the evaluation of the degradation:

- up to 325°C, Au penetrates along the bordes, giving origin to triangular features (typical of solid state diffusion) which modify the space charge region, thus decreasing the current flow; Ti barrier is still intact;

- at 350°C, Au uniformily penetrates through Pt and Ti layers, creating a defect-rich region into the channel (that is the classical gate-sinking phenomenon).

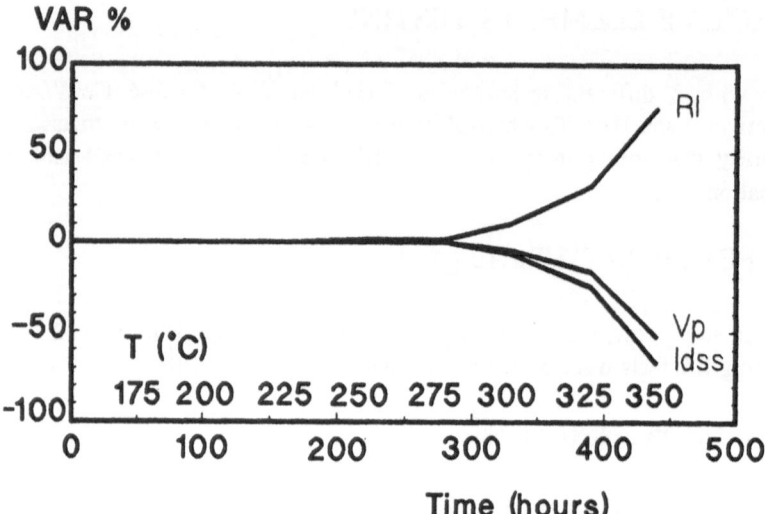

Fig. 9: Mean FETs parameters degradation during thermal step stress.

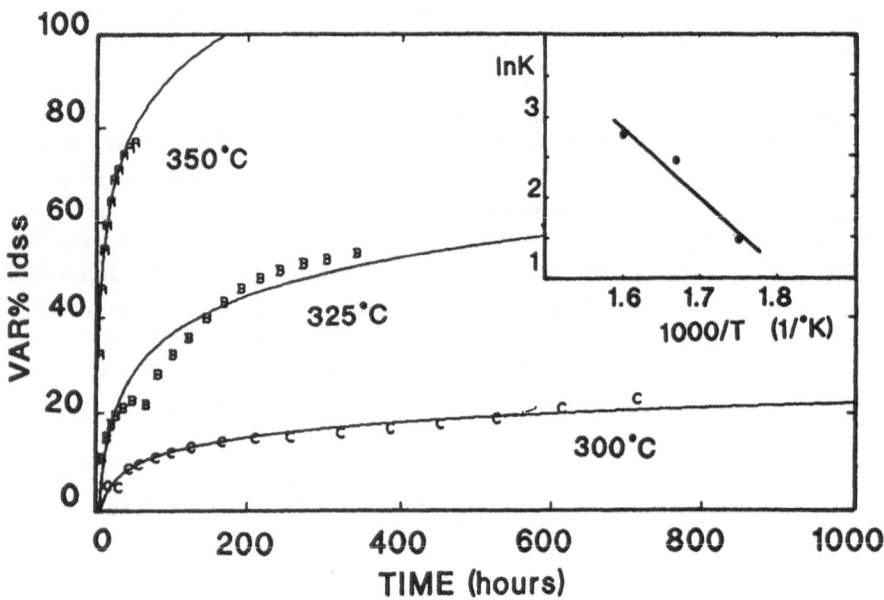

Fig. 10: Mean Idss variation during storage tests at Tamb=300°C, 325°C and 350°C. The best fit is a logarithmic dependance on time: $VAR\%I_{dss} \approx K_0 e^{-Ea/kT}$ lnt. The activation energy is $Ea \approx 0.75$ eV.

The results are summarized in fig. 11, where a TEM image of a sample aged at 300°C is shown, with the two competing phenomena; it is interesting to note that through-barrier interaction happens only at 350°C at very short time (it was detected in one sample after 3 hours only!).

Backgating characteristics also changed at high temperatures: fig. 12 shows how drain current is practically insensitive to substrate bias up to 300°C, while at 325 and 300 °C a sharp Id decrease happens, together with substrate current increase.

5.1.2 High Forward Gate Current and High Temperature Reverse Bias

These two tests were performed at Tamb=160°C, with source and drain grounded; 10^5 A/cm² forward gate current density and -5V gate voltage were kept respectively; several FET structures were used (with different gate length and width) to analyze possible geometrical effects; up to 6000 hours no degradation happened.

5.1.3 Operating Life Tests

Three lifetests were performed, with DC bias, at Tch=175, 200 and 225 °C, making use of both test fixtures previously described.

No catastrophic failure happened, and the parameters degradation are shown in Fig. 13, in terms of mean Idss, Vp and Ron values; at 225°C the results are still consistent with gate-semiconductor interaction, but we focussed our attention in particular on the two devices (out of 12) that degraded at 200°C, in contrast with the excellent stability of the others.

Again we had to use TEM to get significant failure analysis results: Fig. 14 reports a TEM image and a sketch of a gate, where the degradation appears different from the one observed during pure thermal stress; in fact, there is a defective region on the borders, that does not take the form typical of solid state diffusion (temperature was too low) but that could be generated by atomic gold diffusion; that is consistent again with the observed degradation (Au compensate donor density [6]) and it can happen at low temperature.

5.2 FET results (Cr/Au/W/Au gate)

Again in this case, the same tests described before were performed, making use of discrete test structures; the technology is characterized by the same metal structure for gate and ohmic contacts, with double implantation providing the Shottky and ohmic characteristic.

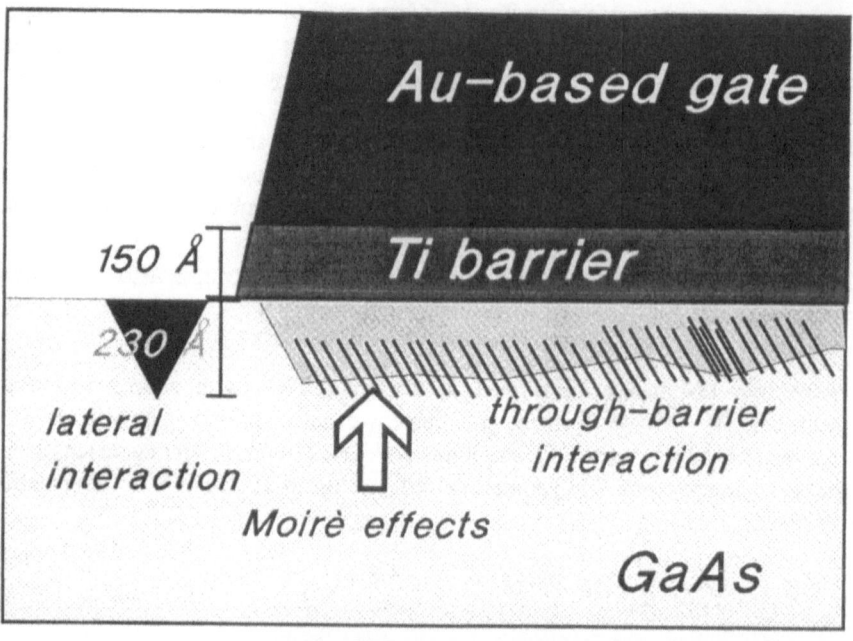

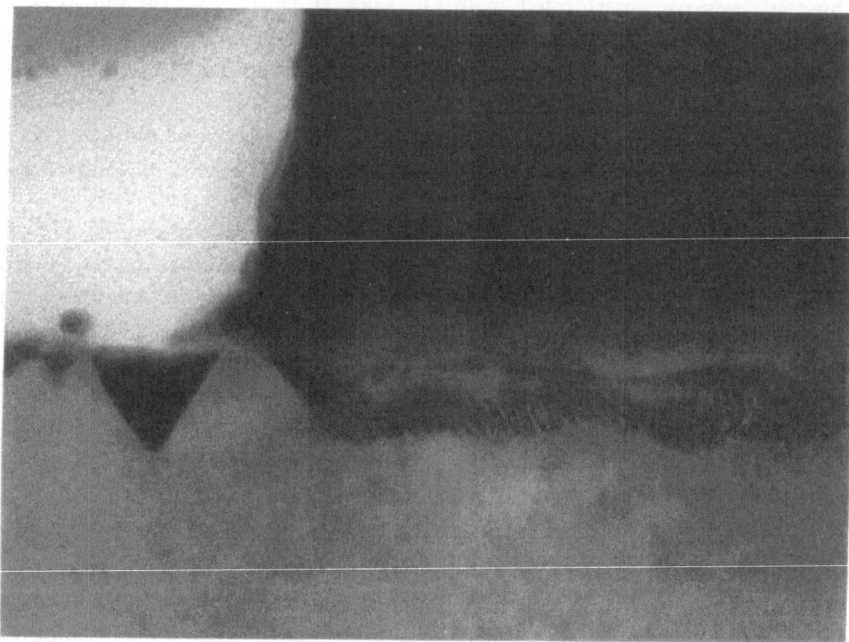

Fig. 11: TEM image with a drawing illustrating the two interconnections during storage at 350°C.

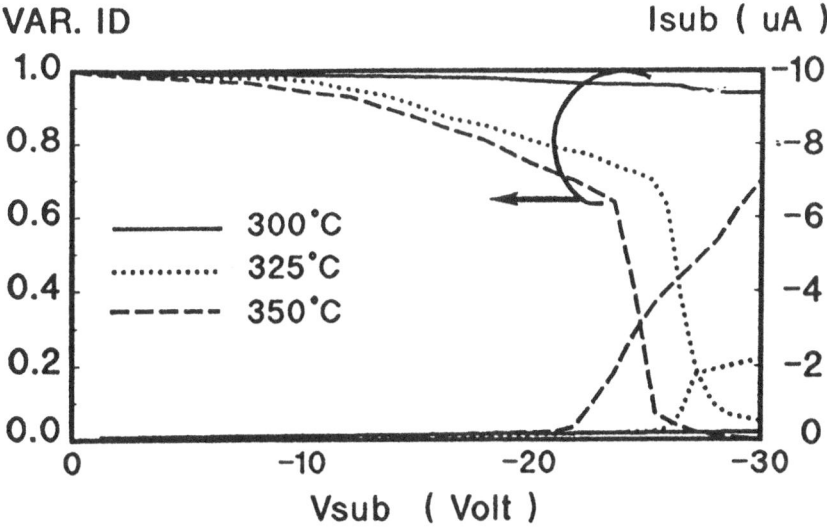

Fig. 12: Backgating characteristics of FETs during thermal step stress.

5.2.1 Thermal step stress

For this test, we used single finger FETs + 2 sidegates, FAT FETs (100 μm gate length) and ungated FETs (to check ohmic contacts stability); the behaviour of classical electrical parameters is summarized in fig. 15: there is a clear difference between FAT FETs, that degraded quite rapidly at relatively low temperature, and the other structures, that are stable up to 350°C, with only minor ohmic resistance increase.

By means of backetching we could understand that metal-semiconductor interaction happens only when electrolytic gold reinforcement is present, (see fig. 16) that is on ohmic contacts and on FAT FET gate; on 0.8 μm gate, without reinforcement, no interaction phenomenon is detectable, as confirmed by electrical measurements. Regarding sidegating characteristics, the aging behaviour is similar to the one previously reported, for backgating, as degradation of threshold voltage for drain and sidegate current is relevant at high temperature only (Fig. 17), the same temperature of metal-GaAs interaction; we define the "threshold" value as the sidegate voltage which causes 20% reduction of drain current and 100% increase of sidegate current.

184

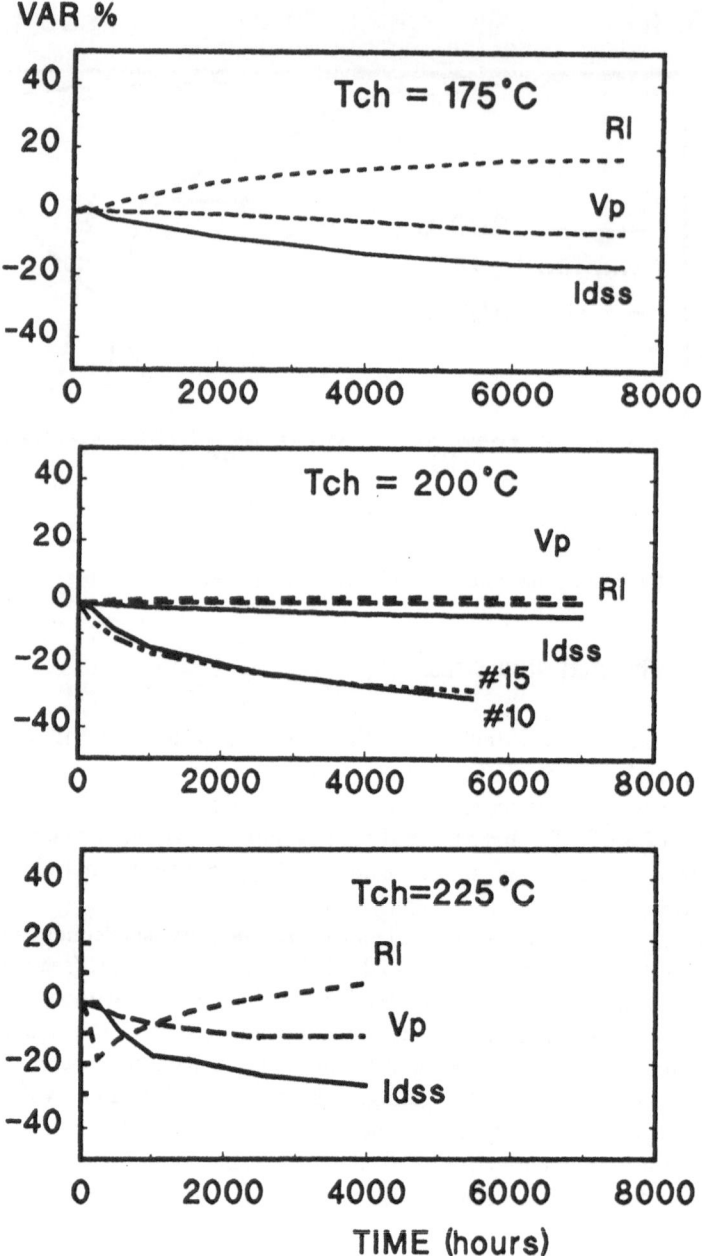

Fig. 13: FET parameters degradation during operating lifetest.

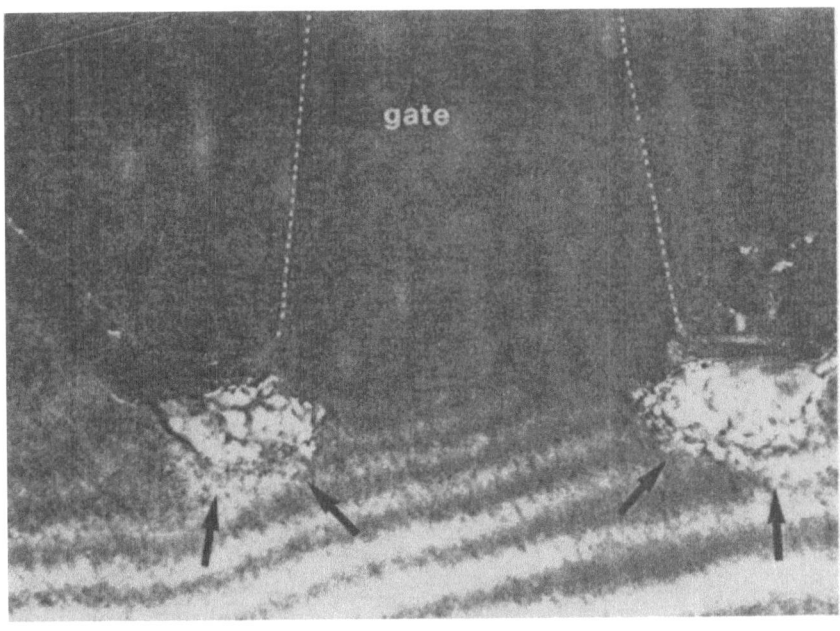

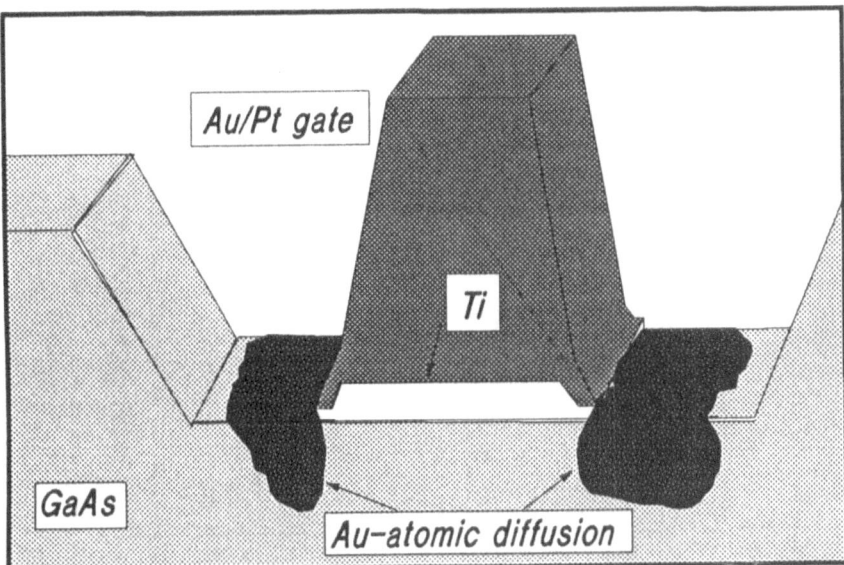

Fig. 14: TEM image of FET degraded during operating lifetest at channel temperature of 200°C.

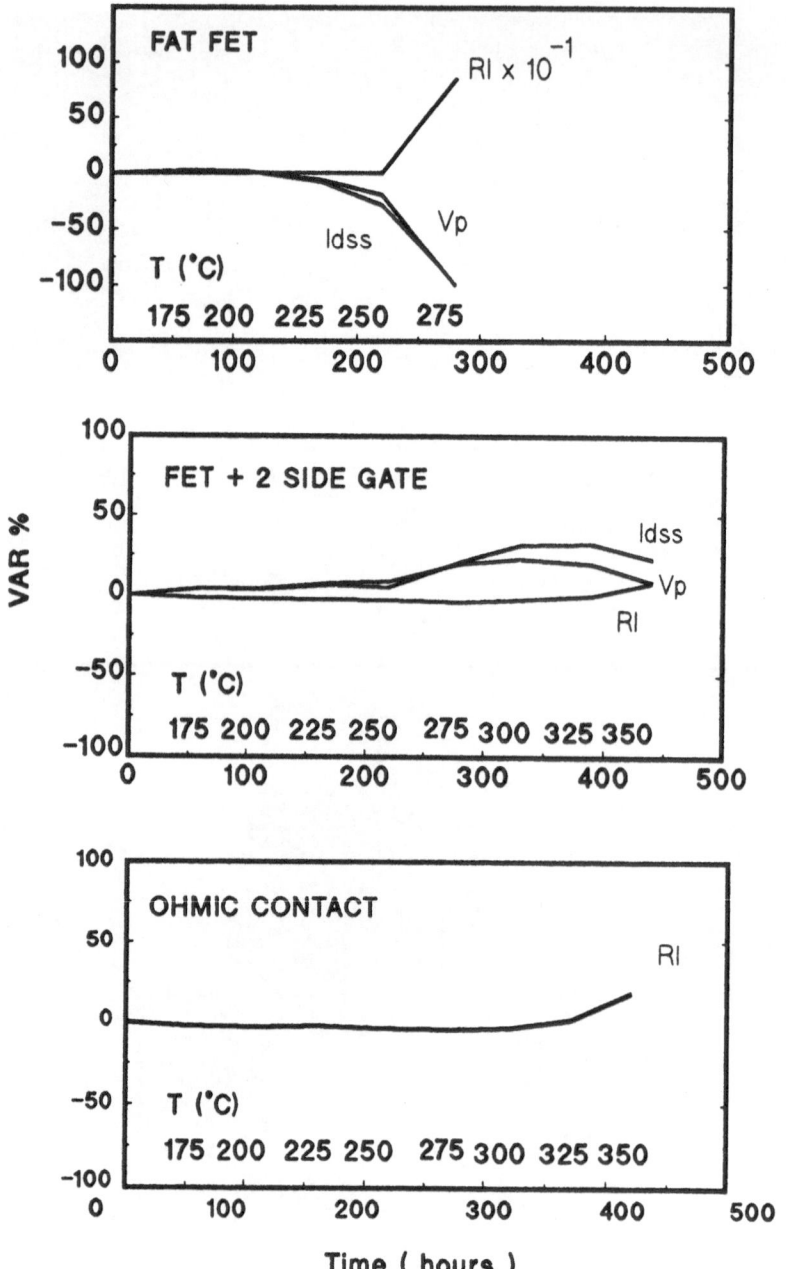

Fig. 15: Mean FETs parameters degradation during thermal step stress for FAT FET (a), FET + sidegate (b) and ohmic contact (c).

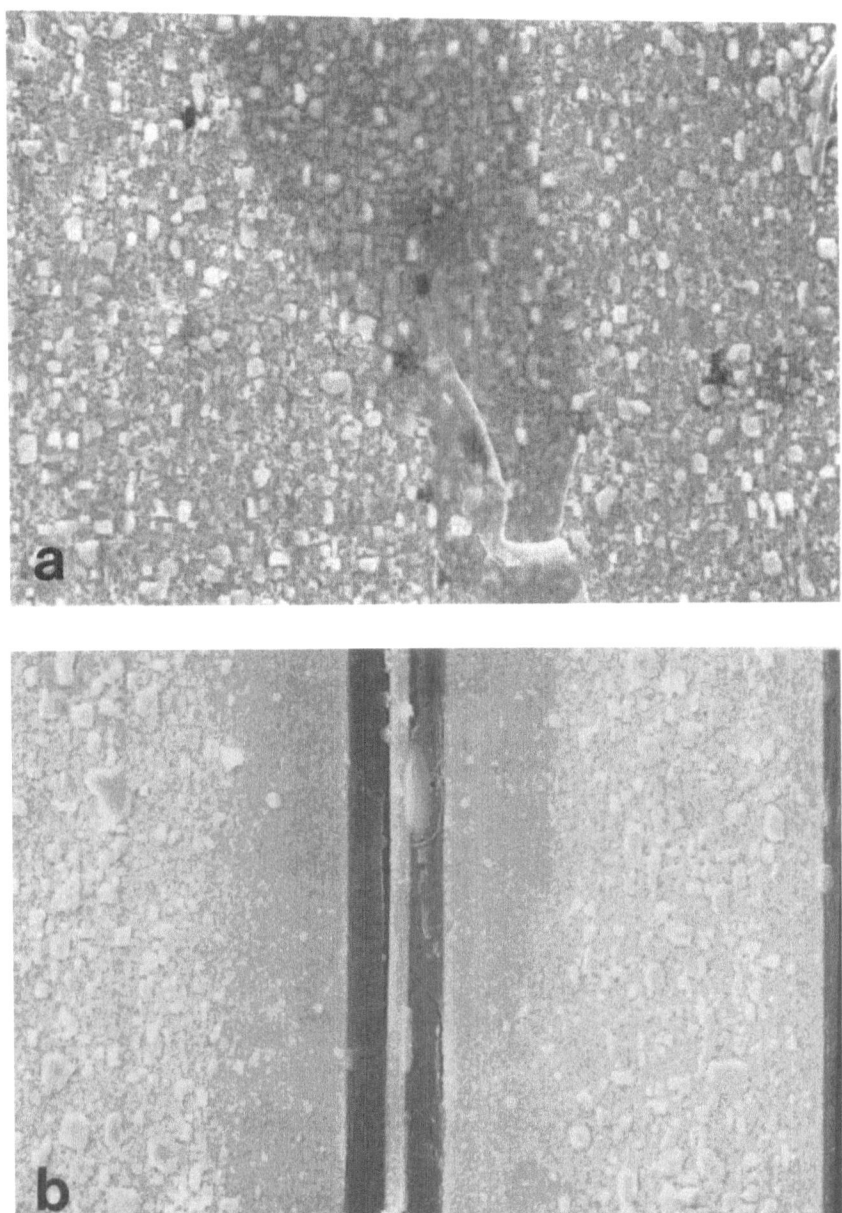

Fig. 16: SEM image of backetching of the FAT FET gate (a) and the FET source-gate-drain (b) degraded during thermal step stress.

5.2.2 High temperature reverse bias and high forward gate current

With the tests performed in the same conditions already described for the other technology, no significant degradation was detected up to 6000 hrs.

5.2.3 Operating life tests

900 μm gate periphery FETs were used, mounted into standard FET packages and biased to reach 175, 200 and 225 °C channel temperature; in the last case, it was necessary to increase the bias voltage, up to 8 Volt, while for the other two temperatures 5 Volt drain voltage was enough to get the proper power dissipation; the aging behaviour during the 3 tests is summarized in Fig. 18: no noticeable degradation happened up to 200°C, while at 225°C some catastrophic failures took place together with parametric degradations; as shown in fig. 19, the degraded samples, before catastrophic failures, were characterized by modification of Id-Vd curves and decrease of gate diode breakdown voltage.

The application of back etching technique allowed to identify the catastrophic failure mechanism as a "classical" burnout, with massive gold interaction including gate, source and drain, while all the parametric failures had indication of gate disuniformities (interaction) always towards the drain (fig. 20); this phenomenon is consistent with the observed electrical behaviour, and it will be discussed later.

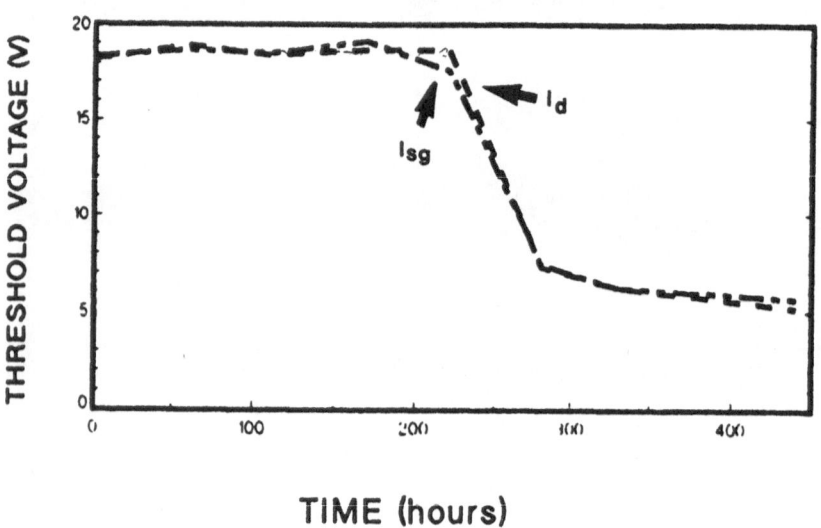

TIME (hours)

Fig. 17: Threshold voltage variation during thermal step stress for FET + sidegate.

VAR%

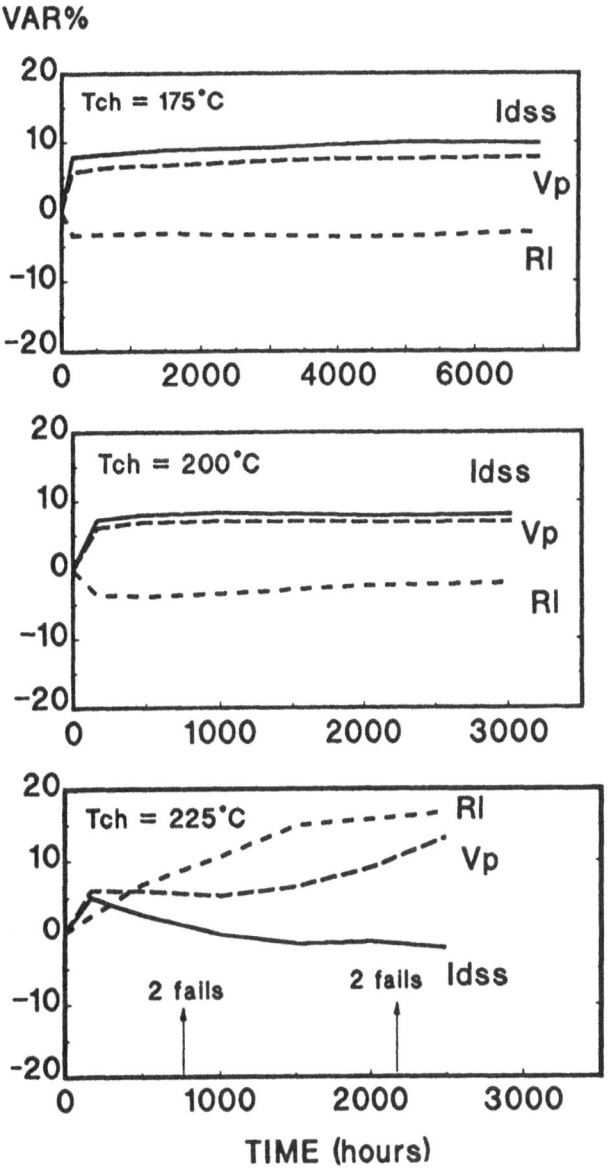

Fig. 18: Means FETs parameters degradation during operating lifetest at channel temperature of:
a) 175°C
b) 200°C
c) 225°C

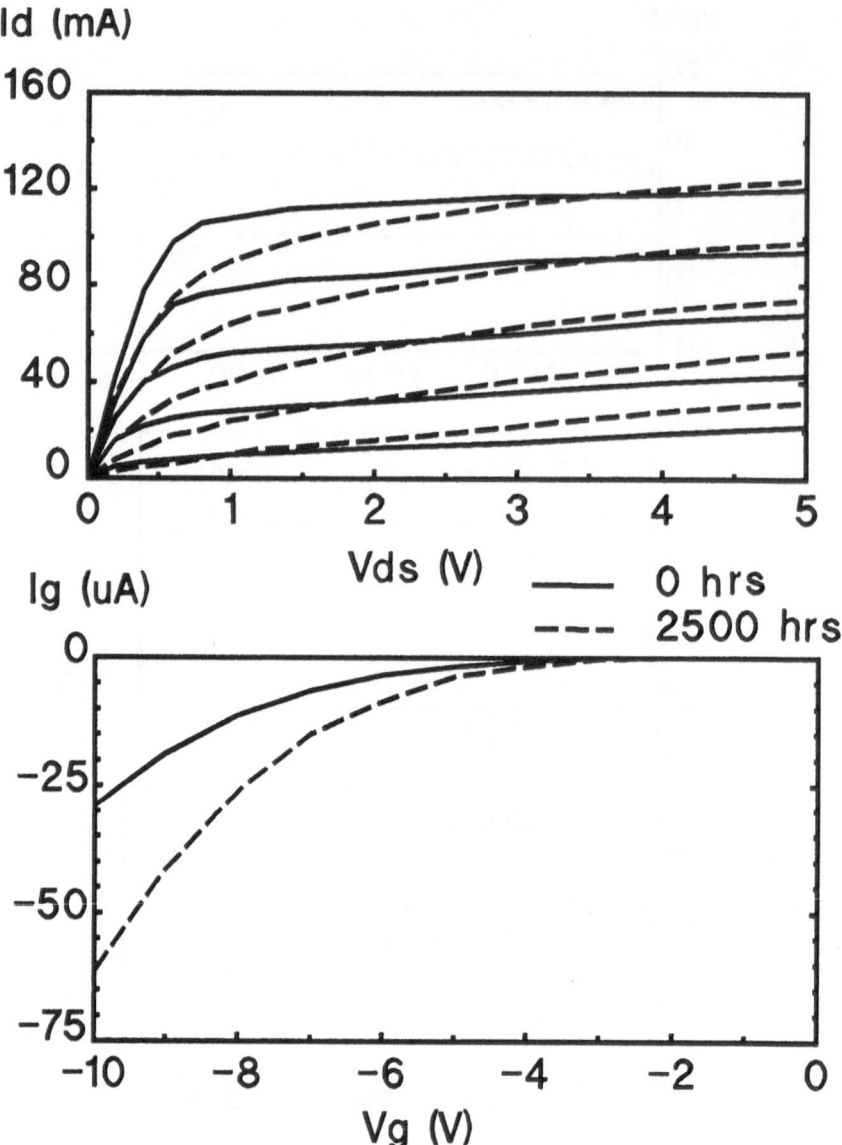

Fig. 19: Variation of Ids vs Vds and Igs vs Vgs reverse curve at different phases of the operating lifetest.

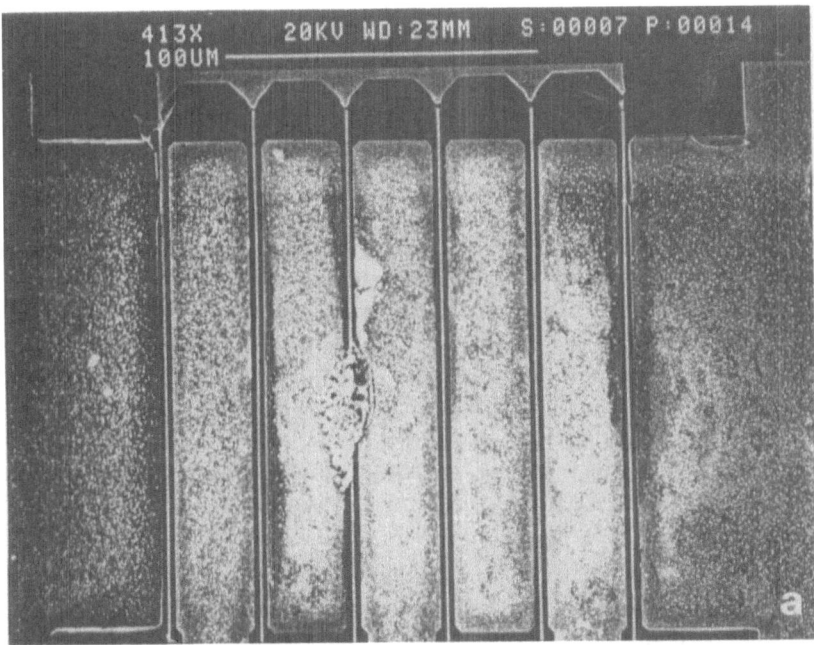

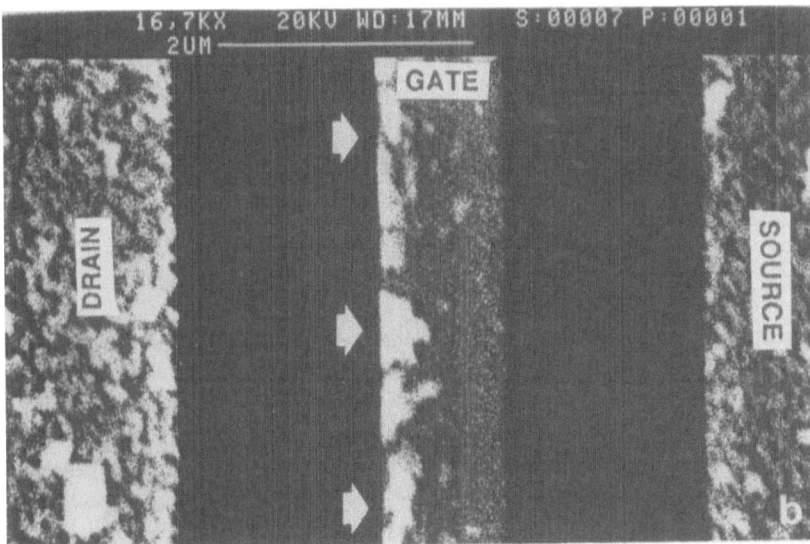

Fig. 20: Backetching image of catastrophic failures (a) and parametric failures (b), with the arrows highlighting the lateral interactions.

5.3 MMICs testing

Simple low level MMICs were tested in order to confirm the results from elements testing; a single stage $0.05 \div 3.5$ GHz feedback amplifier was chosen for Ti/Pt/Au technology, and a two stage $1.5 \div 2.5$ GHz was used for Cr/Au/W/Au technology.

All the device were mounted into commercial hermetic flat-packs, suitable for the defined frequency ranges; electrical characterization included DC parameters (supply current and input leakage current) and S parameters.

To determine the lifetest temperature, we perform IR analysis: in the case of single stage amplifier the hottest region was the FET active area, with an estimated channel to case overtemperature around $33°C$; the case temperatures was then set in order to reach $200°C$ and $225°C$ channel temperature. In the other case, IR analysis gave the information that the hottests regions were not the two FETs (that were around $15°C$ higher than the case) but the two drain bias resistors, that reach $40°C$ overtemperature; anyway the case temperature was set to get $200°C$ and $225°C$ on the FETs.

The results are summarized in figg. 21 and 22, where, for both technologies, it seems that S parameters variation (S_{21} only is reported for simplicity) are almost negligible, despite some degradation of supply current; anyway no catastrophic failure happened during the tests.

5.4 HEMT results

Low noise HEMT devices were developed, featuring AlGaAs/GaAs material, 0.25 μm WSi mushroom gate and AuGeNi ohmic contact; discrete devices were submitted to thermal step stress and operating life tests.

5.4.1 Thermal step stress

The results of this test, that was run in the same conditions of FETs, are shown in Fig. 23: drain current variation at Vds=50mV is reported too, because this parameter is sensitive to ohmic contacts degradation, that is the likely failure mechanism; in fact, by means of microsectioning and TEM, heavy ohmic contact material penetrations were pointed out, in some cases down to the superlattice confirement layers; as several times reported in the past, these interdiffusion features are Au and Ge rich, while As outdiffuses (significant As quantities are infact detected above the ohmic alloy).

Gate metal and GaAs surface looked absolutely stable up to $350°C$.

5.4.2 Operating Life Test

12 devices for temperature were operated at Tch=175, 200 and 225 °C respectively, biased at 3V, 12mA; up to 4000 hrs at least, no significant

193

degradation was detected for any transistor parameters, the maximum Ids decrease being less than 5% at the highest temperature, due to ohmic contact resistance increase.

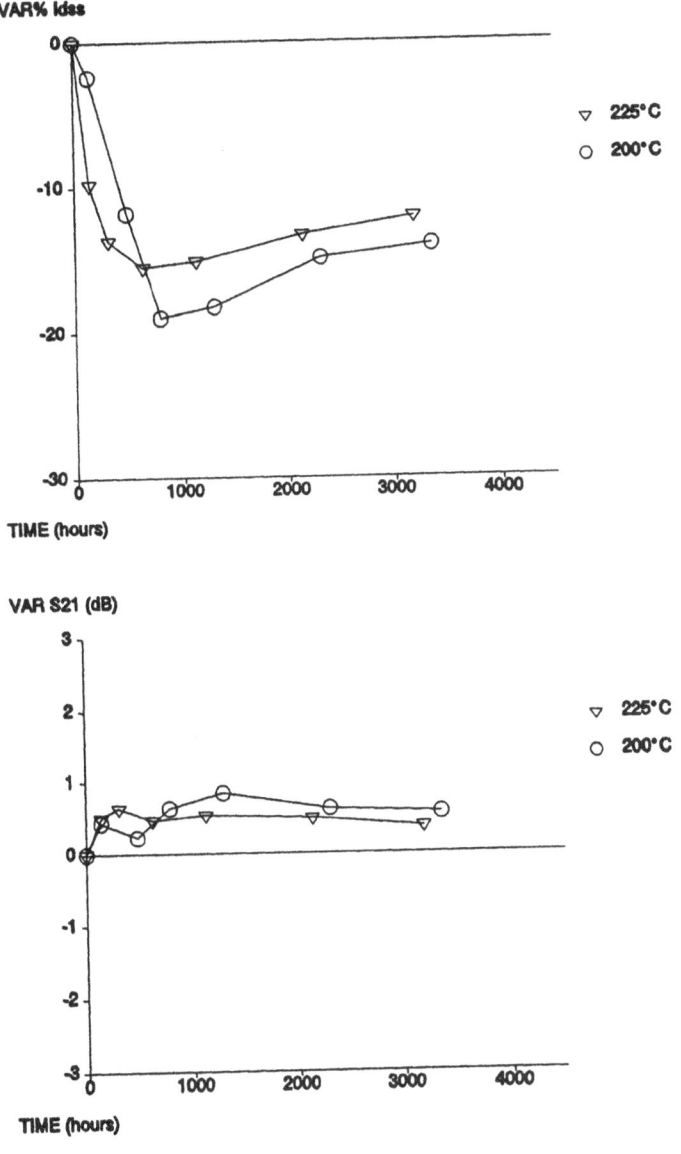

Fig. 21: Single stage MMIC mean results.

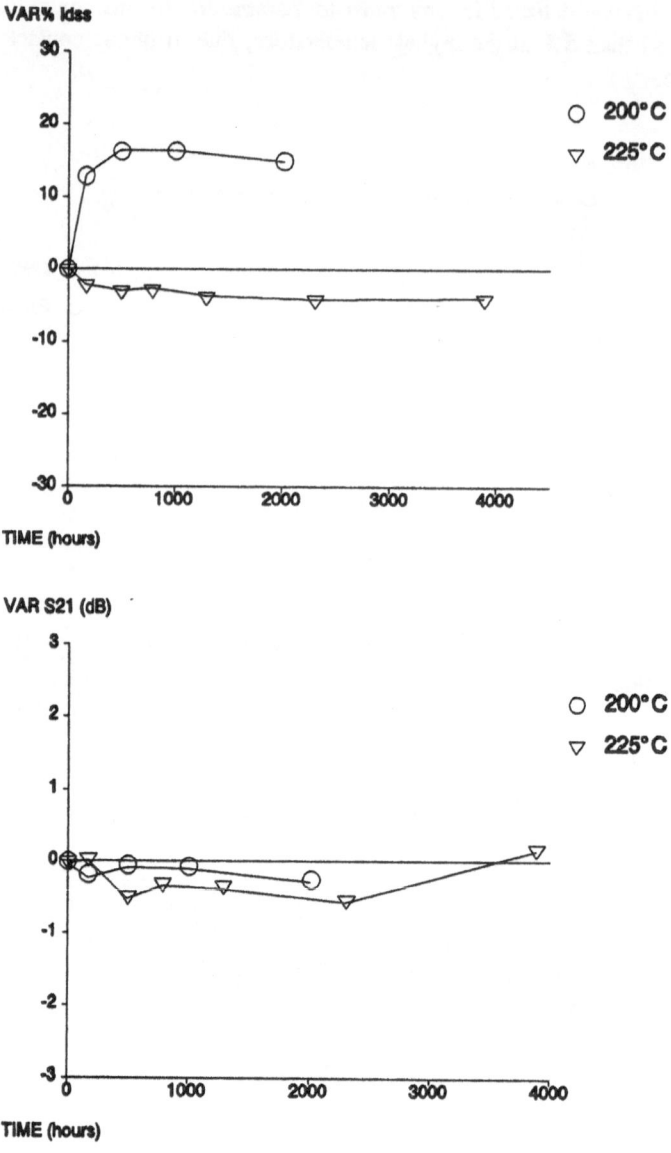

Fig. 22: Two stage MMIC mean results.

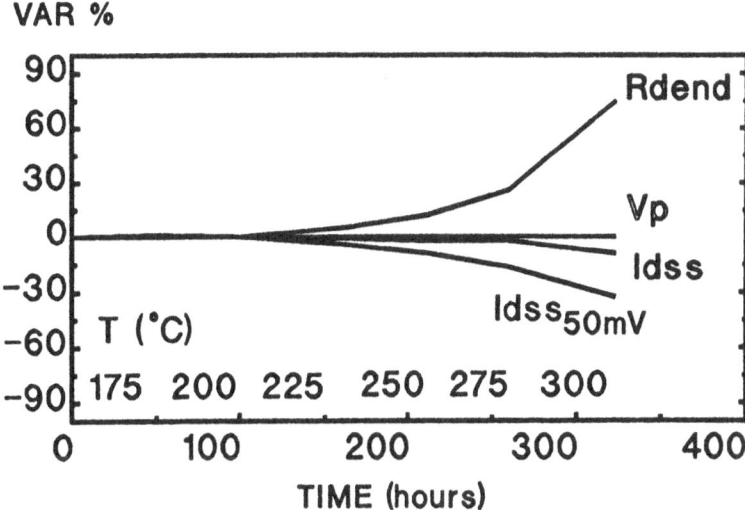

Fig. 23: Mean HEMTs parameters degradation during thermal step stress.

6. DISCUSSIONS

The results presented so far, are now discussed in terms of the main detected failure mechanisms.

6.1 Ohmic contacts electromigration

Some cases of ohmic contacts eletromigration have already been presented (see [6] for a summary) and one of them [8] was demonstrated on the same technology as our work; using the data reported in fig. 5, a tentative calculation of the accelerating factors due to temperature and current density can be made: the results are shown in fig. 24, where the mean parametric variation (after 2800 test hours) is plotted vs local temperature and current density.

The degradation seems to fit quite well the Arshenius law (with Ea=0.5eV) and a power dependance on current density (with n~2.8), after the necessary normalization has been made to take into account the reciprocal influence of the two stresses; minor variations of the power factor can be expected, due to the fact that current density values are calculated into the interconnecting vias ($\sim 12 \mu m^2$), while electromigration actually takes place in the ohmic layer; here, current density levels are expected to be at least ten times higher than those into the vias. With an arbitrary failure criteria on of 20% resistance increase, medium lifetime, at 80°C ambient and maximum allowed current density, exceeds 3×10^6 hrs for the test structure; for actual circuits, this value has to be increased due to the presence of thich ohmic contact reinforcement that lowers the current density.

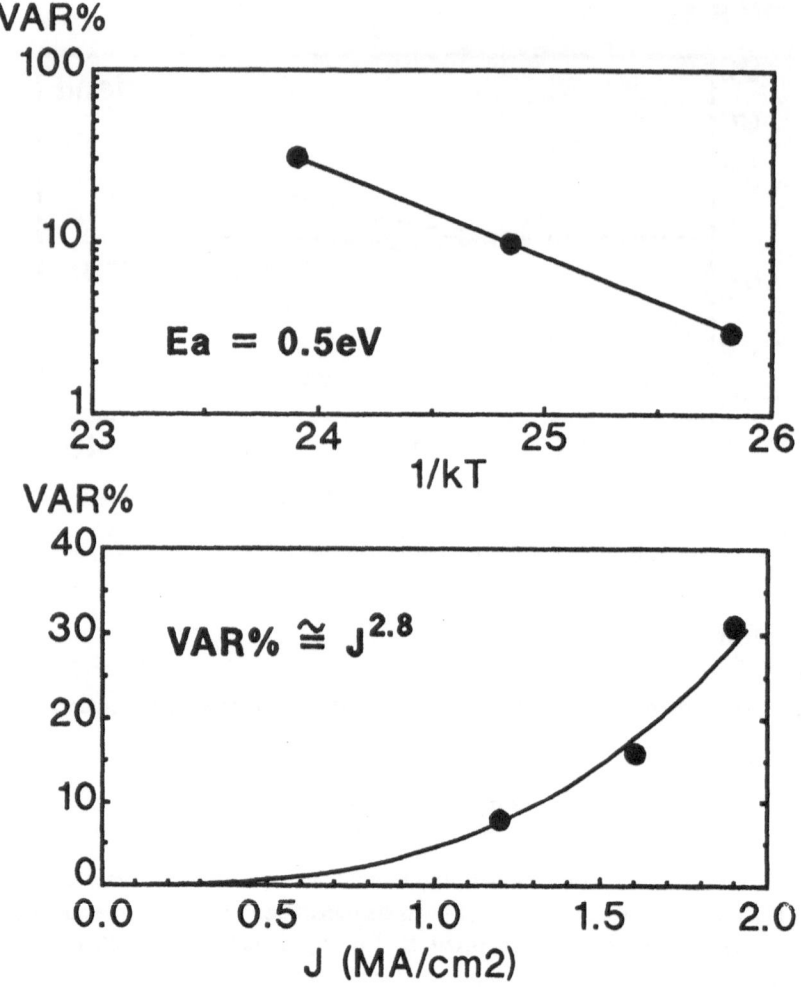

Fig. 24: Dependance to current density and temperature for mean percent variation (after 2800 test hours) of interconnection resistance during lifetest.

6.2 Gate-semiconductor interaction

Several phenomena were observed concerning this mechanism.
Starting from finaly thermal effects on Ti/Pt/Au, two steps have been clearly identified, as already described in parag. 5.1.1; while from the electrical measurements it is not easy to separate the effects, it seems that through-barrier

interaction has a threshold temperature, as it was never found up to 325°C at several hundreds hours; anyway, with the fitting law reported in fig. 10, we can rule out any effect of thermal interaction at actual operating temperature. It is interesting to note that the possibility of interaction starting along the borders was postulated long time ago [9], but to our knowledge it was never experimentally demonstrated.

When performing operating life test, another type of interaction was pointed out, that is different from the previous two, as it does not involve massive gold diffusion; even if the failure analysis results should be carefully considered, atomic gold diffusion seems the most likely failure mechanism, again along the borders; as Au compensates the donors, this phenomenon produces the same macroscopic electrical effects of massive diffusion.

From the observation of the electrical behaviour, it seems that this atomic Au diffusion is typical of "weak" devices; one possible hypothesis is that it happens when Au residues are present on the surface, possibly coming from gate layers misalignement or non-perfect lift-off; these residues can be very small, but atomic Au is enough to modify electrical characteristics, even at low temperature.

Cr-based contacts are thermally stable up to 350°C; interaction with GaAs can be found only when electric field is applied, and it is clearly found towards the drain, where electric field is higher and temperature is hotter; this interaction is consistent with gate-drain leakage increase and drain conductance increase, observed during the test. Even without considering the effects of higher drain voltage, a worst case extrapolation from our data yields an activation energy higher than 1.9eV, with an estimated medium lifetime exceeding 10^8 hrs at Tch = 125°C.

6.3 Ohmic contacts interdiffusion

The degradation of alloyed ohmic contacts is well known since a long time [6]; as it is a high activation energy effect, it can be hidden by other phenomena during operating lifetest at "low" temperature.

In our HEMT work, the excellent stability of gate metal and GaAs surface allowed to clearly identify the effects of ohmic contact degradation, both during thermal step stress and lifetest.

Moreover, in HEMTs, the difference between channel and ohmic contacts temperature is very low (< 10°C), meaning that ohmic contact temperature is enough to cause interdiffusion; in FETs, even for low power devices, ohmic contacts are 30÷40°C colder than the gate [10], preventing any significant degradation in our lifetests.

After the failure mechanism identification, discussed in paragraph 5.4.1, the increase of both Rs and Rd during lifetest was found to follow a linear dependance on the square root of time (that is an other confirmation of interdiffusion effects); as shown in Fig.25, the slope follows the Arrhenius law, with Ea = 1.5÷1.8eV, consistent with other available results.

198

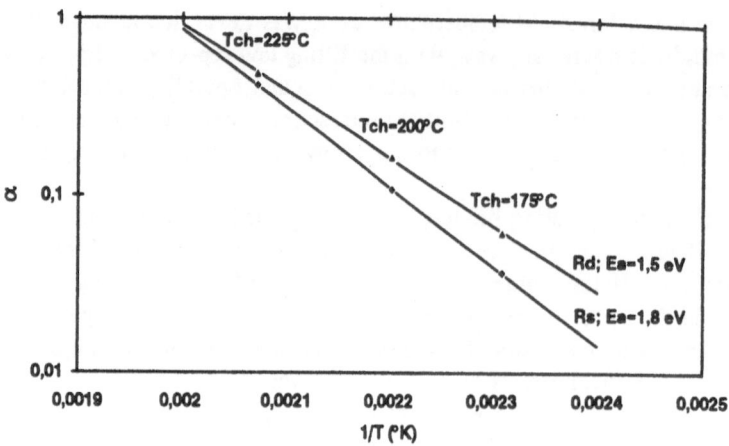

Fig. 25: Activation energy for HEMTs ohmic contact degradation.
VAR% $R_s, R_d \approx \alpha\ t^{1/2}$, where $\alpha = \alpha_0\ e^{-Ea/kT}$.

7. CONCLUSIONS

Several conclusions can be drawn from our work.

- intrinsic FET reliability is high, with ohmic and gate degradation happening in extreme conditions only; no surface related problem has been found;
- passive elements shouldn't be a major problem, anyway the possibility that they can be the hottest MMIC elements has to be carefully addressed;
- electromigration characteristics of Ti/Pt/Au metal are very good, up to ~60 MA/cm²;
- ohmic contact electromigration in actual devices is probably avoided by contact reinforcement;
- backgating/sidegating degradation happens at high temperature only, and it seems to be related to metal-semiconductor interaction;
- MMIC small signal parameters are not dependent on DC parameter variation;
- the developed HEMT technology is absolutely stable up to 225°C Tch, being at least comparable with the best devices available from the market.

REFERENCES

[1]   M. Belfort et al.,"A methodology for the space qualification of GaAs MMICs" CNET Technical Note, 3rd issue, Dec. 92.
[2]   W. Roesch, D, Stunkard "Proving GaAs reliability with IC element testing" presented at MANTECH Conf. Nashville, Nov. 88.
[3]   F. Magistrali "An overview of GaAs reliability" Proc. of GaAs Applic. Symp., Noordwijk, The Netherlands, April 92.

[4]   B. Riccò et al. "Reliability of GaAs MESFETs" in Semiconductor Device Reliability, A. Christou and B.A. Unger eds., pp. 455÷469, Kluwer Academic Publ., 1990.

[5]   C. Canali et al. "Test fixture for MESFET reliability life tests" Microelectronics Reliability, Vol. 27, N°5, pp. 897÷911, 1987.

[6]   E. Zanoni et al. "Metal-GaAs interaction and contact degradation in microwave MESFETs" Quality and Reliability Eng. Int., Vol. 6, pp. 29÷46, 1990.

[7]   F. Fantini et al. "Back-etch: an effective tool for characterization and failure analysis of MESFET devices" Proc. of Int. Symp. for Testing and Failure Analysis, pp. 235÷241, Los Angeles, Nov. 88.

[8]   T. Abbott and J. Turner "Some aspects of GaAs MESFET Reliability" IEEE Journal on Microwave Theory and Techniques, Vol. 24, N° 6, p. 317, 1976.

[9]   S.P. Murarka "High temperature stability of AuPt/n-GaAs Schottky barrier diodes" Solid State Electronics, Vol. 17, pp. 869÷876, 1974.

[10]   C. Canali et al. "Correlation between thermal resistance, channel temperature, infrared thermal maps and failure mechanisms in low power MESFET devices" Microelectronics Reliability, Vol.29, No.2, pp.117÷124, 1989.

AKNOWLEDGEMENTS

The authors would like to aknowledge first of all the support of EEC and the collaboration of COSMIC consortium, especially of GMMT and Siemens; then, we are in debt with all our collegues at Alcatel-Telettra Quality Dept. and GaAs Comp., that took part to the work; special thanks are due to M. Vanzi, now with Univ. of Cagliari, for TEM techinque and failure analyses in general.

4.2 Forward and Reverse Engineering

P. H. Ladbrooke, J. P. Bridge, and D. R. Bowler
GaAsCode Ltd, St John's Innovation Centre, Cambridge CB4 4WS, UK

1 Introduction

The traditional approach to MMIC manufacture is based around the concept of the "fixed foundry" process, reflected in terminology such as the "20GHz process" or "the half micron process". Here, technology is a constant, serving to reproduce the microwave properties of active and passive devices and circuit elements characterised previously by measurement. Design parameters are restricted principally to the dimensions and values of passive components. Innovations are to be avoided unless test circuits can be built and characterised. Although this may be possible for passive elements, the investment necessary to bring about a change in the performance of an active device - in particular, a FET - is prohibitively expensive, and requires the development of a new "process".

The traditional approach is inflexible and may be unworkable, especially as operational frequencies increase beyond 26GHz and device dimensions shrink. The traditional approach may lead to circuits with decreasing yields unless designers are free to alter technology in order to achieve specification and to minimise performance spreads. This approach may be referred to as *technological design centering*. In general, both active and passive devices would be brought into the design. A more restricted approach would be to regard the circuit as most sensitive to the properties of the active devices, and to adjust the active device behaviour alone: this method might be referred to as *FET centering*. These approaches immediately raise two requirements:

(i) It must be possible to predict the outcome of structural changes on the microwave performance of the device and circuit. Without this capability, the approach would be uneconomic.

(ii) The technology used to fabricate components and circuits must be capable of responding to the demands of the device designer with accuracy and precision, without the need for a long sequence of trials.

The MONOFAST consortium set out to demonstrate that software and fabrication tools have evolved or could be developed to the point where high-performance circuits can be designed and fabricated with predictable yields starting from physical descriptions of the components with the minimum of trial and error.

2 Forward and Reverse Modelling - Definitions

2.1 Forward Modelling

Forward modelling consists of forecasting the microwave and millimetre-wave S-parameters, power gains, noise parameters, and RF output power of a FET from the physical description of the device and its bias conditions, where by physical description is meant the doping profile, gate length, inter-electrode spacings, recess depth and shape, surface charge (or potential) and material transport properties.

Forward modelling is used for the following purposes.

1. Designing FET structures and refining FET design.

2. Focusing technological effort by forecasting device behaviour with practical gate recesses and interpreting surface conditions.

3. Identifying the microwave characteristics of the technological mean FET (so circuit design is carried out around a ''centered'' FET).

4. Pin-pointing the technological origin of electrical variations in FETs.

5. Serving as the basis MMIC of sensitivity analysis and yield forecasting.

2.2 Reverse Modelling

Reverse modelling uses the techniques of forward modelling in reverse, and consists of analysing measured FET S-parameters, including S-parameters measured at many bias points, using physics-guided routines which

(a) permit recovery of values for some of the physical parameters of the devices from which the S-parameters came, and

(b) result in an equivalent circuit, the S-parameters of which duplicate the measured S-parameters.

3 The MONOFAST Project

The objective of the MONOFAST project was to demonstrate feasibility of a 44GHz low-noise amplifier for space flight use to the following specification:

> Gain: >15dB   Return losses:<-10dB   Noise figure: circa 4dB
> Bandwidth: 1GHz   Centre frequency: 44GHz

The methods of forward and reverse modelling were employed in the MONOFAST project as follows.

3.1 "Design Limits" Design

"Design Limits" design involves a high-level abstraction of the methods of reverse modelling, resulting in a preliminary specification of the FET and first estimates of the gate length and layer doping needed to meet the overall circuit specification.

3.2 Detailed Device Design

A full forward modelling simulation is used to carry out a detailed design of the FET concurrently with, and iteratively with, initial circuit design and practical development of the FETs. The objective is to develop a specification for the doping profile, the gate recess depth and shape, the inter-electrode spacings and all other structural features of the FET or HEMT required, taking the "Design Limits" FET as a starting point.

3.3 Selection of Circuit Topology and Preliminary Design of Circuit

The selection of a circuit configuration is based on analysis of the sensitivity of trial circuit designs to uncontrolled variations in technology affecting the FETs. A forward modelling simulator is used for this purpose. At this stage in the design process, the need for further changes to the FET structure may be apparent. For example, in the MONOFAST project the gate length of the device required was reduced from 0.25µm to 0.2µm.

3.4 Analysis of Millimetre-wave Characteristics of FETs and Relationship to Technology

Because a reverse modelling S-parameter fitter is physics based, a good fit to measured S-parameters will be obtained only if the S-parameters are those of a FET (or a HEMT) measured in a well calibrated network analyser. Thus the quality of fit obtained can be used as a check of the network analyser calibration. A second way in which reverse modelling is useful is to de-embed finally back to "the MMIC FET" for circuit design. The procedure uses physics-based methods to identify and eliminate parasitic capacitances and inductances present in practical test FETs. Thirdly, reverse modelled technological data, as obtained from physics-based fitting of measured S-parameters, can be used to check that the main technological parameters (metallurgical gate length and channel doping density) in practical FETs have the values intended.

3.5 Sensitivity Analysis of the MMIC

If a forward-modelling FET simulator is linked to a circuit simulator as in Figure 1, the sensitivity of an overall circuit to variations in technology affecting the FETs can be forecast. In a mature technology, random variations in MMIC behaviour are due more to process variations affecting the active devices rather than the passive elements. Sensitivity analysis allows a degree of choice of the circuit configuration and circuit values to minimise the sensitivity to (some) process variations.

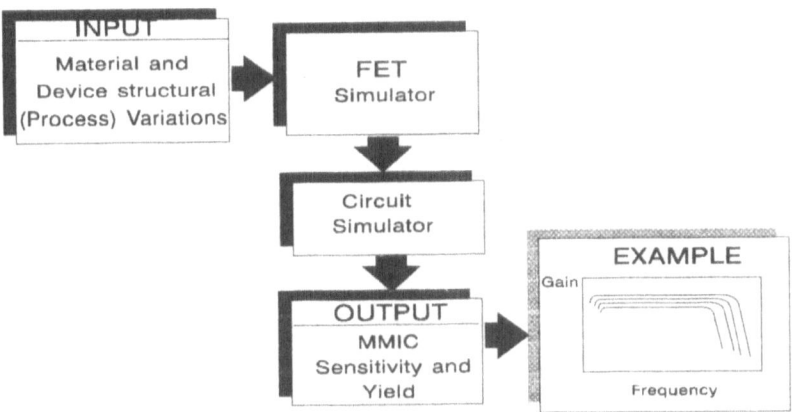

Fig. 1 Basis of MMIC Sensitivity Analysis and Yield Forecasting

3.6 Yield Forecasting

The yield of an MMIC may be forecast using the Monte Carlo method to generate random, practically realistic, varying values for the physical parameters of the FET. The technique automatically generates and keeps all the correct correlations which exist between the equivalent circuit element distributions or S-parameter distributions.

Strictly, spreads or distributions for all the technological parameters should be independently measured to provide the necessary background information. In MONOFAST, "likely representative" spreads were inferred from practical and modelled sources of data, and Gaussian distributions were assumed.

4 Modelling Techniques

4.1 Forward Modelling: Macro-cell Methods

An appreciation of macro-mesh techniques is best gained by contrasting them with the more familiar micro-mesh simulators. Micro-mesh simulators use the same equation set throughout the entire device, and only when the overall solution is obtained does the simulator recognise that some regions are depleted of electrons, others contain cool electrons, and yet others hot electrons. Put another way, they use no ab-initio intelligence about the internal conditions in an operating FET, namely that there are always regions within which certain physical processes dominate. The general rules governing the size, shape, and boundary conditions applying to these distinctive regions can be found by experimental and theoretical methods. Such knowledge allows solvers to be constructed wherein solutions to reduced equation sets are sought within large, spatially non-uniform cells, the solutions within each region being matched at the boundaries. The term "macro-cell solver" usefully distinguishes this type of solution.

A macro-cell solver satisfies equations expressing conservation of electron numbers and current, with the electron motion governed by a generic velocity function in which velocity is related to the local two-dimensional field via electron energy. In these simulators, the nature of the dominant physical processes occurring within each macro-cell is prescribed, and what remains to be found is their magnitude and spatial extent. In contrast to the methods of micro-mesh simulators, where the mesh is fixed and the nature and magnitude of the physical processes occurring within each mesh rectangle, triangle or box are found by the process of simulation, iteration in a macro-mesh solver is required to find the boundary positions of each macro-cell. A macro-cell solver therefore uses a sparse, adaptive mesh, and promises a good balance between speed and accuracy. Fig. 2

gives a typical division of a device into box-like macro cells (the cells need not be rectangular boxes).

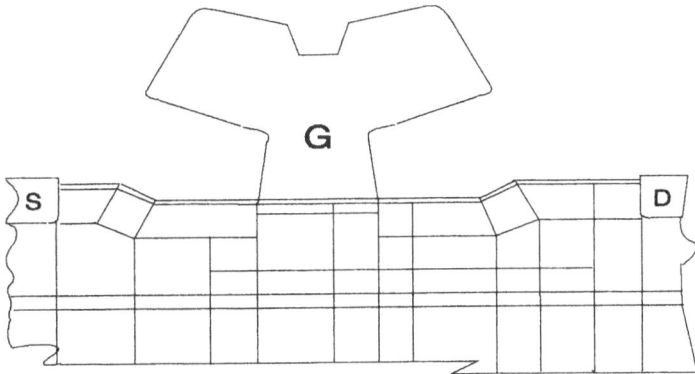

Fig. 2 Division of a FET into Macro Cells

There are practical phenomena occurring in FETs, for instance the detailed bias-dependent behaviour of transconductance, which cannot be explained properly by solvers based on a simple velocity-field characteristic as might apply to a long sample of bulk material. It would appear unnecessary, however, to solve for the dependence of the electron velocity upon two-dimensional field repeatedly, which is what a Monte Carlo device simulator does. There is a generic form for the electron drift velocity as a function of two-dimensional field which has a parallel in silicon device practice, where the term ''mobility model'' is used. Such a model requires much practical data to establish, and hinges on the idea that ballistic motion and velocity overshoot occur to an unimportant extent in real, two-dimensional HFETs. Although to some workers this idea remains unpopular, it appears to be consistent with the work done here.

A macro-cell method also offers an opportunity for calculating the equivalent circuit of a FET or HEMT, and thereby the microwave behaviour, directly. From the charge stored within each macro cell, the magnitudes and phases of the current passing through it and the voltage across it, an equivalent circuit for that macro cell can be determined. The fact that the dominant physical processes within each macro cell are prescribed (because the cell boundaries are chosen that way) infers that any given cell always has a cell equivalent circuit of the same topology. Assembling the cell-level equivalents into an overall equivalent circuit allows the microwave behaviour of the device to be computed using ordinary methods of network theory. The dc characteristics can then be calculated by integrating the incremental (or small-signal) terminal currents with respect to the terminal voltages in the limit of zero frequency.

4.2 Reverse Modelling

The link between FET physical parameters and S-parameters can be viewed in reverse, i.e., instead of trying to predict S-parameters from technological parameters, values for some of the technological parameters can be recovered from measured S-parameters in the process of fitting an equivalent circuit using physics-based optimisation. From measured S-parameters at a single bias point, it is possible only to recover an estimate of the metallurgical gate length and a weighted average figure for the channel doping density. To recover data on the recess depth and width, and on the doping profile, S-parameter data must be available for the space-charge layer extension and depth located at various positions along the recess and in the doping profile - in a sense to "sample" them, i.e. multibias S-parameters must be available. Reverse modelling uses the edges of the depleting region as a two-dimensional "probe" of the device's material and structure, just as a capacitance-voltage profiler uses a one-dimensional depletion edge as a probe of the free-electron profile.

There are three roles for reverse modelling: first, in understanding the physical processes in FETs which determine microwave behaviour, i.e., in helping to develop physical models; second, in diagnosing the structural origin of microwave performance variations as and when they arise in practice; third, in performing fast fits of equivalent circuits to measured S-parameters, and extracting bias-dependent equivalent circuits. Fourteen equivalent circuit elements are found by fitting using physics-based algorithms, with the intrinsic elements correctly related. The series parasitic elements R_s, L_s, R_d and L_d are fitted without recourse to "cold FET" or dc measurements since these elements must be regarded as bias dependent and cannot be held fixed.

Because the fit is found using physics-based algorithms rather than just numerical searching routines, the fit can only be that of a physically meaningful equivalent circuit, and in the process the values of the FET's principal technological parameters are recovered. This property can be used to assess whether any particular set of S-parameter measurements was obtained from a "good" network analyser calibration, and whether they were properly de-embedded. The principle involves making a judgement as to the quality of the fit, and systematically de-embedding until the errors are corrected.

If a fit is poor, it indicates that the measured S-parameters are not those of a FET alone: poor calibration of the network analyser may have distorted the measurements or, in the case of bonded and packaged devices, there are large parasitic reactances associated with the mounting of the chip. These kinds of uncertainty can be detected and systematically eliminated by physics-based analysis of measured S-parameters.

5 Application in "MONOFAST"

5.1 Design Limits Design of FETs

The following first-order expression connects the metallurgical gate length with the channel doping and depletion depth required to achieve a given small-signal power gain:

$$L_G = (v_{sat}/2\,\pi\,f).[R/(R_g + R_i)].[1/G_p] - \varepsilon\,(V_{D'G} + V_{B0})/(qNd)$$

where the symbols are grouped as follows:

Circuit data: f is the frequency, R is the load resistance, G_p is the power gain required, and $V_{D'G}$ is the drain-gate voltage established by the power supply.
FET equivalent circuit data: R_g is the gate resistance and R_i is the intrinsic channel resistance.
Device technology data: N is the channel doping density and d is the depth of depletion in the channel.
Materials data: $v_{sat} = 10^7$ cm sec^{-1} = 10^5m sec^{-1} is the saturated drift velocity of electrons in GaAs, $\varepsilon = 1.15 \times 10^{-10}$ Cvolt^{-1}m^{-1} is the RF permittivity of GaAs, and V_{B0} is the gate Schottky barrier height, approximately 0.8V.
For the MONOFAST circuit, with an overall gain of >15dB, allowing for three gain stages and circuit losses, an estimate of G_p = 7dB per FET stage was arrived at. Assuming:

f = 44GHz	R = 50 Ω	G_p = 7dB	R_g = 1 Ω ;	R_i = 9 Ω ;
$V_{D'G}$ = 3V;	V_{B0} = 0.8V	N = 5 x 10^{17}cm^{-3};		d = 0.05μm

gave
$$L_G = 0.25\mu m$$

as the first estimate of the gate length required (given N = 5 x 10^{17}cm^{-3}). (The consequences of different layer doping can be investigated easily.)

5.2 Detailed FET Design and Practical Realisation

In the MONOFAST circuit, a basic FET structure was preferred to a HEMT for its layer simplicity, and anticipated good reproducibility. A heterojunction FET with an AlGaAs etch stop layer gives good manufacturability and performance. Using a simulator based on the methods of section 4.1, the final technological specification for the MONOFAST FETs was determined as follows:

Channel doping	$5.00 \times 10^{17} \text{cm}^{-3}$
Cap doping	$1.10 \times 10^{18} \text{cm}^{-3}$
Channel thickness	85nm
Cap thickness	15nm
Gate length	0.2µm
Recess depth	44nm

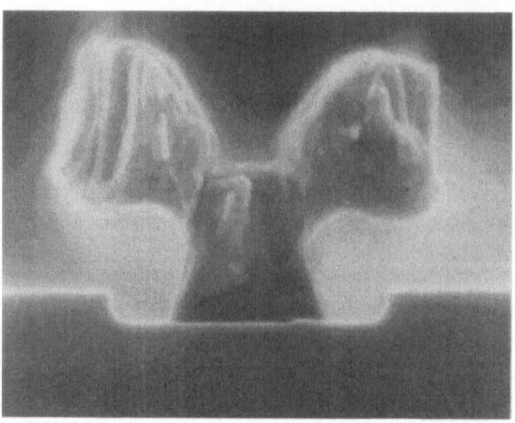

Fig. 3 Cross-section of Practical MONOFAST 0.2µm FET

Figure 3 is a scanning electron micrograph of a typical FET from the MONOFAST programme. The gate, which has a metallurgical length of 0.2µm, is positioned centrally in a shallow recess of depth governed by an AlGaAs etch-stop layer. Molecular beam epitaxy is used to achieve the layer control required for overall reproducibility, with a lightly P-doped buffer giving a sharp tail to the profile.

The FETs have been characterised to 60GHz using a Wiltron 360 network analyser with Cascade 50GHz on-wafer probes. Figure 4 is an example of the measured S-parameters compared with the forecasts for a FET at low-noise bias. The curves marked as "calculated" are technology-based predictions using carefully measured technological data. The agreement is representative of all four S-parameters.

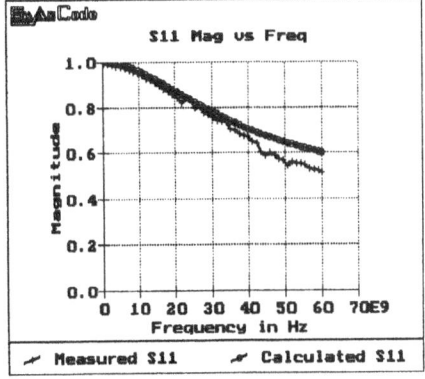 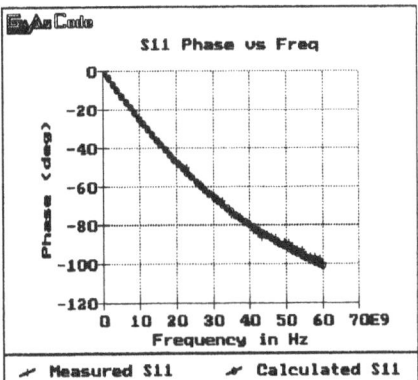

Fig. 4 Example of S-parameters Forecast Using a Forward Modelling Simulator, Compared with Measurements.

Given that the material and structural dimensions of practical FETs are always different from the nominal design values, exact agreement between measured and forward-modelled S-parameters cannot be expected. For the comparison to be exact, the simulator must be able to learn what is specific about a particular FET without having to rely on prior characterisation of the material and cross-sectional dimensions. A reverse modelling simulator has precisely this purpose.

Fig. 5a shows the measured bias dependence of transconductance for a 0.2μm GaAs MESFET compared, in Fig.5b, with the forecast. Both the measured and simulated transconductance exhibit compression, with respect to gate-source bias near open channel, which arises from a strong re-distribution of electric field two-dimensionally, mainly in the corners of the gate recess. At lower values of g_{m0}, there are differences in the gradient of the measured and forecast transconductance with V_{DS}. In this region, the detailed shape of the doping profile tail has a great influence and it is difficult to tell whether the differences arise from model deficiencies, or from an inability practically to measure the exact form of the profile tail.

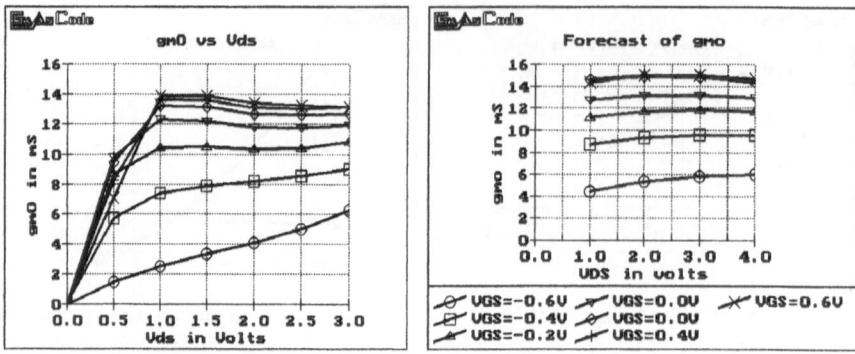

Fig. 5 Measured and Forecast Bias Dependence of Intrinsic Transconductance

5.3 Noise Results (to 26GHz)

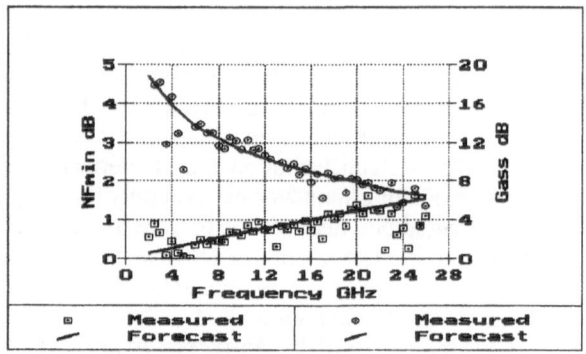

Fig. 6 Measured and Forecast Noise Figure and Associated Gain for the MONOFAST FET

Figure 6 shows the measured and forecast noise figure and associated gain of the MONOFAST FET biased for minimum noise (the maximum frequency to which on-wafer noise measurements were available was 26GHz.) The lines through the measured points are not simply best-fit curves, but are the design forecasts from the simulator based on carefully measured technological data. The results have application at lower frequencies, notably in high-volume manufacturing for low-noise discrete devices and MMICs for direct broadcast satellite (DBS) receivers. At 12GHz, the FETs have 0.75dB noise figure and 11dB associated gain - a performance which is comparable with that of commercial HEMTs at these frequencies.

5.4 Reverse Modelling to 60GHz

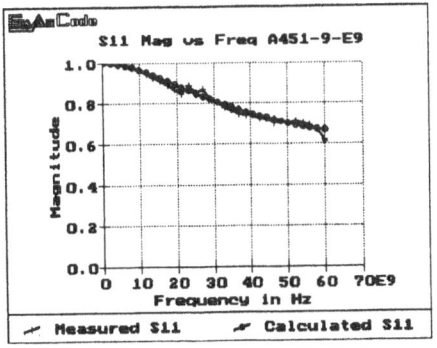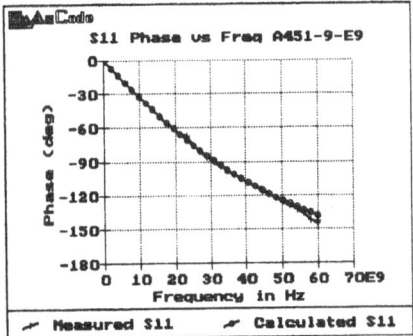

Fig. 7 Practical S-parameters Fitted to 60GHz Using a Reverse-Modelling Fitter

Figure 7 is an example of the fit quality that can be achieved (in this case, for a FET with higher than average transconductance). Figure 8 is a histogram showing the variation in metallurgical gate length for a batch of FETs as recovered by reverse modelling. Typically, the agreement is to within 10% of the value measured using a scanning electron microscope.

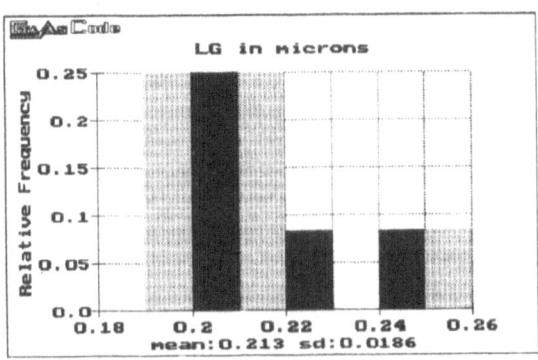

Fig. 8 Spread in Reverse Modelled Gate Length from a Batch of MONOFAST FETs

5.5 Statistical Variation of Equivalent Circuit Elements

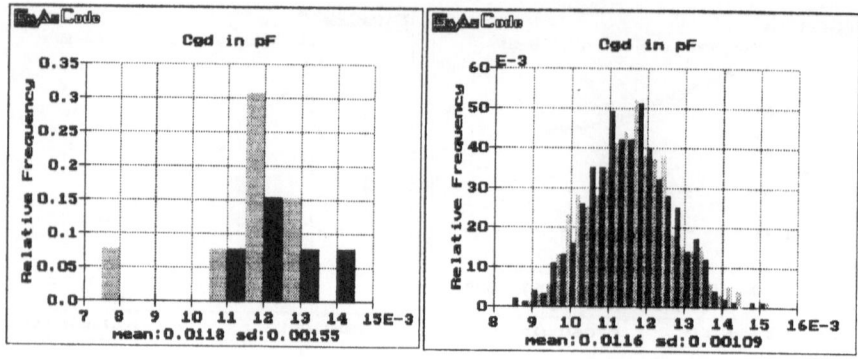

Fig. 9 Example of Measured and Forecast Spreads in Equivalent Circuit Elements.

Figure 9 shows an example of the practical spreads for a sample of devices compared with that forecast from the FET simulator. The Monte Carlo method is used to generate statistical variations in all the FET technological parameters (as in Figure 10), from which the variations in electrical parameters are calculated as in Table 1.

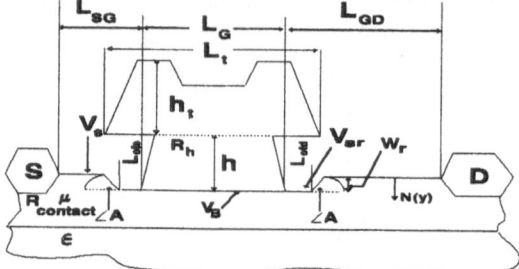

Fig. 10 FET Technological Parameters for a Forward Modelling Simulator

Comparing the practical distributions of all elements with those forecast led to the following conclusions for the MONOFAST FET:

(i)   The practical mean values of C_{gc}, C_{gd}, R_i, R_d, C_{ds}, I_{DSS} agreed well with those forecast,

(ii) the practical mean values of R_g, R_{ds}, g_{m0} and R_{opt} agreed reasonably well with those forecast,

(iii) the practically occurring mean values of τ and R_s were within the forecast distributions, although at the extremes of the range.

Table 1: Forecast Statistics of MONOFAST FET 44GHz Parameters at $V_{DS} = 2V$; $V_{GS} = 0V$

Parameter	Mean	Standard deviation	Parameter	Mean	Standard deviation
C_{gc} (pF)	0.0732	0.00688	C_{gd} (pF)	0.0116	0.00109
g_{m0} (mS)	24.7	1.94	τ (ps)	1.66	0.233
R_d (Ω)	-4.78	3.14	R_g (Ω)	1.5	0.142
R_s (Ω)	4.28	1.51	R_i (Ω)	8.14	0.747
R_{ds} (Ω)	314	38.4	L_g (pH)	1.94	0.197
I_{dc} (mA)	14.2	6.01	V_p (V)	0.892	0.368
Stability (k)	0.675*	0.117	G_{ass} (dB)	5.91	1.04
NFig (dB)	3.04	0.453	MTG (dB)	6.16	0.523
MAG (dB)	8.13	0.43	C_{ds} (pF)	0.0341	0.0003
f_T (GHz)	53.9	3.51	R_{opt} (Ω)	15.4	6.94

* Note: there are several devices with negative k, which could cause circuit instability.

The practical FETs were close enough to the FET performance specification used in the MMIC design for the MMIC to function correctly at the first fabrication attempt (see section 5.6).

5.6 MMIC Results

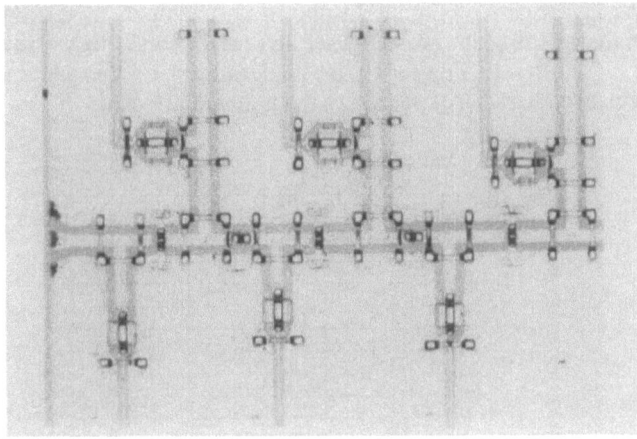

Fig. 11 Practical MONOFAST 44GHz Low-noise Amplifier

Figure 11 shows the practical MMIC. It employs three FETs, co-planar waveguide and metal-nitride capacitors. An electromagnetic simulator was used to investigate some features of the passive network, such as T-junctions and placement of the ground-interconnecting airbridges. Figure 12 shows the design responses (as continuous lines) and the measured responses (as discrete symbols) for S_{21}, S_{11}, S_{22} and S_{12} (no noise measurements at 44GHz have been made on the MMIC). The amplifier met the design specification at the first full processing attempt.

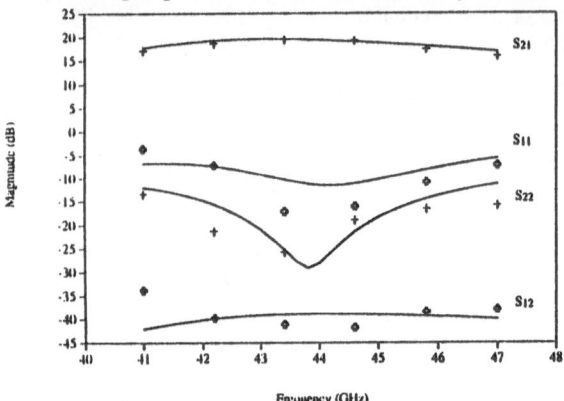

Fig. 12 Design Responses of the MONOFAST MMIC Compared with Measured Responses

The final configuration of the amplifier was chosen from among a number of contenders as that which promised least sensitivity to process variations affecting the FETs (the principal randomly varying element in a mature MMIC technology). The sensitivity of MMIC variants to all the FET technological parameters in Figure 10 was investigated. Figure 13 shows an example, in this case the forecast sensitivity of the gain, noise figure, and return losses to the source-side recess offset, L_{ofs} (refer to Figure 10 for a definition of this parameter). All the performance characteristics except the gain are affected.

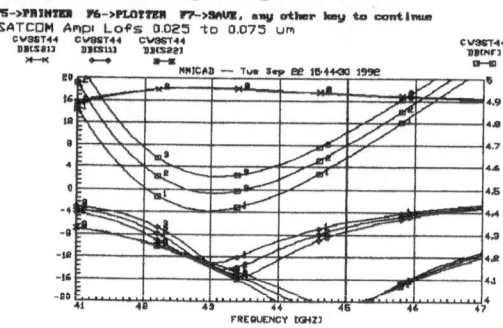

Fig. 13 Simulated Sensitivity of the MMIC to Source-side Recess Offset

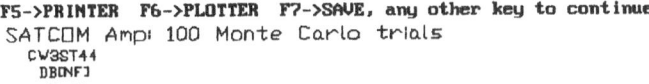

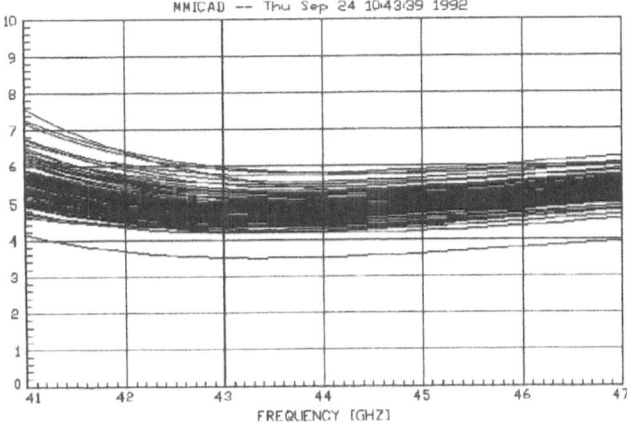

Fig. 14 Forecast Spread in Noise Figure of the MMIC

Figure 14 gives the likely spread in noise figure as forecast for one hundred randomly varying FETs (spreads in the gain and return losses are obtained similarly). Monte Carlo methods have been used in the forward modelling FET simulator to select technological parameters from assumed Gaussian distributions with mean and standard deviation values taken from technological practice. For each set of randomly selected technological parameters, the FET millimetre-wave S-parameters and noise parameters have been forecast, and the amplifier response has been calculated with all three FETs assumed identical. The passive network was unchanged in these simulations.

Figures 15, 16 and 17 show the cumulative yield on S_{11}, S_{21} and S_{22}. The continuous lines are forecasts from the Monte Carlo yield analysis; the practical yield of circuits is represented by the discrete symbols. The agreement between the forecast and the practical yield on S_{11} and S_{21} is good enough for the forecasting method to be judged useful as a design aid for manufacturability. There is a large discrepancy between the measured and forecast S_{22} in Figure 17, the origin of which has not been traced (but, fortunately, the practical return loss is better than forecast).

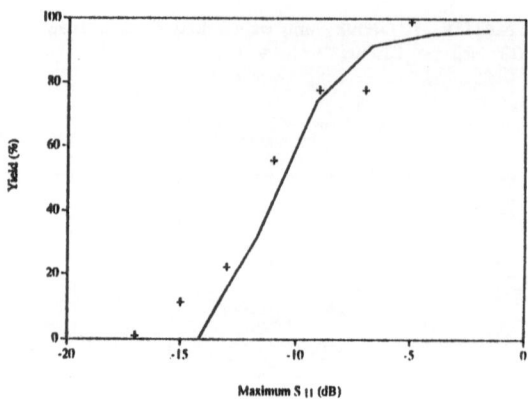

Fig. 15 Forecast (continuous line) and Practical (discrete points) Cumulative Yield on S$_{11}$.

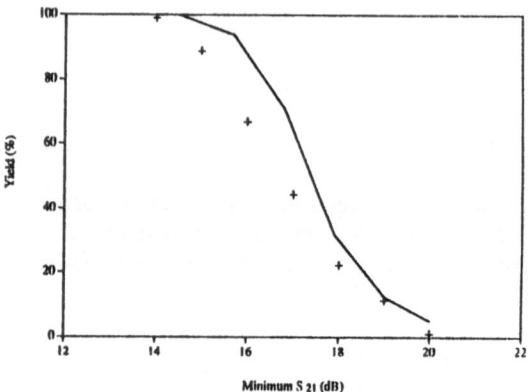

Fig. 16 Forecast (continuous line) and Practical (discrete points) Cumulative Yield on S$_{21}$.

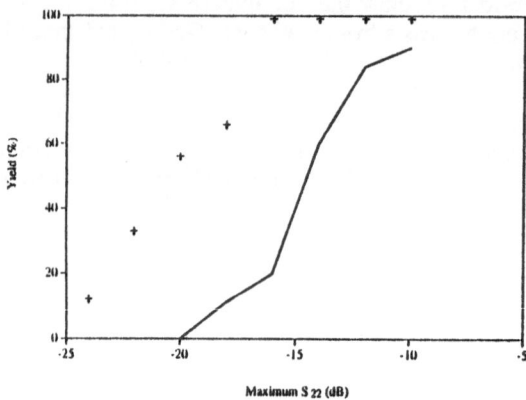

Fig. 17 Forecast (continuous line) and Practical (discrete points) Cumulative Yield on S$_{22}$.

The foregoing cumulative yield curves address the performance parameters one-by-one, without considering the simultaneous behaviour of the others. For instance, for one particular chip S_{11} might be well within specification while S_{21} might not satisfy the specification. What is of ultimate interest is the yield of chips which satisfy all performance parameter specifications at the same time.

The yield forecasts for all performance parameter specifications combined are given in Table 2 for two noise figure specifications (4.5dB and 5dB). Although no on-wafer noise measurement equipment was available to verify the forecasts, it is seen that the noise figure which can be accepted is likely to be a prime factor governing the yield of the 44GHz MONOFAST MMIC.

Table 2: Compound Yield Forecasts

For the combined specification		Forecast Yield
Gain	>15dB	5%
Return losses	<-10dB	
Noise figure	< 4.5dB	
Gain	>15dB	45%
Return losses	<-10dB	
Noise figure	<5dB	

6 Conclusions

The role of forward and reverse modelling in the MONOFAST programme may be summarised as follows:

1. The methods were validated to 60GHz for S-parameters, and to 26GHz for noise and associated gain.

2. MONOFAST demonstrated that technology can be responsive to the need for "process centering".

3. "Design Limits" design of the active devices led to a FET which was close to the final need.

4. Design of the MMIC around a "centered FET" led to first-time working circuits.

5. The worst of the sensitivities of the MMIC were engineered-out by concentrating on critical features of the FET technology.

6. The circuit itself was not "design centered" for maximum yield.

Acknowledgements

The participants in the MONOFAST project were:

The Nanoelectronics Research Centre, University of Glasgow; the National Microelectronics Research Centre, University of Cork; Alcatel Espace, Toulouse; the Microelectronics Research Centre, University of Cambridge; Farran Technology, and GaAs Code Ltd. The work described in this paper represents the collective efforts of all participants.

4.3   On-wafer Noise Parameter Measurements

T. Sporkmann
Institut für Mobil- und Satellitenfunktechnik, D-47475 Kamp-Lintfort, Germany

Introduction. Noise parameter measurement systems have been used for many years now. In the early years, the noise parameters were determined by tuning the input of the DUT until the noise figure was reached. Today, a number of at least four input impedance states is used and, based on this information, the four noise parameters can be calculated. This procedure is clearly much more time efficient and also, due to possible measurement redundancy the results are more accurate than they used to be. However, since noise parameter measurements are difficult and the verification of the measurements is not possible up to now, a few people have pronounced their doubt in the results of these measurements. Several publications treated the measurement accuracy as well as accuracy improvements of state of the art systems in the last years /1//3/. Unfortunately, these investigation treat normally only some of the aspects in this scope. Based on the open questions concerning these measurements, an measurement system investigation was started under the umbrella of the ESPRIT-COSMIC project. Within this project, a state of the art noise parameter measurement system was installed and a detailed investigation of the measurement system uncertainties was conducted.

Measurement System. Noise parameter measurement systems are widely used by semiconductor manufacturers or device modelers. There are two US-companies which offer quite similar systems covering the frequency band from 2GHz to 26GHz. Despite the fact that noise parameter measurements are by at least one order of magnitude more difficult than S-parameter measurements, most people believe in the results achieved by these systems. A typical measurement set-up is depicted in figure 4.1.1. However, it was a task within the ESPRIT-COSMIC project to investigate in detail which accuracy's can be expected by utilising this kind of measurement system. The theory of operation as well as the calibration for these systems is described in /5/ and shall not be explained here. It should only be pointed out, that a complete vector error correction is typically use to correct for the input loss e.g. The true second stage noise will be determined and utilised during the calculations. Many sources of measurement uncertainties can be expected even with a state of the art measurement system as used for this investigation. To investigate the system uncertainties, three methods were used in parallel. First of all, parameterical studies were conducted.

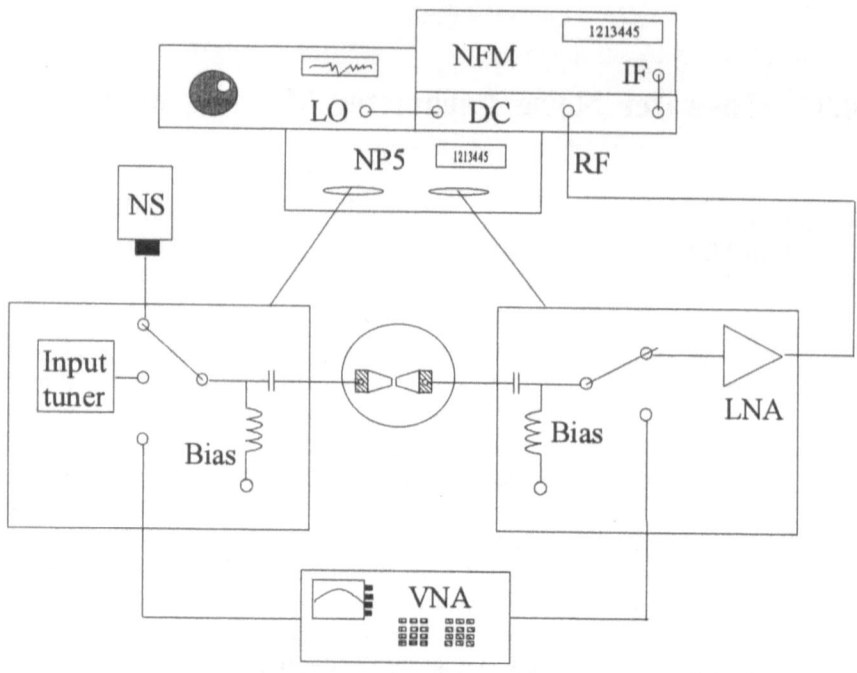

Figure 4.1.1: State of the art noise parameter measurement system (2GHz-26GHz)

In the second step, a sensitivity analysis of the most critical parameters was performed and last but not least, a complete statistical analysis concluded the investigations. The practical results for the following investigations were achieved by using a conventional On-Wafer probable HEMT device. This device features an F_{min} of about 1.3dB at 10GHz, and an associated gain of about 8dB. The magnitude of the Γ_{opt} varies from 1 @ 2GHz down to about 0.6 @ 6GHz. The typical low noise bias point was applied to determine this information. It must be pointed out that the results given below are strongly device dependent. This is reflected in the results of the sensitivity analysis for instance. This investigation shows, how the uncertainty changes as function of F_{min} or G_{ass}.

Parametric Investigations: For the parametric investigation of the measurement system all the above listed uncertainties were varied within a certain range, and the new noise parameters were then calculated. Some of the results are shown in figure 4.1.2. Figure 4.1.2a depicts F_{min} as function of the variation in the ENR value of the noise source. The results in this graph are quite remarkable since conventional systems translate errors in ENR directly into the measurement errors. In our case, the measurement uncertainty due to the uncertainty in the noise source is reduced by almost 50%. This is due to the fact that several noise power measurements are made instead of only one hot- and one cold-measurement. Figure 4.1.2b shows the "sensitivity" of the measurements in respect to the error in the measured S-parameters. In other words, if the uncertainty of the S-parameters would be about 4%, the resulting uncertainty in F_{min} of our device would be about 30% at 10GHz. It is clear that the other noise parameters are also effected by these variations. A general overview on the various results is given in two comprehensive COSMIC reports.

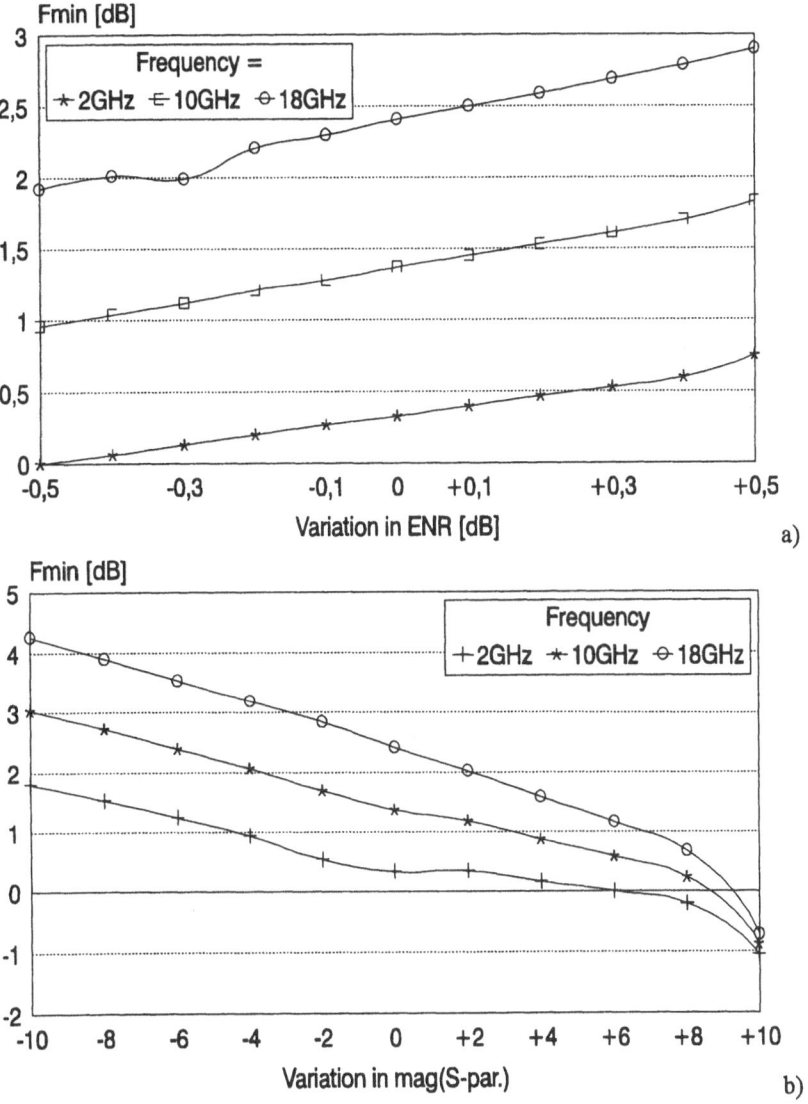

Figure 4.1.2: parameterical analysis
a) F_{min} as functions of the variation in the noise sources ENR values
b) F_{min} as function of the variation in the mag. of the DUT's S-parameters

Sensitivity Analysis: Some of the results of the sensitivity analysis are reported in figure 4.3. The investigations for this figure concerned the post receiver gain and noise parameters as well as the reflection coefficient of the source impedance's. It can be seen that only the source gamma sensitivity features a strong frequency dependency. However, at the edges of the frequency band large sensitivity values are reached. Among these curves, the sensitivity of the post receiver gain is the largest with a value of about -6. The table in figure 4.1.3 depicts the sensitivity value for F_{min} in respect to the individual S-parameters. The critical parameters in this case are S_{11} and S_{21}. An example is also given for a real case.

Example at 12GHz

$\dfrac{\Delta F_{min}}{\Delta S_{11}} = 7.0$	$\Delta S_{11} = 0.003$	$\Delta F_{min} = 0.021dB$
$\dfrac{\Delta F_{min}}{\Delta S_{12}} = -1.0$	$\Delta S_{12} = 0.002$	$\Delta F_{min} = 0.002dB$
$\dfrac{\Delta F_{min}}{\Delta S_{21}} = -5.0$	$\Delta S_{21} = 0.017$	$\Delta F_{min} = 0.085dB$
$\dfrac{\Delta F_{min}}{\Delta S_{22}} = 1.0$	$\Delta S_{22} = 0.008$	$\Delta F_{min} = 0.008dB$

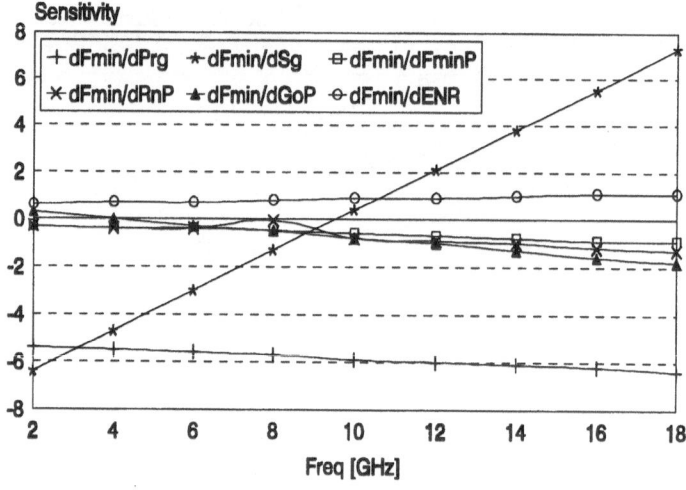

Figure 4.1.3: Sensitivity of F_{min} with respect to some system parameters vs. the frequency

223

The given variations in S_{11} represent changes due to different calibrations techniques, LRM or SOLT, for instance. The resulting change in F_{min} is also listed. The resulting uncertainty for F_{min} is thus 0.116dB. In real measurements the uncertainty was 0.15dB for the above given example.

Statistical Analysis: The results of the statistical analysis were already described /4/ and shall not be repeated here.

Device Measurements. It is clear that device results look quite similar from device to device. For this reason, only some typical results will be shown here. Figure 4.1.4 depicts the F_{min} and the G_a of a HEMT and in addition, the noise and gain data of a 2 stage amplifier constructed out of such HEMT's. It can be seen, that the F_{50} and F_{min} for the amplifier are almost identical in the frequency band of interest. In other words, the amplifier is well matched to achieve the minimum noise figure. The min. noise figure of a single device goes from 0.2dB at 2GHz up to 2.5dB at 18GHz. The F_{50} of the HEMT is about 2dB higher than F_{min} over the complete frequency band.

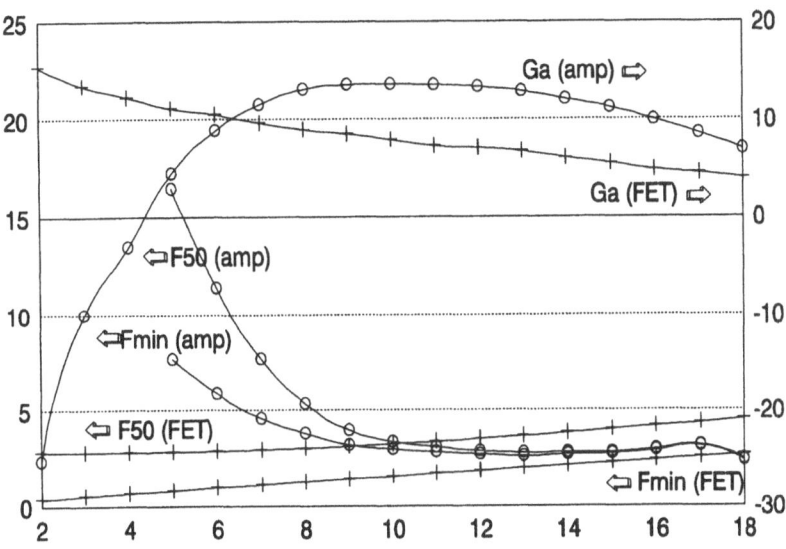

Figure 4.1.4: F_{min} and G_a of a typical HEMT and a 2 stage amplifier

Figures 4.1.5 and 4.1.6 depict some typical data for a 4x75µ FET. While figure 4.1.5 depicts the variation in F_{min} of a conventional FET over a number of 14 samples figure 4.1.6 shows the minimum noise figure of these devices as function of the drain current. It can be seen, that the spread in F_{min} is less than 0.2dB for all the tested devices. Also, the best F_{min} for these devices can be achieved at roughly 20% of I_{dss}.

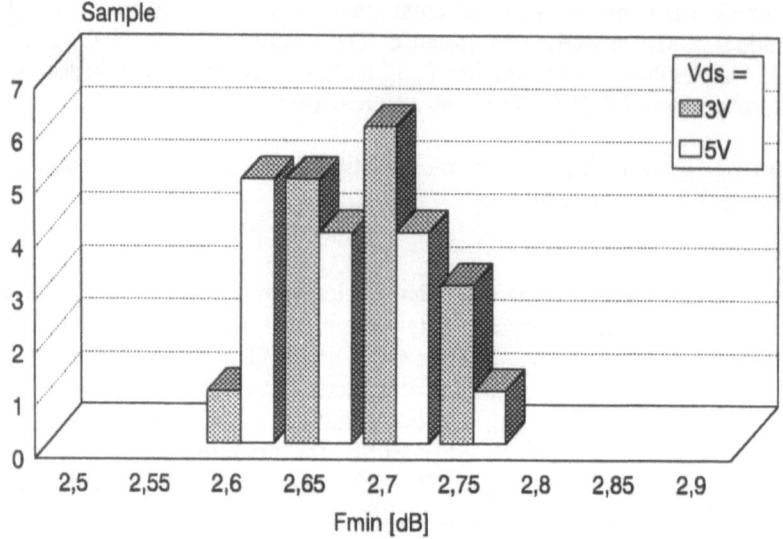

Figure 4.1.5: Spread of F_{min} of a 4x75µ FET over over 14 samples

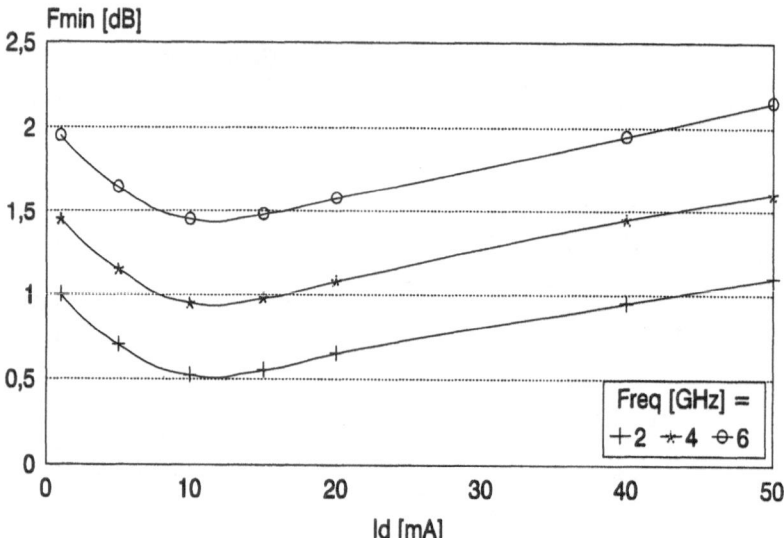

Figure 4.1.6: F_{min} data of a 4x75µ FET as function of I_d with frequency as parameter

Amplifier Measurements. In the case of amplifier measurements there are two remarkable points about the work in COSMIC. First of all it should be pointed out, that all noise parameters of amplifiers, including the F_{50} values, may be measured with the introduced measurement system. The typical results of four different amplifiers are depicted in figure 4.1.7. The curves of F_{min} as well as for

G_{ass} for all four amplifiers are shown in figure 4.1.7. The frequency band from 2GHz to 26GHz is covered.

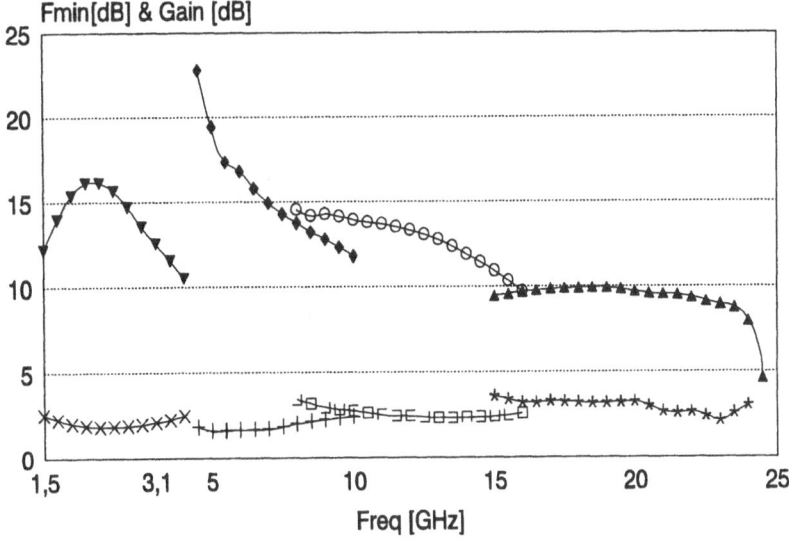

Figure 4.1.7: Typical F_{min} and G_{ass} results of four different amplifiers

Besides the noise figure in 50Ohms and the gain of the amplifiers, the four noise parameters are also part of the measured results. In the case of the amplifier it is not needed to know the Γ_{opt} or the F_{min}, however, if F_{min} and F_{50} are known, or if the magnitude of Γ_{opt} is known, conclusions on the quality of the noise match can be made. For instance, if F_{50} and F_{min} are close in values or even identical, this means that the noise match is quite good. Results of this nature are shown in figure 4.1.4. In practice, values for Γ_{opt} around 0.1 and smaller should be achieved for good noise performance.

Another topic on amplifier noise measurements is the application of various measurement systems. Due to the facts that amplifiers are often matched to 50Ohm and that gain is achieved in the band of interest one might draw the conclusion that expensive measurement systems are not needed. However, the numerous error sources within the scalar measurement systems may cause extremely large errors during such measurements. For this reason, scalar and vectorized measurement system have been applied to amplifiers measurements. Results of this comparison are depicted in figure 4.1.8. Even though the results in 4.1.8 look quite similar, there might be a difference of up to 0.3dB in F_{50} for instance. Thus, an error of more than 10% can be expected if scalar measurement systems are utilised. It is typical that the curves measured with the vectorized measurement system are smoother that the traces determined by utilising scalar systems. Such comparisons have been conducted on three types of amplifiers, single ended, travelling wave amplifiers and balanced amplifiers. In the band where the amplifiers are well matched to 50Ohms the agreement is usually quite good. However, outside this band the, the noise figures determined by the various methods show a large spread. Our investigations show, that even for amplifier measurements, vectorized measurement systems are mandatory.

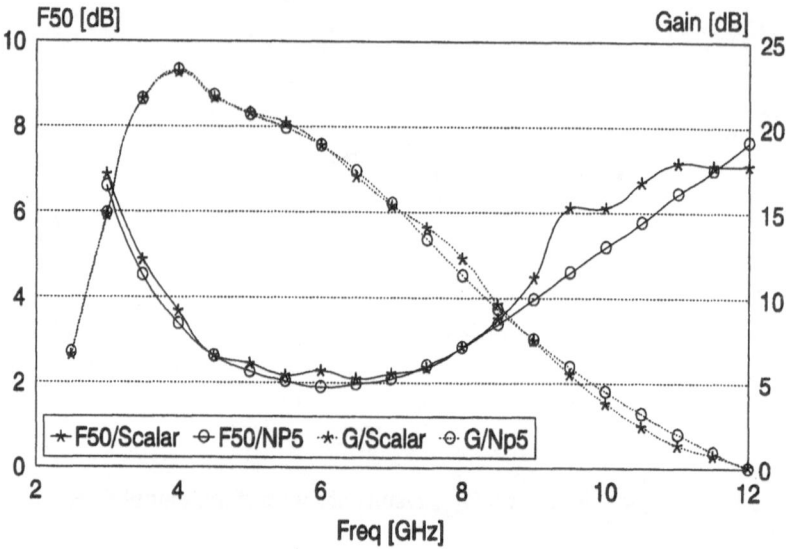

Figure 4.1.8: Comparison between scalar and vectorized amplifier noise measurement results

Conclusion. The work reported here was conducted under the umbrella of the ESPRIT II-COSMIC project. For the first time, a comprehensive investigation on uncertainties in noise parameter measurement systems was shown. The types of investigations range from parametrical-, over sensitivity-, to statistical investigations. The investigations have shown that the uncertainties depend on the device as well as on the system components. Errors due to the uncertainty in the ENR of the noise source for instance are reduced by about 50%. Nevertheless, the total uncertainty is much larger than expected. Thus it is not uncommon to have 0.2dB of error in the measurement of F_{min}. This is especially true if a shielded room is not available. This fact is somewhat frightening since device manufacturers are already dealing with noise figures of 0.5dB at 12GHz. Since the sources of the uncertainties have now been defined, it seems possible to improve the measurement accuracy.

References

/1/   Davidson, A., Leake, B., Strid E., "Accuracy Factors in Microwave Noise Parameter Measurements", IEEE MTT-Digest, pp. 827-830, June 1990.

/2/   Adamian, V., Uhlir Jr., A., "A novel procedure for Receiver Noise Characterization", IEEE Trans. on Instrumentation and Measurement, Vol. IM-22, No. 2, pp. 181-182, June 1973.

/3/   Davidson, A. C., Leake, B. W., Strid, E., " Accuracy Improvements in Microwave Noise Parameter Measurements", IEEE Transactions on Microwave Theory and Technique, Vol. 37, No. 12, pp. 1973-1978, December 1989.

/4/   Moniz, J. M., Adamian V., Wandrei D.,"Noise Parameter Measurement Accuracy and Repeatability Considerations", atn Application Note AN-001.

/5/   atn, "NP5 users manual".

4.4 Noise Modelling in Semiconductor Devices

G Ghione, F Bonani, M Pirola and C U Naldi
Politecnico di Torino, I-10129 Torino, Italy

1 Introduction

Gallium arsenide field-effect transistors in low-noise amplifiers for microwave communication systems have shown, during the past years, outstanding noise performances (*i.e.*, low noise figure and high associated gain) at increasingly high frequencies. State-of-the-art 0.25 μm gate length GaAs MESFETs exhibit noise figures around 1 dB at 20 GHz with associated gain in excess of 10 dB, while even superior performances are attained by conventional GaAs-AlGaAs and pseudomorphic high electron mobility transistors (HEMTs). Such devices are now commercially available in both discrete and integrated form, and performances are being steadily improved through the use of more advanced doping profiles, new epitaxial structures, and shorter gate lengths.

The optimization of low-noise GaAs FETs traditionally requires the technological process to be iterated until a satisfactory tradeoff between gain and noise performances has been achieved. A physics-based computer-aided tool for device design and optimization clearly is a convenient alternative, particularly for integrated devices. Although the physics-based device simulation is now a routine tool in the design of silicon devices, its use in optimizing the noise performances of GaAs FETs has been up to now sporadic, despite the excellent work carried out in the field during the last few years by some European research groups. In particular, a quasi-2D approach to the DC, AC and noise simulation of MESFET devices [8], later extended to HEMTs [15] has been proposed by the Lille University group. The work described here follows a similar path, but proposes for the MESFET a more general, two-dimensional approach to the AC and noise analysis. Concerning the HEMT, the developed model is an analytical implementation of a quasi-2D approach, computationally efficient and accurate enough to carry out parametric studies on a given technology in order to analyze performance limits and trends.

This contribution is structured as follows. Sec. 2 gives a short overview on the noise physics and modelling approaches. Sec. 3 is devoted to the discussion of the MESFET noise model developed within the framework of the COSMIC project and the related results. In Sec. 4 the HEMT noise model is described in detail, with particular attention to the scaling study carried out on the SIEMENS conventional HEMT technology; references to further work fostered by the COSMIC project are also given. Finally, some conclusions follow in Sec. 5.

2 An overview on noise modelling

The electrical behaviour of electron devices, circuits or communication systems is described in terms of electric signals, which can be voltages and currents, or, at a microscopic level, electromagnetic fields and charge distributions. However, such quantities actually are random processes characterized by a mean or expected value and a superimposed, small-amplitude random *fluctuation*. Such fluctuations are a consequence of the microscopic processes on which the operation of electron devices and circuits are based; they also are the fundamental cause of electrical noise in electronic circuits, *i.e.*, of an unwanted, small-amplitude signal superimposed to the signal carrying information.

The basic cause of noise in electron devices and circuits can be found in the *transport processes*, whereby electron and holes are set into motion by an applied field or a concentration gradient. At a microscopic level, the carrier motion is not a coherent drift, but rather a random walk, where *free flights*, performed at a very high velocity, take place between *scattering events*. During free flights, each carrier reacts to the applied external forces; scattering events, *i.e.* collisions with lattice vibrations, impurities, photons, or other carriers, abruptly change the carrier velocity, thus causing the average velocity of a carrier ensemble to fluctuate. Since an extremely high number of carriers is present, the huge velocity fluctuations of a single carrier only induce extremely small fluctuations in the average velocity of the ensemble. As the random-walk behaviour of carriers is the very cause of carrier diffusion in the presence of a concentration gradient, the microscopic noise mechanism associated to carrier transport is known as *diffusion noise*. This is by no means the only fluctuation mechanism present in a semiconductor, since the average velocity of a carrier ensemble can fluctuate not only because the velocity of each carrier fluctuates, but also because electron-hole pairs are generated or recombine, thus causing a carrier to appear or to disappear from the ensemble. The resulting noise mechanism is called *generation-recombination noise*, and plays a minor role in high-frequency devices and circuits.

Confining our attention to diffusion noise, semiconductor theory allows the power spectrum of the current density fluctuations to be expressed in terms of the carrier density and of their *diffusivity*. To translate this theoretical, microscopic result into manageable information at the circuit level, one must be able to evaluate the effect of a current density fluctuation occurring within the bulk of an electron device on the voltages and currents measured at the device terminals. Two basic ideas can help in performing this task. Firstly, current fluctuations due to microscopic noise have a small amplitude with respect to the average value of this parameter: thus, they can be viewed as a small-signal excitation, and the fluctuation of external voltages as corresponding system response. In other words, the power spectrum of the external fluctuations induced on the device terminals by an internal, microscopic current fluctuation localized in a point of the device is simply proportional to the power spectrum of the latter, through a frequency-dependent transfer function. Secondly, random carrier fluctuations taking place in different points of the device are statistically uncorrelated; thus, the power spectrum of the overall external voltage or current fluctuations simply results by summing up all contributions arising from the different device areas. The correlation spectrum of fluctuations at different device ports can be estimated in a similar way. This approach to the noise characterization of electron devices is known as the *impedance-field method* [22]. Apart from giving obvious results in simple conditions (for instance, the well known expression for the r.m.s. voltage fluctuations arising from thermal noise in a resistor, thermal noise being a particular case of diffusion noise at or near thermodynamic equilibrium), the impedance-field method allows the noise behaviour of electron devices to be modelled from a physics-based standpoint in any operating conditions. The impedance-field method can be implemented either numerically through finite-element based techniques [12] or analytically, by means of suitable approximations; see *e.g.* [23] for the case of the MESFET.

3 The MESFET noise model

The physics-based MESFET noise model used within COSMIC is based on a two-dimensional implementation of the impedance field method (IFM) [22, 25] in the framework of the MESFET simulator MESS; further details on MESS can be found in [11]. The IFM implementation, based on an efficient technique akin to the so-called *adjoint method* for the noise analysis of electrical networks [20], is described in [12].

The experimental validation of the model was carried out on a GEC-MARCONI 0.6 μm F-20 foundry MESFET. In order to have a more direct insight on the model outcome, comparisons were performed not only on overall, system parameters like the minimum noise figure, optimum source impedance and noise resistance (NF, Z_o, and R_n, respectively), but also on the power and correlation spectra of the short-circuit current and open-circuit voltage fluctuations. These were derived, on the frequency band 1-18 GHz, from the scattering and noise system parameters, measured for $V_{DS} = 1, 2, 3, 4$ and 5 V for several values of the drain current.

In agreement with theoretical estimates [23], the power spectrum of the short-circuit drain current fluctuations $S_{i_D}(f)$ is almost constant with frequency, while the power spectrum of the short-circuit gate current fluctuations $S_{i_G}(f)$ has a f^2 behaviour. Similarly, the correlation spectrum between the drain current and the gate voltage fluctuations, $S_{i_{D}e_{G}}$, is nearly frequency-independent in magnitude and its phase linearly depends on frequency. Since S_{i_D}, S_{e_G} and $S_{i_D e_G}$, besides being almost white, appear to be less affected by parasitics than other spectra (although the parasitic input capacitance somewhat affects S_{e_G} and $S_{i_D e_G}$), these parameters have been chosen for comparison with the simulation, rather than other possible linearly dependent sets of noise spectra.

The diffusivity-field model (see *e.g.* [10]) has been found to play an important role. In fact, the power spectrum of the microscopic current density fluctuations in a n-type material is equal to $4q^2 nD$, where n and D are the DC electron density and differential diffusivity, respectively. In most discretization schemes for the continuity equation (*e.g.*, the Scharfetter-Gummel scheme) the low-field Einstein relationship is unphysically enforced also at high fields. In FET modelling, this assumption has a minor impact on the DC and small-signal behaviour, but implies a microscopic noise source proportional to $\mathcal{E}^{-1}$ at high electric fields, thereby underestimating the noise in the velocity saturated part of the channel. In the present approach, a modified diffusivity-field model is used whereby the Einstein relation $D = D_0 = (kT/q)\mu_0$ holds up to a critical field $\mathcal{E}_c$ for which $D(\mathcal{E}_c) = D_\infty$; for $\mathcal{E} > \mathcal{E}_c$ the diffusivity is constant and equal to the high-field value D_∞. Although the high-field diffusivity [17] is not easily estimated in doped materials from first principles [16], its value can be assumed as a technology-dependent tuning factor. Indeed, the best fit between measurements and simulation is achieved with $D_\infty/D_0 \approx 0.2 - 0.3$ [13], in agreement with theoretical considerations [16].

The measured and simulated S_{i_D}, S_{e_G} and $S_{i_D e_G}$, as a function of the drain current, are shown in Fig. 1 - 4, respectively, for $V_{DS} = 5$ V and for several values of the D_∞/D_0 ratio. The operating frequency is $f = 5$ GHz; however, apart from the phase of the correlation spectrum, all spectra are almost frequency independent. The error bars refer to the standard deviation resulting from an average measurement uncertainty of $\pm 7.5\%$ on Z_o and R_n and of $\pm 10\%$ on NF (in dB), and has been numerically generated through Monte Carlo simulation. The uncertainty of the measured S-parameters has been taken as negligible.

The behaviour of measured and simulated data *versus* bias is now discussed. The drain current spectrum is almost linear *vs.* the drain current and is nearly independent from V_{DS} after saturation. The best agreement with measured data is obtained with D_∞/D_0 between 0.2 and 0.3 (see Fig. 1). The open-circuit gate voltage fluctuation spectrum (Fig. 2) is almost constant *vs.* I_{DS}, apart from the pinch-off region where, however, the experimental uncertainty is high. This can be qualitatively explained by observing that the channel resistance increases with decreasing drain current, but, at the same time, the capacitive coupling with the gate electrode becomes lower. Although the spectrum is dominated by the ohmic region of the channel, a fairly good agreement is achieved again for D_∞/D_0 between 0.2 and 0.3. (A parasitic input capacitance of about 0.15 pF has been included in the simu-

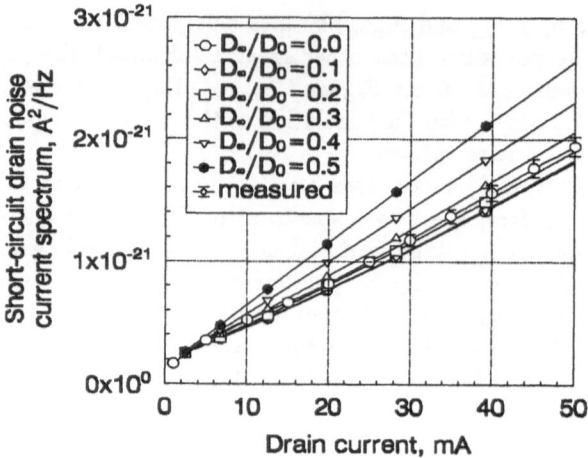

Figure 1. Measured and simulated spectra of drain current short-circuit fluctuations *versus* the drain current in full saturation (V_{DS}=5 V). The frequency is $f = 5$ GHz.

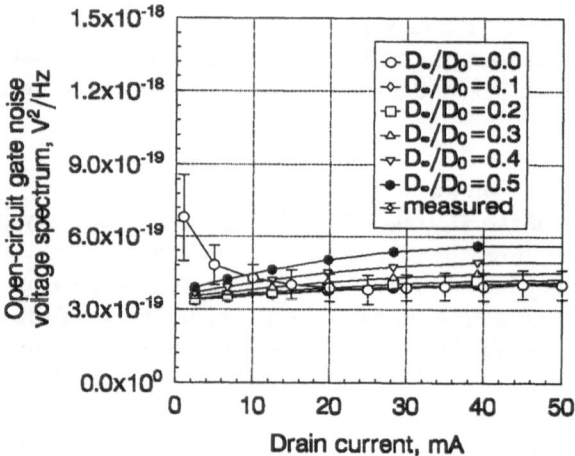

Figure 2. Measured and simulated spectra of gate voltage open-circuit fluctuations *versus* the drain current in full saturation (V_{DS}=5 V). The frequency is $f = 5$ GHz.

lated results shown.) The same comments apply to the correlation spectrum shown in Fig. 3 and Fig. 4. The poor agreement on the correlation phase can be ascribed to the choice of reference planes in the experimental set up, which is not accurately reproduced in the simulation.

Finally, an example of measured and simulated frequency behaviour of the noise figure, optimum source resistance and reactance for the low-noise bias

233

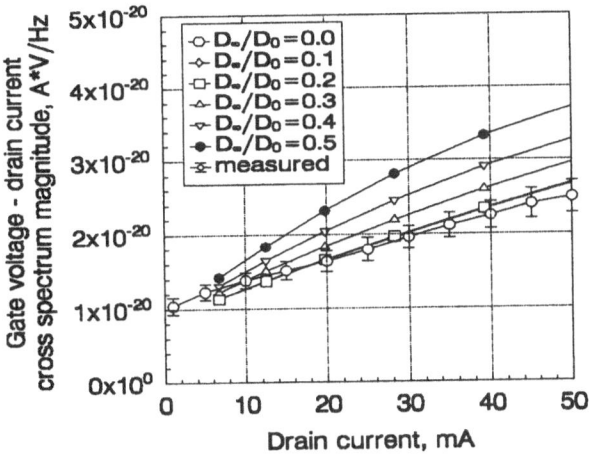

Figure 3. Measured and simulated magnitude of cross spectrum between drain current short-circuit and gate voltage open-circuit fluctuations *versus* the drain current in full saturation (V_{DS}=5 V). The frequency is $f = 5$ GHz.

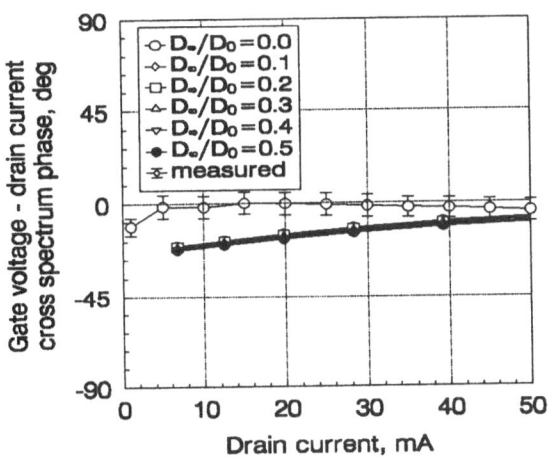

Figure 4. Measured and simulated phase of cross spectrum between drain current short-circuit and gate voltage open-circuit fluctuations *versus* the drain current in full saturation (V_{DS}=5 V). The frequency is $f = 5$ GHz.

point ($I_{DS} \approx 10$ mA) are shown in Fig. 5, Fig. 6 and Fig. 7, respectively. The simulated and measured optimum reactances ($X_o \approx 1/\omega C_{GS}$) differ since the parasitic gate-source external capacitance has not been included into the simulation.

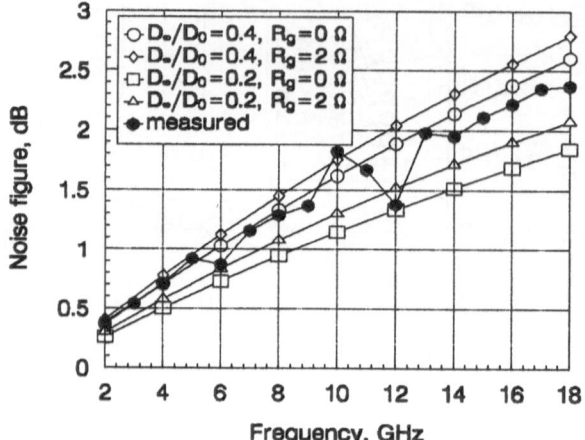

Figure 5. Simulated and measured minimum noise figure for the low-noise bias point ($V_{DS} = 5$ V, $I_{DS} \approx 10$ mA) *versus* the operating frequency.

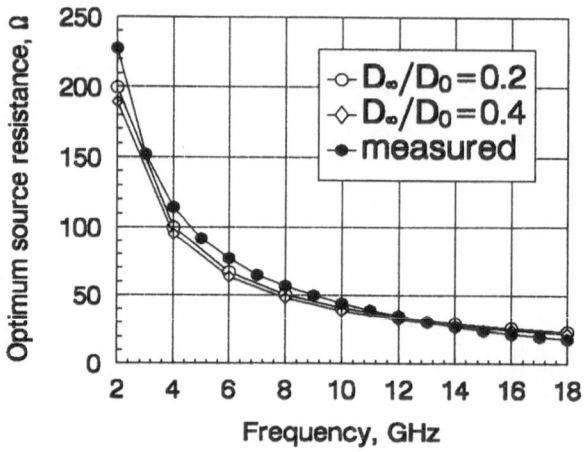

Figure 6. Simulated and measured optimum input resistance as a function of frequency for the low-noise bias point.

Further validations of the 2D noise model were carried out against literature data, and comparisons were performed with other models, both analytical and numerical, with satisfactory results. From the standpoint of numerical efficiency, the MESFET simulator despite being based on a finite-element like discretization strategy, is only moderately CPU intensive. As an example, the DC, AC and noise analysis typically requires a few CPU minutes for each working point on a medium-power workstation.

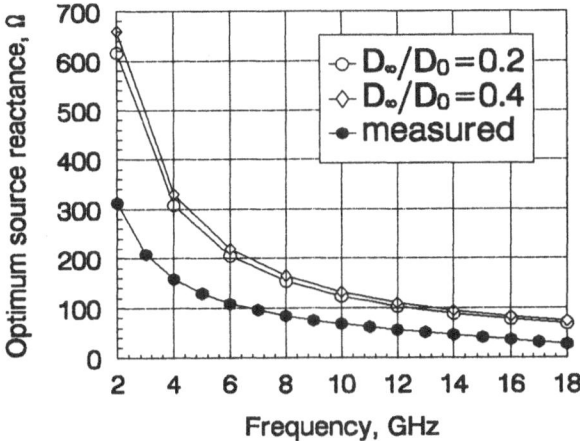

Figure 7. Simulated and measured optimum input reactance as a function of frequency for the low-noise bias point.

4 The HEMT noise model

The HEMT noise behaviour was extensively investigated by Cappy *et al.* (see [7] and references therein) through a 1D numerical implementation of the impedance field method (IFM) based on a quasi-2D energy-transport physics-based model. An analytical IFM implementation was proposed for the HEMT by Brookes [6], who extended the MESFET noise model of Statz, Haus and Pucel [23]. Some improvements were introduced by Ando and Itoh [1] on the basis of a more accurate nonlinear charge control model. The model developed within COSMIC aims at removing some of the limitations of Ando's approach; on the basis of the validation, carried out against experimental data taken from devices built on the same epitaxial structure, a partly empirical model for short-gate effects was derived, and a scaling study was performed on the noise and small-signal performances of the process for gate lengths in the range 0.5-0.1 μm.

The Ando model [1] makes use of an analytical relationship describing the sheet density of the channel mobile charge (called the *two-dimensional electron gas*, 2DEG), as a function of the gate bias. This *charge control model* assumes fully ionized donors in the supply layer, and is based on the approximate model by Shey and Ku [21] for the quantum behaviour of the 2DEG in an AlGaAs/GaAs modulation doped structure. In this work, a numerical charge control model allowing for deep and shallow (incompletely ionized) donors in the supply layer (see [19]), was implemented; the 2DEG quantum model was derived from [9] and validated through a self-consistent Poisson-Schrödinger 1D simulator.

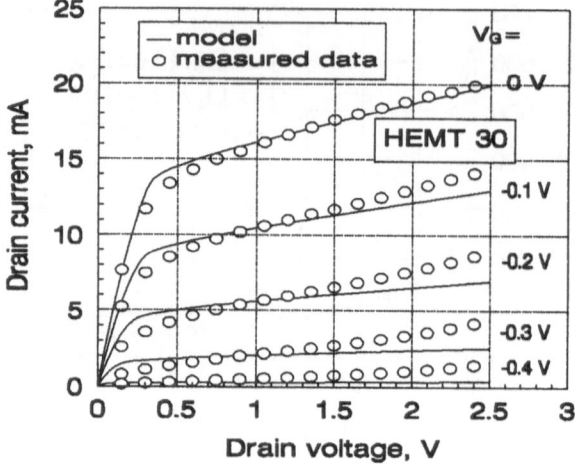

Figure 8. DC curves for 0.5 μm SIEMENS HEMT.

To preserve the analytical framework of the approach, the resulting numerical sheet charge control relationship was fitted to the analytical model in [1]; for a more detailed discussion, see [5, 3]. To take into account the substrate leakage behaviour of real devices, the DC and AC models were coupled to the semi-empirical substrate conductance model proposed by Pavlidis [18]. Moreover, a better agreement with the experimental behaviour of g_m near pinch-off was obtained by means of the initial mobility model proposed by Drummond and Sherwin [9].

The model performances were validated against experimental results concerning two SIEMENS HEMTs of different gate length built on the same epitaxial structure. The first device considered [3] was a 0.5 μm X-band SIEMENS HEMT intended for second stage applications in DBS preamplifiers. The epitaxial structure is grown using MBE and consists of an undoped buffer layer (GaAs/AlGaAs), an undoped AlGaAs spacer layer and a homogeneously doped supply layer. By adjusting the thickness of the supply (AlGaAs) layer around its nominal value good agreement was found on the DC curves (Fig. 8); the output conductance in saturation was simulated through the Pavlidis model. To match the drain saturation current with the given pinch-off voltage, an equivalent saturation velocity $v_s \approx 1.5 \times 10^7$ cm/s was used. Fig. 9 shows a comparison between the in-package measured noise figure and the simulation, as a function of the drain current; the operating frequency is 12 GHz. Package parasitics increase the noise figure of about 0.1 dB with respect to the on-chip value; for the associated gain, however, the degradation caused by external parasitics is much more significant, see [3].

Then, a short-gate 0.25 μm SIEMENS HEMT was considered, built on the same epitaxial structure as the 0.5 μm HEMT described above. To account for non-stationary transport effects, the equivalent v_s was estimated through the empirical model by Graffeuil [14], whereby the equivalent satu-

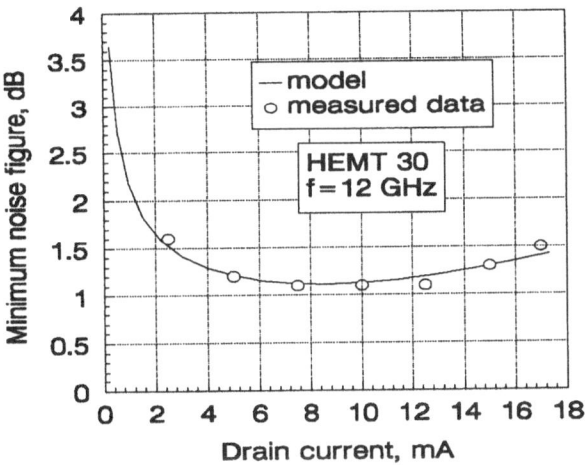

Figure 9. Minimum noise figure *vs.* I_D for the 0.5 μm SIEMENS HEMT.

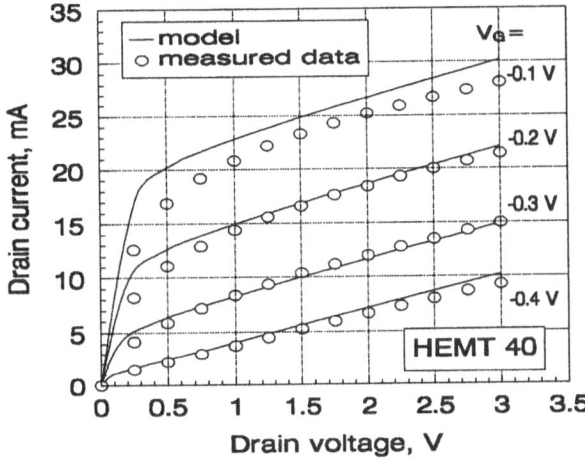

Figure 10. DC curves for 0.25 μm SIEMENS HEMT.

ration velocity of the 2DEG approximately depends on the gate length. The model was also modified to match the v_s value for the 0.5 μm device (which is close to the bulk 2DEG value). The simulated DC curves agree well with the experiment, see Fig. 10, and so does the simulated associated gain (with external parasitics added), which is compared in Fig. 12 to the on-chip measurements at 12 and 20 GHz; finally, the measured and simulated on-chip noise figure are shown in Fig. 11. The agreement obtained is satisfactory at both frequencies considered; the flat behaviour of *NF* with respect to I_{DS} (see Fig. 11) allows the device to be operated at high gain without any significant degradation of the noise figure.

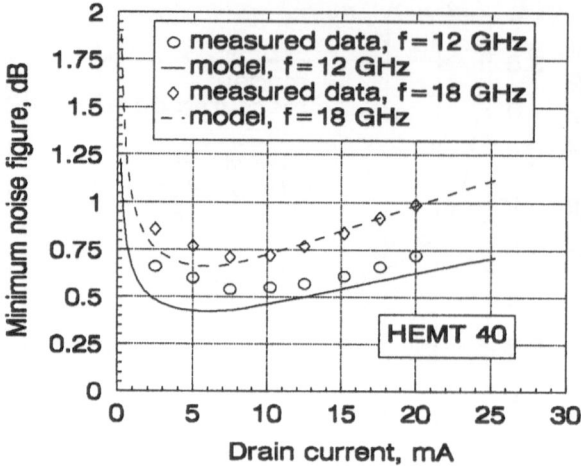

Figure 11. Minimum noise figure *vs.* I_D for the 0.25 µm HEMT.

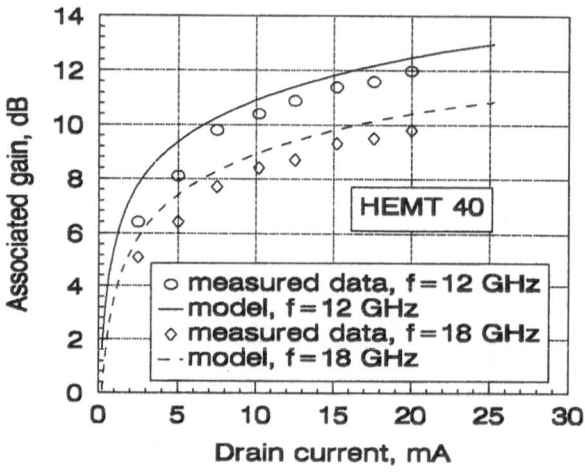

Figure 12. Associated gain *vs.* I_D for the 0.25 µm HEMT.

To investigate the performance limits of the "conventional" AlGaAs-GaAs HEMT technology, a gate length scaling study was carried out on devices having the same epilayer structure and overall layout. The simulated noise figure at $f = 12$ GHz, the transconductance g_m and intrinsic f_T are shown in Fig. 13, Fig. 14, Fig. 15, respectively, as a function of the gate bias (with $V_{DS} = 3$ V) and for different gate lengths. The resistive parasitics were estimated as an average value for the 0.5 and 0.25 µm devices as $R_S = R_D = 3 \Omega$, $R_G = 0.5 \Omega$. The transconductance g_m was chosen as a figure of merit for the device gain instead of the associated gain, since the latter is heavily influenced by the external parasitics; the transit frequency was estimated according to the definition in [24].

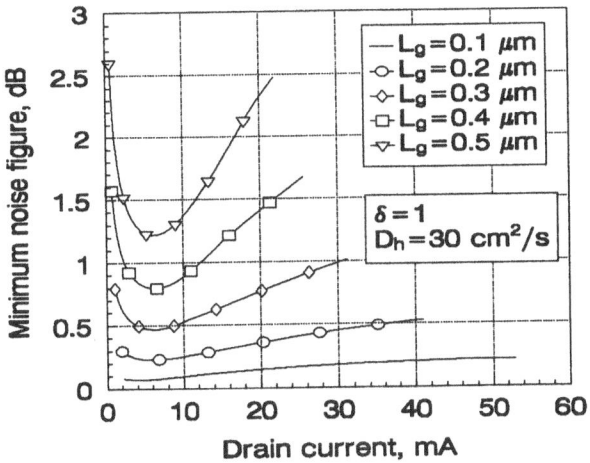

Figure 13. Scaling behaviour of minimum NF at $f=12$ GHz *vs.* I_D as a function of L_g.

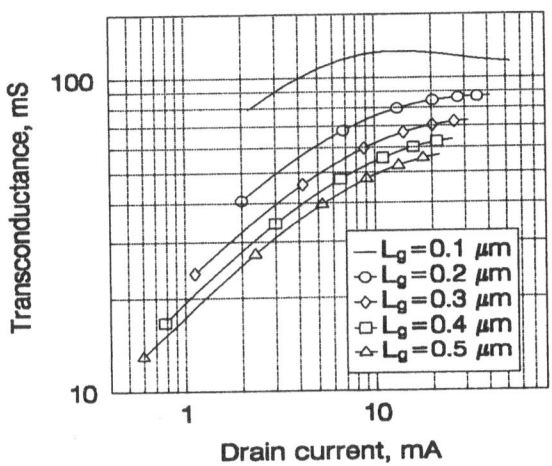

Figure 14. Scaling behaviour of g_m *vs.* I_D as a function of L_g.

The shortest gate length for which a long-gate behaviour prevails, thereby enabling to use a two-zone channel formulation, is around 0.1 μm for the device structure under examination. With the Graffeuil model for v_s, a minimum intrinsic NF of 0.1 dB at $I_{DSS}/10$ is achieved for the 0.1 μm device. The scaling study was also performed with constant $v_s = 1.5 \times 10^7$ cm/s; for the shortest gate length, a $\approx 20\%$ higher noise figure and a $\approx 30\%$ lower g_m and f_T were obtained, but the same qualitative trends were found for the noise and AC parameters *vs.* the gate length.

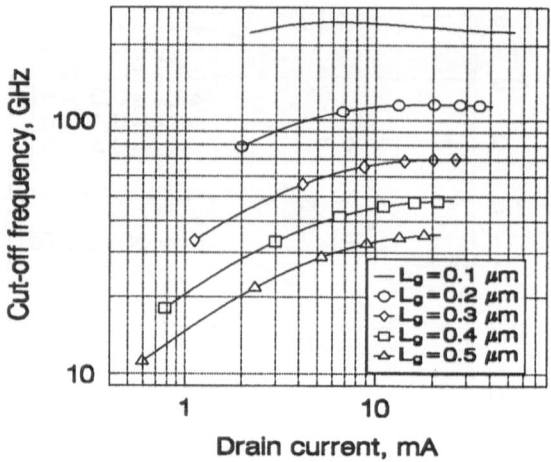

Figure 15. Scaling behaviour of f_T vs. I_D as a function of L_g.

Owing to its quasi-analytical nature, the HEMT model is very efficient from a computational standpoint, and runs on a PC with negligible CPU times. Further comparisons, not shown here, were also carried out with other models, namely with the HELENA simulator from Lille University [15]. The DC results obtained are in fair agreement, while for the noise figure the HELENA model foresees a substantially similar minimum noise figure, but a sharper increase of NF with increasing drain current. Further extensions of the HEMT model developed within COSMIC allow for the simulation of devices realized on arbitrary epitaxial (pseudomorphic or lattice-matched) structures.

5 Conclusions

A set of CAD tools for the design and optimization of low-noise microwave FETs, developed within the COSMIC project, has been briefly described. The physical models have been validated through a systematic comparison with experimental data concerning the GEC-MARCONI foundry MESFET process and the SIEMENS low-noise HEMT process. Finally, the models have been exploited for device optimization, scaling, and parametric analysis.

References

[1] Y. Ando and T. Itoh, "DC, small-signal, and noise modelling for two-dimensional electron gas field-effect transistors based on accurate charge-control characteristics", *IEEE Trans. on Electron Devices*, Vol. ED-37, No. 1, pp. 67-78, Jan. 1990.

[2] W. Baechtold, "Noise behavior of GaAs field-effect transistors with short gate lengths", *IEEE Trans. Electron Devices*, vol. ED-19, pp. 674–680, 1972.

[3] F. Bonani, G. Ghione, C. U. Naldi and F. Ponse, "A CAD-oriented quasi-physical HEMT noise model for device design and optimization", *Proceedings of GAAS92*, Noordwjik, April 1992.

[4] F. Bonani, G. Ghione, C. U. Naldi, R. D. Schnell and H. J. Siweris, "HEMT short-gate noise modelling and parametric analysis of NF performance limits", *Proc. IEDM 1992*, pp. 581–584, San Francisco, 13–16 Dec. 1992.

[5] F. Bonani and G. Ghione, "Noise modelling of HEMTs", *Alta Frequenza – Focus on GaAs electronics: toward zero dimensional structures*, Vol. 5, No. 2, pp. 28–36, March-April 1993.

[6] T. M. Brookes, "The noise properties of high electron mobility transistors", *IEEE Trans. on Electron Devices*, Vol. ED-33, pp. 52–57, 1986.

[7] A. Cappy, "Noise modelling and measurement techniques", *IEEE Trans. on Microwave Theory & Techniques*, Vol. MTT-36, No. 1, pp. 1–10, Jan. 1988.

[8] B. Carnez, A. Cappy, R. Fauquembergue, E. Constant and G. Salmer, "Noise modeling in submicrometer-gate FETs", *IEEE Trans. on Electron Devices*, Vol. ED-28, pp. 784–789, 1981.

[9] T. J. Drummond and M. E. Sherwin, "Analytic MODFET models beyond the triangular well approximation", *Solid State Electr.*, Vol. 33, No. 7, pp. 885–891, 1990.

[10] R. Fauquembergue, J. Zimmermann, A. Kaszynski, E. Constant and G. Microondes, "Diffusion and the power spectral density and correlation function of velocity fluctuation for electrons in Si and GaAs by Monte-Carlo methods", *J. Appl. Phys.*, Vol. 51, p. 1065, Feb. 1980.

[11] G. Ghione, C. Naldi, F. Filicori, M. Cipelletti and G. Locatelli, "MESS - A Two-Dimensional Physical Device Simulator and its Application to the Development of C-band Power GaAs MESFETs", *Alta Frequenza*, pp. 295–309, Sept. 1988.

[12] G. Ghione and F. Filicori, "A computationally efficient unified approach to the numerical analysis of the sensitivity and noise of semiconductor devices", *IEEE Trans. on CAD*, Vol. CAD-12, N.3, March 1993.

[13] G. Ghione, F. Bonani and M. Pirola, "High-field diffusivity and noise spectra in GaAs MESFETs", submitted for pubblication to *Journal of Physics D: Applied Pysics*.

[14] J. Graffeuil and P. Rossel, "Semiempirical expression for direct transconductance and equivalent saturated velocity in short-gate-length MESFETs", *IEE Proc.*, pp. 185–188, 1982.

[15] H. Happy, "HELENA: un logiciel convivial de simulation des composants a effet de champ", *Ph.D. Thesis*, University of Lille, 1992.

[16] M. de Murcia, D. Gasquet, S. Elamri, J-P. Nougier and J. Vanbremeer-sch, "Diffusion and noise in GaAs material and devices", *IEEE Trans. on Electron Devices*, Vol. ED–38, No. 11, pp. 2531–2539, Nov. 1991.

[17] J. P. Nougier, "Noise and diffusion of hot carriers", in *Physics of nonlinear transport in semiconductors*, D. K. Ferry, J. R. Barker, C. Jacoboni eds., Plenum Press 1980, pp.415–465.

[18] M. Weiss and D. Pavlidis, "The influence of device physical parameters on HEMT large-signal characteristics", *IEEE Trans. on Microwave Theory & Techniques*, Vol. MTT–36, pp. 239–249, Feb. 1988.

[19] F. Ponse, W. T. Masselink and H. Morkoc, "Quasi-Fermi level bending in MODFETs and its effect on FET transfer characteristics", *IEEE Trans. on Electron Devices*, Vol. ED–32, pp. 1017–1023, Jun. 1985.

[20] D. Rohrer, L. Nagel, R. Meyer and L. Weber, "Computationally efficient electronic-circuit noise calculation", *IEEE Journal of Solid-State Circuits*, Vol. SC–6, No. 4, pp. 204–212, Aug. 1971.

[21] A. J. Shey and W. H. Ku, "On the charge control of the two-dimensional electron gas for analytic modeling of HEMTs", *IEEE Electron Device Lett.*, Vol. 9, pp. 624–626, Dec. 1988.

[22] W. Shockley, J. A. Copeland and R. P. James, "The impedance field method of noise calculation in active semiconductor devices", in *Quantum theory of atoms, molecules and solid state*, P. O. Lowdin ed., Academic Press 1966, pp. 537–563.

[23] H. Statz, H. A. Haus and R. A. Pucel, "Noise characteristics of gallium arsenide field-effect transistors", *IEEE Trans. on Electron Devices*, Vol. ED–21, No. 9, pp. 549–562, Sep. 1974.

[24] P. J. Tasker and B. Hughes, "Importance of source and drain resistance to the maximum f_T of millimeter-wave MODFETs", *IEEE Electron Device Letters*, Vol. 10, No. 7, pp. 291–293, July 1989.

[25] K. M. van Vliet, A. Friedman, R. J. J. Zijlstra, A. Gisolf and A. van der Ziel, "Noise in single injection diodes. I: A survey of methods", *J. Appl. Phys.*, Vol. 46, No. 4, pp. 1804–1816, April 1975.

4.5   Non-linear Modelling

Y Crosnier and G Salmer
Institut d'Electronique et de Microelectronique du Nord, 69655 Villeneuve d'Ascq,
France

1. Introduction

The continuous progress of heterostructure transistors offers more and more
possibilities of development in the order of high power or non linear applications
over larger and larger frequency bands from 1GHz up to 40 and even 60GHz. In
parallel, the need of a strong CAD environment naturally arises which has to be
particularly suited to the specific aspects of the wide variety of existing devices,
i.e. single and multi-heterojunction field effect transistors (HFETs) realized in
the AlGaAs/GaAs system or in the AlGaAs/InGaAs and in InAlAs/InGaAs
systems on GaAs or InP substrates, and also hetero bipolar transistors (HBTs)
realized in the AlGaAs/GaAs and GaInP/GaAs systems. Characteristics required
for operating these devices under large signal conditions depend obviously on
the aimed application. From this point of view, two main classes of applications
have to be distinguished. The former deals with power amplification. The main
requirement is then a maximized linearity of the device behaviour up to the
extreme current and voltage excursions. The latter comprises all the non linear
functions whose the most famous are mixers and frequency multipliers. In such
applications, it is necessary that one or several elements of the device exhibit a
maximized non linearity. It is clear that whatever the application is, linear or
non linear, designing the active device and its functional circuit needs accurate
identification and mastering of the physical aspects of the non linearities. As
concerns the HFET, the physical causes of non linearity can be summarized as
follows .

   In its general form, the structure of the HFET is a more or less complicated
pile of alternating narrow and wide band gap semiconductor layers surmounted
by the Schottky gate and the ohmic source and drain contacts (fig. 1). Such a
structure despite the fundamental capacitance-like behaviour of the electron gas,
with respect to the gate voltage, has inherent reasons for being non linear. The
first one is linked to the fact that the charge instead of being confined in a
unique quantum well is distributed in several quantum wells, and, in some
extent, in adjacent supplying layers, so that the gate control appears as a
succession of separate commands across the whole pile [1]. The second one lies
in non uniformities resulting either from differences of electron mobility among

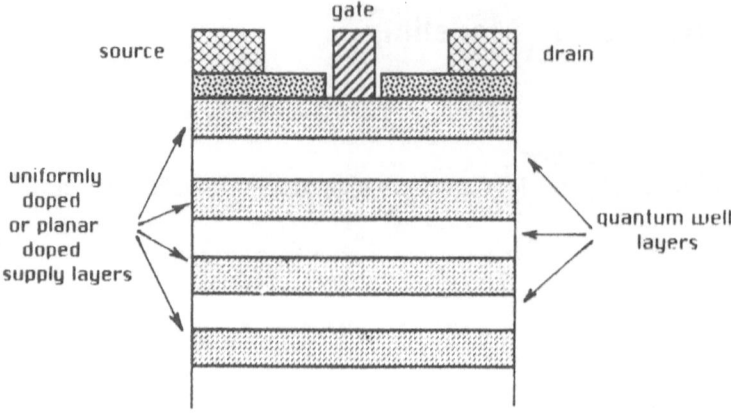

Fig. 1 - General representation of a multiheterojunction HFET.

the various materials constituting the structure layers or from the presence of traps. In particular, AlGaAs is well known as a low mobility material compared to GaAs or InGaAs. On the other hand, AlGaAs is also well known for its propensity to exhibit traps when uniformly doped [2] (the so called "DX centres") and in a certain extent when planar doped [3]. Beside these effects which are essentially related to the vertical command, there are additional effects which are more especially linked to the longitudinal electric field resulting from the application of the drain-source voltage and which are influenced by the gate geometry, the recess geometry and the gate to drain spacing. In particular, most important are the overshoting and hot carrier phenomena. And finally, the last causes of non linearity to be considered relate to the conductive and breakdown operating conditions, i..e;. when the gate is forward biased or largely reverse biased. In fact these conditions correspond practically to the most important non linearities and constitutes the ultimate limits of the device. All these effects are responsible of changes both in the internal spatial distribution of electrons and in their dynamic flow and, consequently, greatly affect the non linear behaviour of the device DC and AC characteristics as functions of the gate-source and gate-drain voltages.

The non-linear modelling implies a double strategy including both a physical approach and an electrical approach. The physical model is necessary to determine what technological features or parameters of the structure have to be especially tailored to minimize or maximize a particular type of non linearity. It is basically the prediction tool which allows to establish a hierarchy among different structural parameters as concerns their relative influences. Such a model, has to be able to describe the device behaviour whatever the bias conditions are. For this reason, it has to account for the maximum of physical phenomena. But for problems of computer time and size it must be restricted to the essential part of the device, i.e. almost the intrinsic part. So, the electrical model drawn from DC and AC characterizations, constitutes a complementary tool to know parasitic elements and get a practical and quantitative determination of the device intrinsic part. A relatively standard topology is now adopted everywhere for the large signal FET equivalent circuit (fig. 2) which is simply an extension of the usual small signal equivalent circuit. But a wide variety of modelling equations are existing whose the most popular are those of Curtice, Curtice-Ettenberg, Statz, Materka-Kacprzak and Tajima. The major distinction between these models is found in the drain-source current expression. All of them were initially developed for MESFETs and for that reason are somewhat limited for the description of HFETs and more particularly for multiheterojunction devices. Recently, several improvements have been propose. One of them consists in modelling the I_{DS} current by two non linear sources.. The former is a bias point dependent non linear equation and the latter represents the differences between DC and dynamic (derived from pulsed measurements) characteristics [4] [5]. An other one, more radical, consists in replacing analytical functions by tabulated ones depending on Vgs and Vds and directly extracted from characterizations [6].

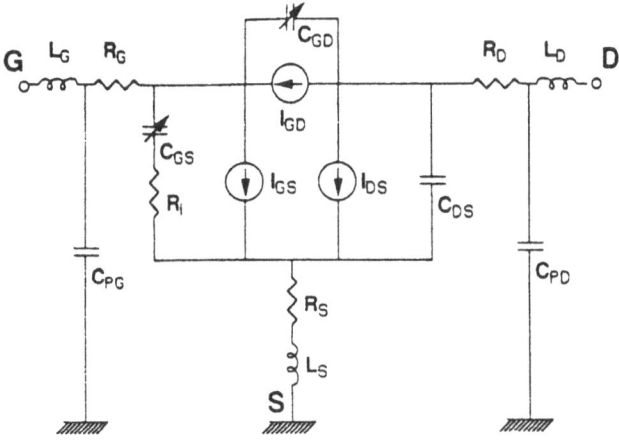

Fig. 2 - The standard non linear HFET equivalent circuit.

Constant progresses in modelling do not exempt the circuit designer from a clear strategy and definition of criteria when facing a given application. For instance, efficiency in high power amplification is more particularly related to the constancy of the cut-off frequency f_{max} versus V_{gs} and V_{ds}. On the other hand, the transconductance profile versus Vgs is a dominant parameter for hot FET gate mixer behaviour. Whereas FET frequency multiplication is dominated by the pinch-off steepness of the I_{DS} current ... As a consequence, it is obvious that the choice of a model and the refinement devoted to the determination and accuracy of a given non linear element depends strongly on the aimed application.

2 Main Simulation Methods

Aid expected from simulation tools is mainly :
• to improve the understanding of their physical problems
• to optimize devices by studying the dependance of their performances on technological parameters
• to design monolithic integrated circuits in which devices are included.

   Among the main requirements that such tools have to satisfy, one can mention
- validity for a large range of technological parameters and d.c. bias conditions.
- capability to simulate complex device structures
- accuracy allowing to include device models in circuit simulators.
- capability to consider large signal conditions.

   As pointed out in the introduction, it is not possible to find a unique simulation tool or device model that satisfies all these requirements. The present section describes briefly the typical set of models which is usually developed in many laboratories. It consists of three physical approaches based on :
- two dimensional solutions of balanced equations, so called 2D hydrodynamic models, such as SIMFET tool.
- quasi 2D solutions of semiconductor equations, such as HELENA software.
- one dimensional charge control model.
and of a more or less in house electrical approach comprising D.C. and microwave characterizations joined to appropriate parameters extraction.

2.1. Two Dimensional Solutions of Balanced Equations also called Hydrodynamic Energy Model [7;18].

These methods are based on the two-dimensional solutions of the balanced semiconductor equations [currents continuity, carriers momentum and average energy conservation equations] and Poisson equations. The conservation equations are deduced from Boltzmann transport equation by imposing several appropriate assumptions, mainly the relaxation time approximation and by considering an equivalent single valley model. Finally, the basic idea is to postulate that all the characteristics of the carriers distribution [electronic temperature, relaxation times, equivalent carrier mobilities] are only dependant on carriers average total energy, these dependances being evaluated by using purely physical models such as Monte Carlo simulations.

These tools initially conceived for MESFET's simulation [7, 8, 10, 11] have been considerably modified in order to treat conventional HFET's [12, 13, 14, 15] ; more recently [17, 18] ; several important physical effects [impact ionisation, quantization and screening] have been introduced allowing to accurately simulate pseudomorphic AlGaAs/InGaAs HFETs with very complex epilayer structure.

The whole system of equations is discretised using a finite difference scheme with a variable mesh size ; it is linearized with respect to time using a dynamic time step. This tool not only gives a good insight of the device physics through isolines of different quantities [carrier concentration and energy, potential, velocities] but also allows to determine electrical parameters and to predict device performance [cut-off frequency, gain, breakdown voltage].

Among the advantages of these models such as the SIMFET software [partially developped in the context of the ESPRIT project AIMS], we can mention that they are able to account for the majority of physical effects that occur in submicronic gate PM HFETs and they need less computational time [by a factor of 10] than particular models such as Monte Carlo simulations. But they need to use RISC stations or powerful computers and computational time may remain long. For instance, with an IBM 3090 600 E equiped with parallel vector processors, for a $0,2\mu m$ gate δ doped PM HFET, it needs 10 CPU min for 1ps real time of transient simulation [18] and 3-4ps are needed for reaching steady state.

2.2. Quasi 2D Models : One Dimensional Solution of Semiconductor Equations [19, 24]

In order to reduce computational time, several models have been proposed, based on the one dimensional solution of semiconductor equations, such as balance and Poisson equations. The main assumption is that the carrier displacement can be considered as one dimensional ; it is assumed that the equipotential lines in the "equivalent" channel are roughly perpendicular to the source to drain axis. Several other assumptions have been used for MESFET simulation [19, 20]. For conventional and pseudomorphic AlGaAs/InGaAs HFET's, total carrier concentration in the "equivalent" channel is calculated by using for each abscissa a charge control model [see 2.3] along the direction perpendicular to the source to drain axis. Despite of these assumptions, such models can be very accurate, as it has been shown by comparison with experiment or more sophisticated simulation tools. In fact, they take into account the main effects that occur even in submicrometer gate HFETs.

These equations are solved numerically by using a one dimensional discretized scheme along the source to drain axis. The main advantage of these tools, such as HELENA software [partially developped in the context of the Esprit project

GIANT], is that they are simple enough to be implanted on desk computer and are very fast : three minutes are necessary to perform a complete AC, DC and Noise analysis of a 0.2µm gate HFET with a 486/50MHz personal computer. Moreover, noise performance can be deduced by using the impedance field method. On the other hand, the main limitations come from the present difficulty to account for two dimensional effects, such as real space charge transfer and carrier injection that directly influence the output conductance and then the available gain, but such limitations will be overcome in the near future. Despite of these drawbacks, these models are very suitable for device optimization : they allow to determine very quickly the dependance of electrical characteristics and expected performance on the main technological parameters [gate length, recess configuration, epilayer characteristics...].

2.3. Charge Control Models [25, 34]

By assuming that the carriers are drifting with an average velocity $<v>$ in the channel, it is possible to predict the drain current density Id and the transconductance g_m if we know the dependance of total carrier concentration n_s in the channel as a function of Vgs, that constitutes the charge control law.

$$Id \; \# \; qnId_s <v> \qquad \text{and } g_m \; \# \; q <v> \frac{dn_s}{dV_{gs}} ;$$

in fact,$<v>$ is a fitting parameter that depends on the gate length.

The linear dependance $n_s = f(V_{gs})$ proposed by Delagebeaudeuf [25] and Lee [26] for conventional AlGaAs/GaAs HFETs does not remain valid for complex δ doped heterostructures. For such epilayer structures, only simulations based on a self consistent solution of Poisson and Schrödinger equation [27, 29, 31, 33, 34] allows to predict accurately the dependance of carrier concentration on technological parameters [epilayer characteristics] and gate to source voltage. However, it has been shown recently [34] that even in double planar doped heterostructures, the energy levels of the first two subbands in the channel are linearly dependant on the carriers density n_s :

$$E_i = A_i + B_i n_s$$

A_1 and B_i being semi-empirical parameters that depend on the epilayer characteristics. This result allows to write a quasi analytical simplified charge control law in the channel that is in very good agreement with experiment and more sophisticated models [34].

In conclusion, these kinds of simulation tools are very fast, but due to the assumptions introduced, they are not able to predict accurately the device performance. However, they can be used for comparative studies and then for computer aided design, even of complex heterostructures.

2.4. Electrical Model

It is generally accepted that a wide variety of measurement techniques is necessary to get a valid non linear electrical model for a given HFET. Such measurements comprise full D.C. characteristics, pulsed D.C. characteristics around various bias points to account for traping and heating effects, C.W. micro ave S parameters with the FET successively put in cold biasing configuration (V_{DS} = 0) then in normal biasing configuration for deembedding the extrinsic and intrinsic elements respectively [35]. Recent progresses accomplished in the field of network analyzers allow presently to complete these measurements with pulsed S parameters together with a pulsed biasing [36] in a pulse duration as short as 100ns. Extraction of the equivalent circuit elements from all these measurements has resulted in the development of sophisticated optimizers and numerical methods. It seems that the main remaining problem lies essentially in the ability of the element relationships to describe accurately very non linear behaviours. This question has been treated in the frame of the ESPRIT program with the purpose to design highly performant millimeter HFET mixers [37]. It has been demonstrated that high degree power series in subtle combination with hyperbolic tangent functions are able to fit with a high flexibility very sinuous or angular non linearities, in particular those dealing with submicronic gate multiherojunctions HFETs.

3. TYPICAL RESULTS

We will give now some typical results allowing to point out the main capabilities of simulations tools that have been developped partially in the context of ESPRIT Programme.

3.1. - 2 D and Quasi 2D solutions of S.C. equations

As it has been previously mentioned, these models are used on one hand to have a better insight of device behaviour and on the other hand to allow device optimization by determining the dependance of device performance on technological parameters.

As a first example, fig. 3 represents the electrons concentration, energy and potential contours for the intrinsic part of a double planar doped pseudomorphic AlGaAs/InGaAs HFET deduced from the 2D hydrodynamic energy model (SIMFET software). The structural parameters of the device are given in the figure. For instance, we can easily make the following observations :
- the rear doping plane is not fully depleted. This device is not well optimized, because low mobility carriers remain in this plane.

- despite of carriers confinement in the well, hot carriers injection occur just after the gate exit.
- a high energy and high electric field domain are located at the same place.

Fig. 4 then gives an example of dependance of device performance on technological parameters deduced also from hydrodynamic energy model. It represents the variations of intrinsic current gain cut-off frequency f_c ($g_m/2\pi$ C_{gs}) versus Vds for a 0,3μm gate length δ doped pseudomorphic FET. The three devices considered here differ by the values of the offset distance between the gate and the edge of the recessed zone (0μm, 0.1μm, 0.2μm). This distance R constitutes a very important technological parameter, because for instance breakdown voltage increases with R. However, as it can be seen in the figure, f_c decreases more quickly with V_{ds} as R is large, and then the large signal behaviour becomes worse and worse. A trade-off between large signal behaviour and high breakdown value must be checked.

As a third illustration, fig. 5b represents the evolutions of current gate cut-off frequency with gate to source voltage following the epilayer characteristics of a double planar doped 0.15μm gate HFET. Such results have been obtained by means of HELENA software (Quasi 2D model). Four epilayer structures have been considered and are represented fig. 5a. They differ mainly by the doping concentration in the doped planes. From this figure, it can be easily seen that the structure case 1 with two doped planes ($\delta_1 = 4.510^{12}$cm^{-2}b and $\delta_2 = 1.510^{12}$cm^{-2}) is the most suitable for power applications, because it presents the flatest variation of f_c as function of Vgs.

3.2. Charge Control Models

As pointed out previously the interest of such models is to allow fast predictions of HFET charge control features and particularly in the case of multiherojunction devices for which 2D and even quasi 2D models are very heavy to use. This interest has been successfully tested in the ESPRIT AIMS project [37] with the realization of special double planar doped, double quantum well heterostructures for mixer applications. The basic idea was that this kind of heterostructure, when commanded by a Schottky gate, could exhibit a two peaks transconductance profile versus the gate voltage, and that, depending on the quantum wells spacing and positions and charges of the doped planes, this profile could be easily tailored. Owing to this tailoring flexibility, it was expected that two kinds of operating mode, both of them in gate configuration, could be possible : the former in fundamental mode, i.e; with the bias gate voltage at pinch-off and the R.F. and L.O. frequencies close to each other, the latter in doubler mode, i.e. with the bias gate voltage corresponding to the transconductance valley and the RF frequency being equal to about twice the L.O. frequency [38] (fig. 6). Advantages hoped were, from the former, the possibility to satisfy easier than with a usual single quantum well HFET the

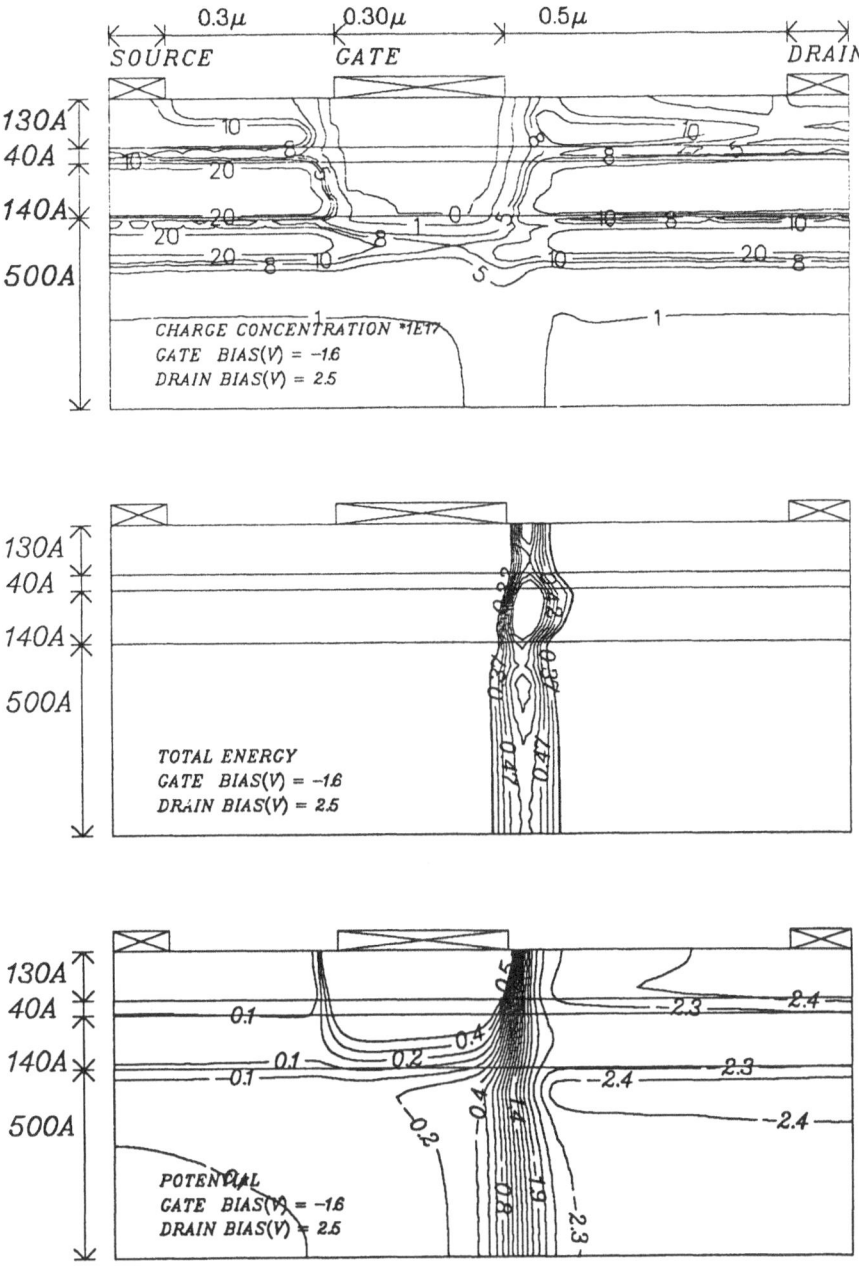

Fig. 3 - Carriers concentration, energy and potential contours in a 0.3μm gate length HFET with two doped planes ($N_1 = 3.10^{12}/cm^2$ and $N_2 = 2,5 \ 10^{12}/cm^2$).

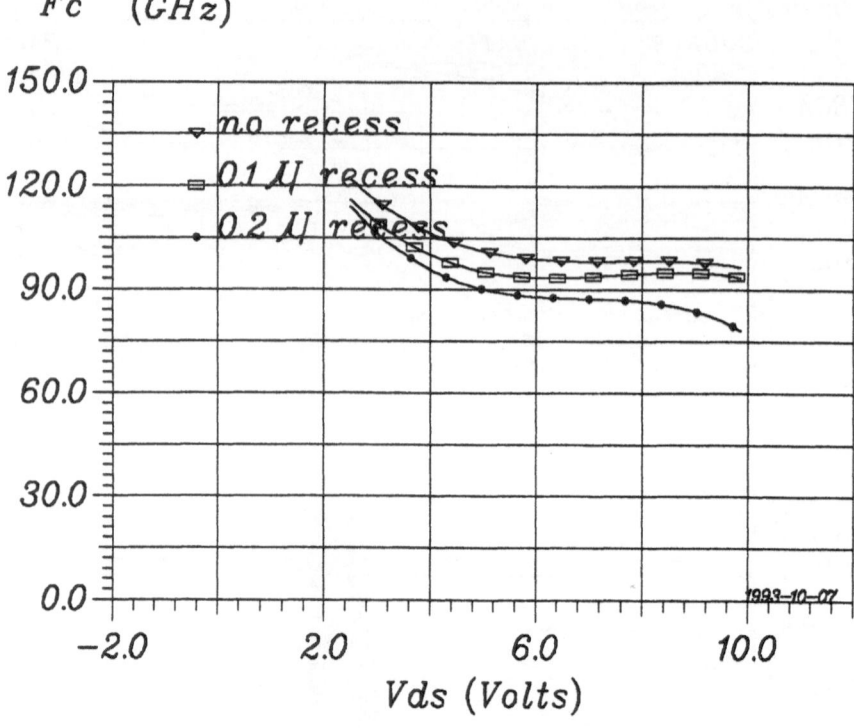

Fig. 4 - Variations of current cut-off frequence f_c = gm/2πCgs versus Vds for 0.3μm gate pseudomorphic HFETs with different gate offset R distances (R = 0, 0.1μm, 0.2μm).

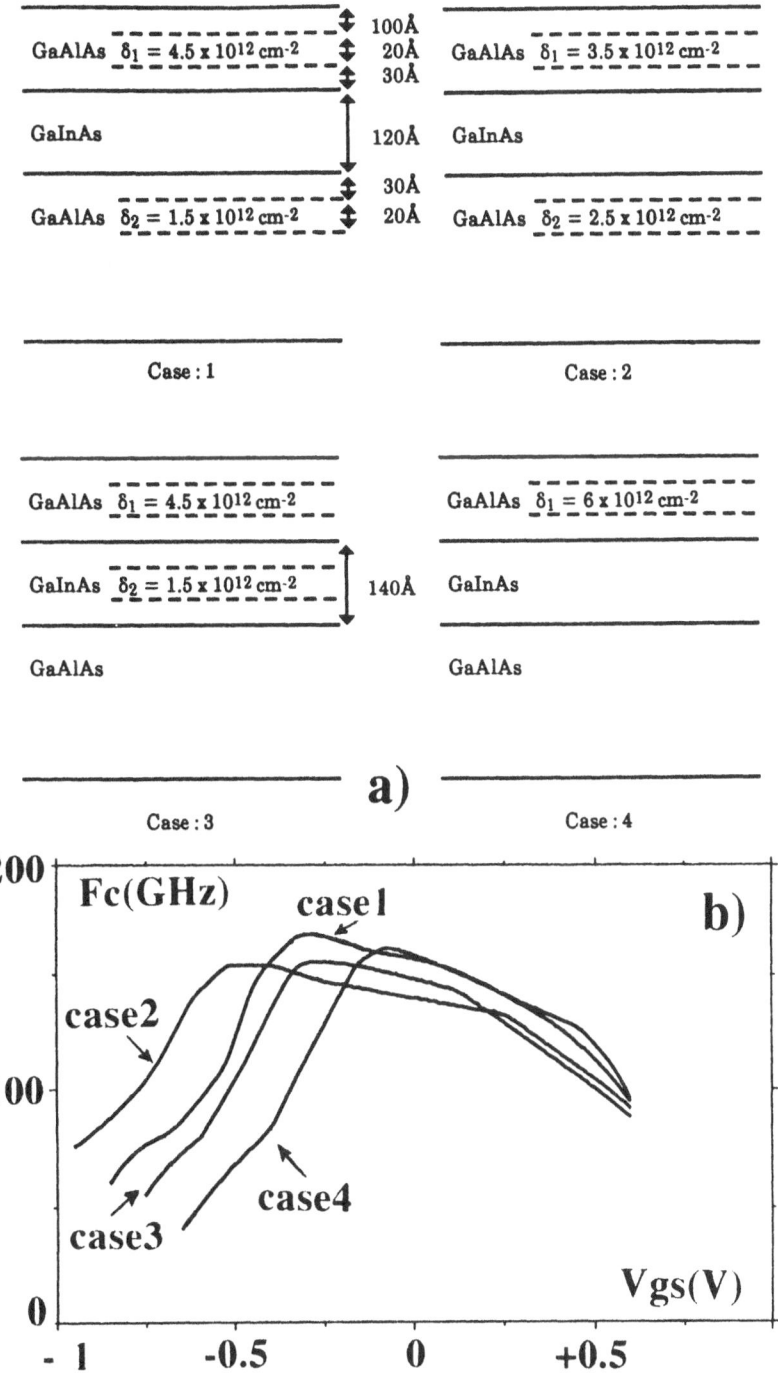

Fig.5 - Characteristics of the four epilayer considered (5a). Corresponding variations of current gain cut-off frequencies versus Vgs (5b).

254

double requirement of high conversion gain and low intermodulation, and, from the latter, the possibility to have a lower frequency for the local oscillator.

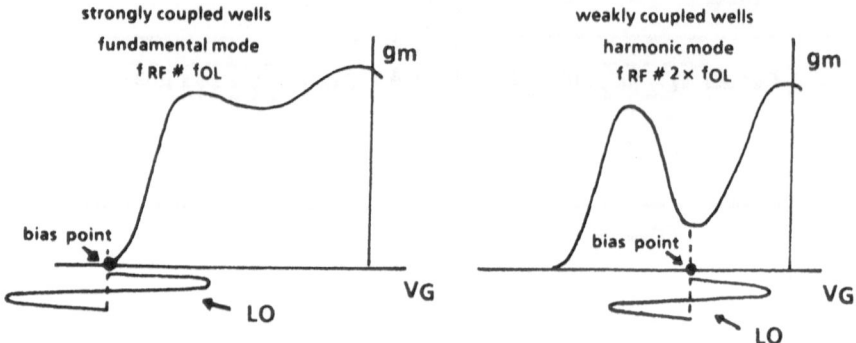

Fig 6 - The two potential mixer operations, in fundamental mode or in doubler mode,

Given the complexity of the needed heterostructures and the fact that many runs should be necessary before getting suitable results, it was patent that a 1D charge control model was particularly appropriate. Experimenting with this approach the influence of the various structural parameters showed the determining role played by the amount of charge of the supply layer separating the two quantum wells and the spacing between them. A global agreement between modelling and experiment [39] was obtained on a large range of variations of the sheet density of this charge, that is up to $4.10^{12} \mathrm{cm}^{-2}$, with about a 10nm spacing (fig. 7a). The important difference obtained in experiment between the quasi static and microwave transconductance profiles was explained on the basis of traping with a classical Fermi statistics and a single donor energy level below the conduction band, scheme somewhat questionable in the case of planar doping. Finally a record non linearity was demonstrated by separating the two quantum wells with a double barrier of uniformly doped AlGaAs of $5.10^{12} \mathrm{cm}^{-2}$ equivalent sheet density and about 30nm thickness (fig. 7b). The 50% valley to peak ratio of microwave transconductance reached with this structure is undoubtly very promising for a future practical use based on a more stabilized technology and a shorter gate length (presently 0.35μm).

3.3, Electrical Model

As pointed out in section 2.4 simulation of HFETs based strongly non linear circuits require often specific in house electrical models especially adapted to the HFET under test, mainly when this one exhibits particularly sinuous and angular

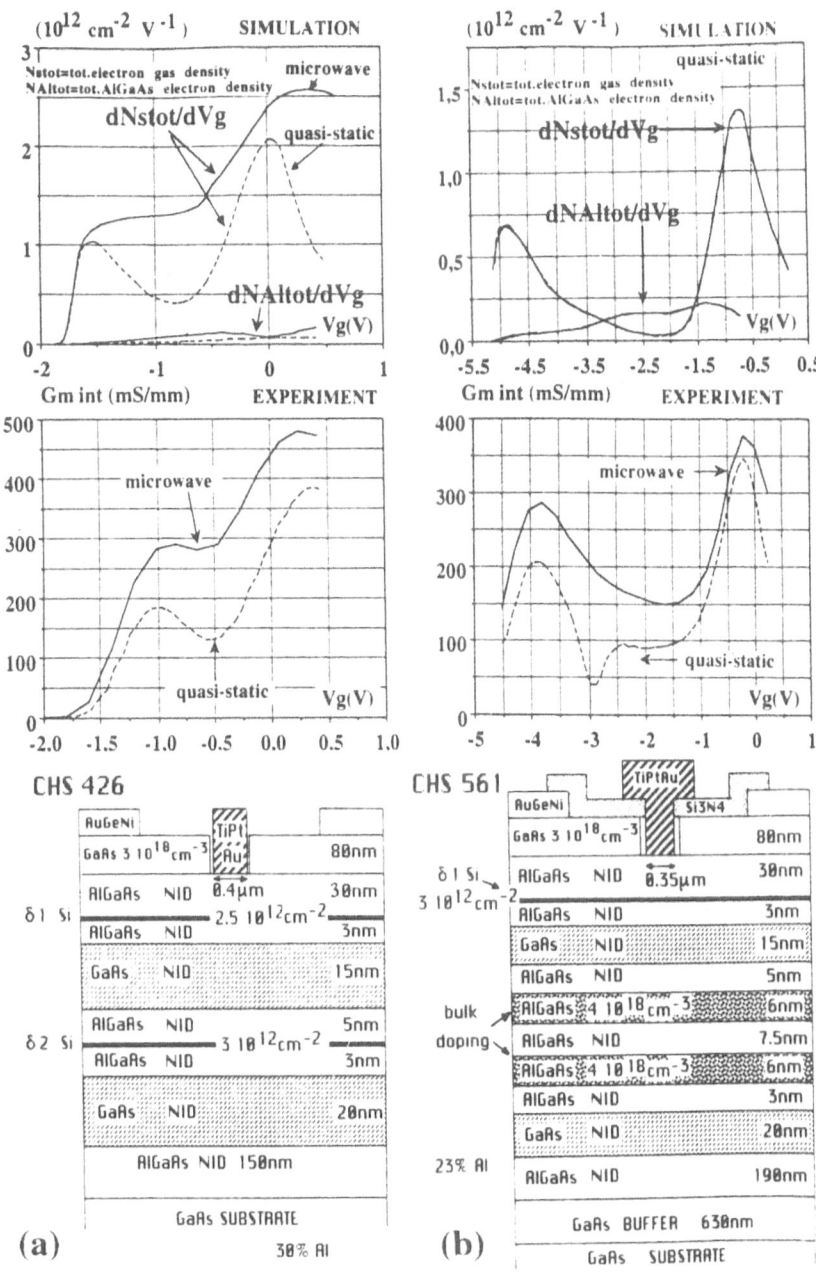

Fig. 7 - Examples of AlGaAs-GaAs double quantum well HFETs designed using the 1D charge control model. The two quantum wells are strongly coupled in case a (CHS 426) and weakly coupled in case b (CHS 561).

non linearities. As an illustration of what can be provided by such simulations the functional analysis of mixers developed in the ESPRIT AIMS project appears exemplary [37]. The core of the HFET model built at Lille University for this purpose consists essentially in polynomial fitting (up to 15th degree) and integration of the device transconductance and gate source capacitance extracted both from D.C. pulsed and S parameters measurements. It takes also into account of the forward and breakdown gate currents. In total it comprises 54 non linear coefficients and 12 linear elements, and it is implemented in the MDS software. After having performed simulations and measurements on a wide variety of single and double heterojunction devices the main teaching was the possibility to predict the conversion gain behaviour with simply looking at the transconductance profile due to the very close correlation between the variations of these two quantities, versus the gate source voltage, for the transconductance, and versus the L.O. power, for the conversion gain. Compared to the transconductance dependance all the other non linearities, in particular the gate-source capacitance one, were found by far less important. As a first example figure 8 shows results corresponding to a AlGaAs-GaAs double quantum well HFET realized on the same epitaxy as the one presented in section 3.2, (figure 7a), that is with two GaAs quantum wells supplied by two doped planes of 2.5 10^{12}cm^{-2}sheet density and 3.10^{12}cm^{-2} sheet density, respectively. The mixer operation was performed using a desk measurement set up with RF and LO frequencies of 20 and 18GHz, respectively. One can notice, first, the remarkable agreement between experimental and simulated conversion gains and, second, the striking similarity of the transconductance and conversion gain profiles, both of them exhibiting the typical two peaks shape. As a second example, figure 9 shows results obtained with a hybrid gate mixer realized on an alumina substrate using a Daimler single quantum well pseudomorphic HFET. In this application RF and LO frequencies are 28.5 and 24.5GHz respectively. Despite the higher frequency range and the increasing possibility of deviations inherent to the parasitic elements uncertainties, the agreement between the simulated and experimental conversion gains is again very good. On the other hand, one can notice that, as expected, due to the use of a single quantum well HFET, the conversion gain profile like the transconductance exhibits a bell shape profile, that is a single peak and not two as in the previous example.

Acknowledgment

The authors would like to thank EEC for supporting, in the framework of ESPRIT Programme, a large part of the works developped in this chapter. They also would like to thank J. Alamkan, A. Cappy, Th. Coupez, J.C. De Jaeger, K. Sherif for their valuable modelling inputs used in this review, D. Theron for multiheterojunction epilayer growth and non linear HFETs processing, R. Allam and C. Kolanowski for the design, fabrication and analysis of HFET microwave mixers.

257

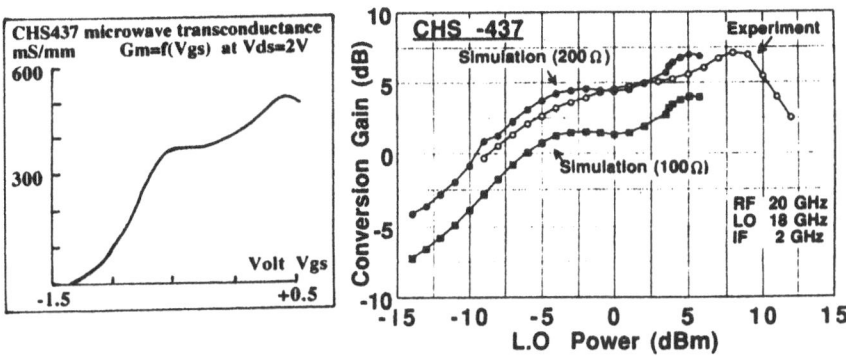

Fig. 8 - Experimental and simulated (for two different load resistances : 100Ω and 200Ω) conversion gains of a double quantum well HFET, with, for comparison, the experimental transconductance profile. Mixer operation was performed using a desk setup measurement.

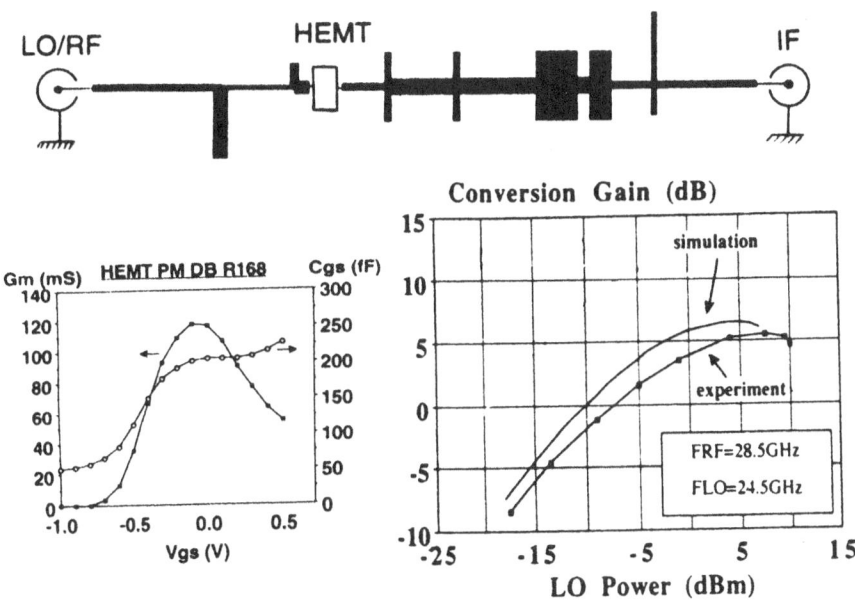

Fig. 9 - Experimental and simulated mixer conversion gains of a single quantum well pseudomorphic HFET with, for information, its transconductance and gate-source capacitance profiles. Mixer operation was performed with the hybrid circuit whose layout is shown.

References

[1] N.H. SHENG et al, "Multiple-channel GaAs/AlGaAs High electron mobility transistors", IEEE E.D.L., n° 6, June 1985.

[2] E.F. SCHUBERT, K. PLOOG, "Shallow and deep donors in direct-gap n type AlGaAs : Si grown by molecular-beam epitaxy", Phys. Rev. B, vol. 30, n° 12, Dec. 1984.

[3] B. ETIENNE, V. THIERRY-MIEG, "Reduction in the concentration of DX centers in Si-doped GaAlAs using the planar doping technique", Appl. Phys. Lett. 52 (15), April 1988.

[4] T.J. BRAZIL, "A universal large-signal equivalent circuit model for the GaAs MESFET", 21st European Microwave Conference, pp. 921-926, 1991.

[5] T. FERNANDEZ et al, "Modelling of operating point non linear dependence of I_{DS} characteristics from pulsed measuments in MESFET transistors", 23th European Microwave Conference, pp. 518-521, 1993.

[6] D.E. ROOT, "A measurement-based FET model improves CAE accuracy", Microwave Journal, Septembre 1991

[7] CURTICE W.R., YUN Y.H., A temperature model for the GaAs MESFET, IEEE Trans. Electron Devices (1981), 28, p. 954.

[8] COOK R.K., FREY J., Two-dimensional numerical simulation of energy transport effects in Si and GaAs MESFET's. IEEE Trans. Electron Devices (1982), 29, p. 970.

[9] WIDIGER D.J., KIZILYALLI I.C., HESS K., COLEMAN J.J., Two-dimensional transient simulation of an idealized high electron mobility transistor. IEEE Trans. Electron Devices (1985), 32, p. 1902.

[10] EL-GHAZALY S., SALMER G., LEFEBVRE M., IBRAHIM M., EL-SAYED O., Two-dimensional simulation of submicronic FET's, 13th European Solid State Device Research Conference. Inst. Phys. Conf. Series (1983), 60, p.127.

[11] EL-SAYED O., EL-GHAZALY S., SALMER G., LEFEBVRE M., Performance analysis of submicron gate MESFET'S. Solid State Electron. (1987), 30, p. 643.

[12] LORET D., Two-dimensional numerical model for the high electron mobility transistor. Solid State Electron. (1987), 30, n° 11, p. 1197.

[13] SHAWKI T., SALMER G., EL-SAYED O., Two-dimensional transient simulation of submicron-gate MODFET's. GaAs and Related Compounds Symposium, Heraklion (1987).

[14] SHAWKI T., SALMER G. and EL-SAYED O., "MODFET 2D Hydrodynamic Energy modeling : optimization of subquarter micron gate structures", IEEE Trans on El Dev, Vol. 37, n° 1, p. 21, 1990.

[15] SHAWKI T., SALMER G., EL-SAYED O., 2D simulation of degenerate hot electron transport in MODFET's including DX center trapping, IEEE Trans CAD of ICAS, nov. 90.

[16] SHAWKI T., SALMER G., Structural optimization of millimeter wave subhalf micron gate MODFET's, European Trans. on Tel. and Rel. Technologies, Vol. 1, n°4, July-Aug 1990.

[17] SHERIF K., SALMER G., REFKY A., EL-SAYED O., Recent results of 2D hydrodynamic energy modelling of MODFET's, Int. Symp. on Signals Circuits and Electronics, URSI, Paris, 1992.

[18] SHERIF K, SALMER G. and EL-SAYED O., "An enhanced 2D hydrodynamic energy model for transient time simulation of complex heterostructure field effect transistors. Simulation of semiconductor devices and processes", 1993, vol. 5, p. 457, Springer Verlag.

[19] CARNEZ B., CAPPY A., KASZINSKI A., CONSTANT E., SALMER G., Modeling of submicrometer gate field effect transistor including effects of non stationary electron dynamics. J. Appl. Phys. (1980), 51, p. 784.

[20] WROBLEWSKI R., SALMER G., CROSNIER Y., Theoretical analysis of the DC avalanche breakdown in GaAs MESFET's, IEEE Trans. Electron Devices (1983), 30, p. 154.

[21] CAPPY A., VANOVERSCHELDE A., SCHORTGEN M., VERSNAEYEN C., SALMER G., Noise modeling in submicrometer-gate two-dimensional electron-gas field-effect transistors. IEEE Trans. Electron Devices (1985), 32, p.2787.

[22] SANDBORN et al, Quasi two-dimensional modeling of GaAs MESFET's, IEEE Trans. on El. Dev., Vol. ED-34, n° 5, May 1987.

[23] SNOWDEN et al, Quasi two dimensional MESFET simulation for CAD, IEEE Trans. on El Dev., Vol. ED-36, n° 9, sept. 1989.

[24] HAPPY H., DAMBRINE G., ALAMKAN J., DANNEVILLE F., KAPTCHE-TAGNE F., CAPPY A., HELENA : a friendly software for calculating the DC, AC and noise performance of HEMTs, Int. Journ. of Microw. and Mill. Wave CAD, vol. 3, n° 1, 14-28, 1993.

[25] D. DELAGEBEAUDEUF and N.T. LINH, Metal n AlGaAs/GaAs two dimensional electron gas field effect transistor, IEEE Trans. on Electron Devices, Vol. ED 29, pp.955-960, June 1982.

[26] LEE K., SHUR M.S., DRUMMOND T.J. and MORKOC H., Current voltage and capacitance voltage characteristics of MODFETs, IEEE Trans. on Elect. Devices, Vol. ED30, pp. 207-212, 1983.

[27] VINTER B., Subbands and charge control in a two-dimensional electron gas field effect transistor, Appl. Phys. Lett., Vol. 44, p. 307-309, 1984.

[28] PONSE F., MASSELINK W.T. and MORKOC H., Quasi Fermi level bending in MODFET's and its effects on FET transfer characteristics, IEEE Trans on Elect. Devices, Vol. ED32, p. 1017-1023, 1985.

[29] YOSHIDA Y., Classical recesses versus quantum mechanical calculation of the electron distribution in the n AlGaAs/GaAs heterointerface, IEEE Trans. on Electron Devices, Vol. ED33, p. 154, Jan 86.

[30] SHEY A.J. and KU W.H., On the charge control of the two-dimensional electron gas for analytic modeling of HEMT's, IEEE Electron Device Letters, Vol. 9, n° 12, Dec. 1988.

[31] ANDO Y. and ITOH T., Analysis of charge control in pseudomorphic two-dimensional electron gas field effect transistors, IEEE Trans. on Elect. Devices, Vol. 35, n° 12, Dec. 1988.

[32] WANG G.W. and EASTMANN L.F., An analytical model for IV and small signal characteristics of planar doped HEMT's, IEEE Trans. on M.T.T., Vol. 37, n° 9, Sept. 89.

[33] PATIL M.B. and MORKOC H., Self consistent calculation of electron density in a two-channel modulation doped structure, Solid State Electronics, Vol. 33, n° 1, pp. 99-104, 1990.

[34] ALAMKAN J., HAPPY H., CORDIER Y. and CAPPY A., Modeling of pseudomorphic AlGaAs/InGaAs/GaAs layers using self consistent approach, Eur. Trans. Tel. Rel. Technologies, Vol. 5, n°4, July 1990.

[35] G. DAMBRINE et al, "A new method for determining the FET small-signal equivalent circuit", IEEE MTT, Vol. 36, n° 7, July 1988

[36] J.P. TEYSSIER et al, "A pulsed S-parameters measurement setup for the non linear characterization of FETs and bipolar power transistors", 23rd European Microwave Conference, pp. 489, 493, Madrid, Septembre 1993.

[37] ESPRIT 5032 AIMS contract, Lille University and Alcatel-Espace contributions.

[38] R. ALLAM, T. COUPEZ, C. KOLANOWSKI, D. THERON, Y. CROSNIER, "Designing and modelling multi-channel HEMT gate mixer", 21st European Microwave Conference, Stuttgart, Sept. 1991.

[39] D. THERON, B. BONTE, C. GAQUIERE, E. PLAYEZ, Y. CROSNIER, "Characterization of GaAs and InGaAs double quantum well heterostructure FETs", IEEE Trans. El. Dev., to be published.

4.6   Power MESFET Modelling

F Gianinni, University of Tor Vergata, Roma, Italy
J L Garcia, University of Cantabria, Santander, Spain

Power MESFET modelling may be addressed in several ways. In this chapter we will consider the possibilities offered by the large-signal equivalent-circuit model, extracted from measurements and suitable for non-linear CAD, by the load/source-pull measurements, which allow the evaluation of the performances of the component directly under large-signal operating conditions, and by physical models, which link the technological parameters of the device to the device performances.

I - LARGE-SIGNAL EQUIVALENT-CIRCUIT MODEL

I.1) - Introduction

The current civil and military applications have allowed the HEMT and MESFET to be incorporated in an increasing number of nonlinear hybrid and MMIC circuits. To design these circuits, nonlinear software programs using harmonic balance or time-domain algorithms are used. These simulation tools are very powerful but they need accurate large signal device models to improve the performance of these circuits and to minimize the number of design and fabrication cycles required. Therefore, foundries need to have more advanced non-linear models than the traditional quasi-static approximation.

An important problem in dealing with the nonlinear modelling of these devices is their anomalous low frequency behaviour: the quasi-static approach is not fulfilled. This behaviour leads to the frequency dependence of transconductance and output resistance and its origin is due to trapping, surface state, etc, [1] [2] and implies that the DC characteristics are an over estimation of the device performance [3]. Furthermore, the dynamic characteristics and the transconductance and output resistance are functions of the operation point. Consequently, changes in the equations and/or topology of the nonlinear circuit of the device must be carried out in order to have an accurate description of the transistor behaviour [4] [5].

The typical way of modelling the nonlinear equation of the channel current is extracting the parameters of this equation from the small signal S parameters in the frequency band of interest at several operation points (quasi-static approximation)

[1] [6] [7]. Another way of obtaining the parameters of this current is using pulsed characteristics [8] [9]. Up to now, this last method is only valid at a given operating point and represents a better simulation of the real function of the transistor because it takes into account its real heating. These methods do not predict a reliable behaviour of low frequency dispersion and DC operation. Recently, several authors have proposed models to solve these problems using a second nonlinear current source [10] [11]. These models are able to predict the low frequency dispersion more or less exactly but do not solve the problem of the operating point dependence of the model parameters nor pulsed measurements [4].

I.2) - DC Measurements

The experimental DC measurements permit us to calculate the access resistors R_g, R_s, R_d and the typical Schottky current parameters I_{ns} and α_s. For these purposes, an automatized experimental set up along with a fitting program, capable of obtaining these kind of measurements by three different methods [4] [5], have been developed. The advantage of this experimental set up, comparing with other authors, is the current source at gate to protect the device and obtain more precise measurements (on carrier/on wafer). In this case, to obtain R_s and R_d, a method that maintains the voltage source and drain ports has been presented. The R_s measurements are made injecting current through gate port sweeping V_{ds} and measuring V_{gs} and I_{ds} (Figure 1). The same method is used to measure R_d interchanging source for drain. In figures 2 and 3, experimental results of R_s and R_d measurements for a GEC-Marconi chip transistor are presented. The R_g measurements are based on the Fukui method [12]. Likewise, the Schottky junction current source is measured injecting current through gate port in common source configuration and the open-circuit drain terminal. The experimental results of figure 4 permit the extraction of the typical nonlinear equation of the Schottky junction current I_{ns} and α_s.

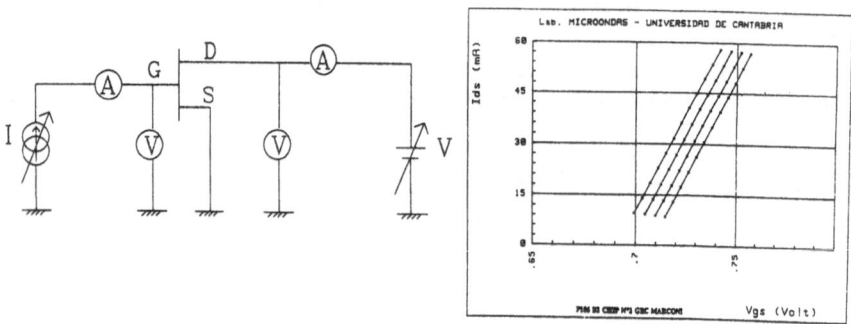

Fig.1 - R_s measurements   **Fig.2 - R_s measurements**
  configuration

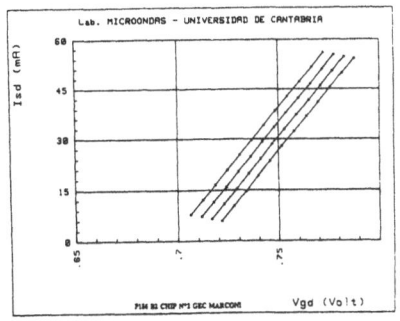

Fig.3 - R_d measurements

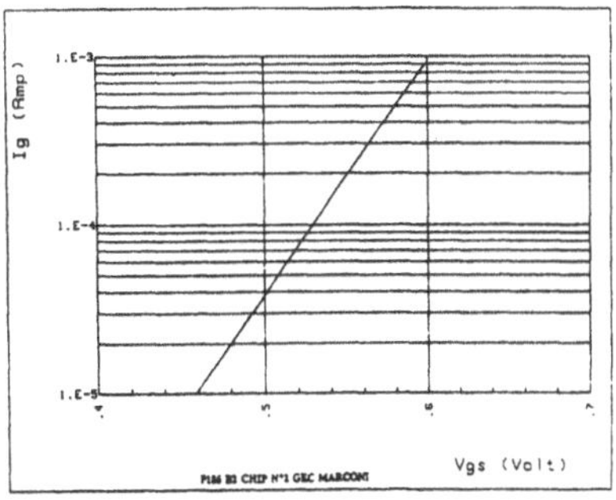

Fig.4 - I_{gs} measurements

I.3) - Pulsed measurements

To characterize the behaviour of a MESFET device it is very important to know the DC and pulsed experimental characteristics. The DC curves allow us to know the DC behaviour and the temperature of the quiescent operating points in RF operation, and the pulsed measurements take into account the trap effects.

In order to measure the continuous and pulsed characteristics of the transistors we have developed an automatic experimental set up along with a control program. The pulsed I/V Measurement System (PIVMS) is a high speed automated set up controlled by two internal microprocessors [13] as we can see in figures 5 and 6. The technical characteristics of the experimental system are:

- It is possible to control the frequency of the pulses.
- The range of the pulse width varies from 300 ns to 2000 ns.
- It has 16 bit resolution in 300 ns pulse width
- It is possible to vary the pulse rise time.
- Voltage and current ranges are:

Gate : +/-10 volts in +/- 200 mA.
Drain: -5 volts to +30 volts in +/-2 Amp.

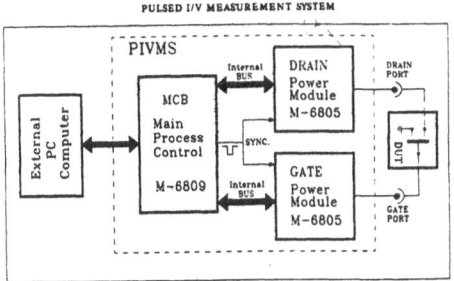

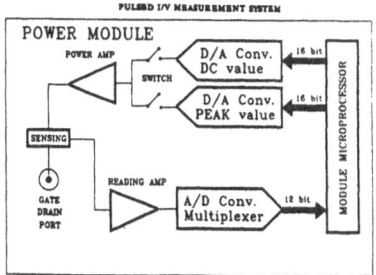

Fig.5 - PIVMS Block Diagram

Fig.6 - Power Module Description

Figure 7 shows the experimental DC and pulsed characteristics of a 6000 μm D07A PML process chip transistor. The V_{gs} range is from -3.5V to -0.25V and the quiescent operating point for the pulsed characteristics is V_{gs}=-2.5V, V_{ds}= 3V.

I.4) - Scattering parameters measurements

These kinds of measurements are very important for knowing the RF behaviour and the low and high frequency dispersion problems. For this purpose we have developed a program capable of controlling the network analyzer to obtain multibias scattering parameters. The measurements can be on wafer or on carrier, with TRL calibration, taking into account the bias way resistors. For power transistor measurements, a special power resistant bias T has been developed. In figure 8, we can see the problems when the bias resistors have not been considered, and figures 12 and 13 show examples of experimental results.

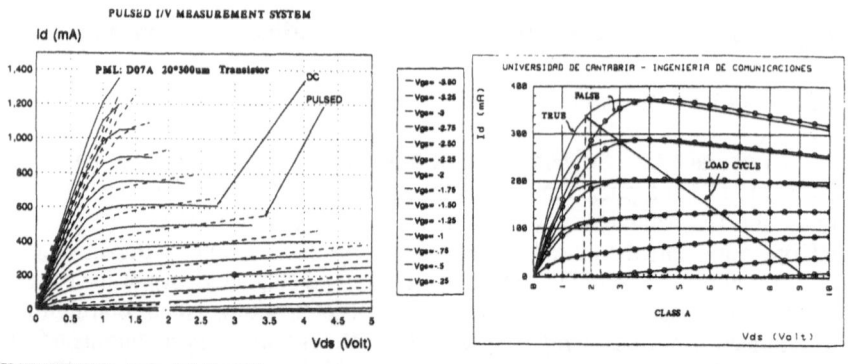

Fig.7 - DC and pulsed characteristics (6000 μm chip transistor)

Fig.8 - Bias resistor influence in DC characteristics

I.5) - Linear extraction

The multibias measurements of the scattering parameters permit us to obtain the parameters of the linear equivalent circuit of the transistor (Figure 9), valid at each operating point in the interest frequency band. This kind of measurements will permit us to obtain not only the linear parameters but the dynamic characteristic curves of the transistor using the quasistatic approximation [1] [6] [11]. For this purpose, we have developed a linear extraction program based on the work of several authors [12] [14] [15] [16], modifying the algorithms to reduce the optimization error and also modifying the equivalent circuit to permit the calculation of packaging elements. This method uses direct extraction techniques to obtain both intrinsic and extrinsic parameters of the linear equivalent circuit of figure 9 from experimental [Y] and [Z] values.

The inductors L_g, L_d, L_s, packaging capacitors C_{pgi}, C_{pdi} and the access resistors R_g, R_s, R_d are obtained from scattering parameter measurements at hot ($V_{ds}=0$, $V_{gs}>0$) and cold ($V_{ds}=0$, $V_{gs}<0$) operating points. We can observe that it is possible to do comparisons between these access resistors and the obtained resistors from DC measurements.

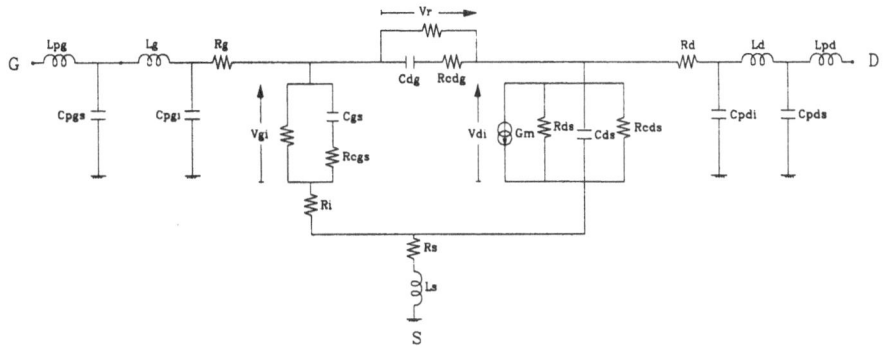

Fig. 9 - GaAs-MESFET Linear Equivalent Circuit

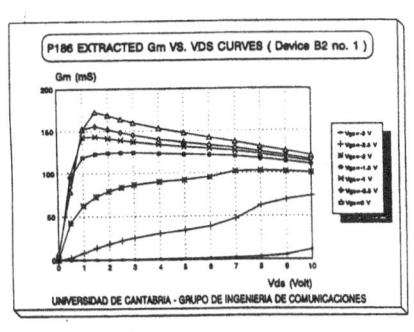

Fig.10 - G_m curves

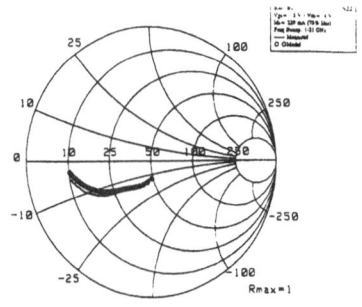

Fig.11 - G_{ds} curves

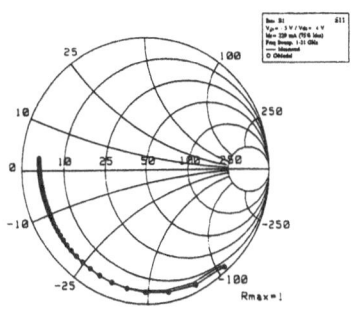

Fig.12a - S11 parameters

Fig.12b - S22 parameters

Taking into account that R_i, τ and C_{ds} are approximately constant, we can do several analytical extractions around the $I_{dss}/2$ and $V_{ds}=3$ or 5 Volts and we can extract the average values. C_{ds} average value is obtained from low and medium frequencies and R_i and τ from high frequencies in the band. After that, multibias

266

extractions permit the calculation of the rest of the parameters of the linear circuit at every operating point.

The algorithms of interpolation using splines and the optimization method with constraints permit us to reach errors 50% less than the above mentioned methods. The error function is calculated comparing the extracted and measured scattering parameters.

Figures 10, 11 show, as an example, the extracted G_m and R_{ds}, functions of external V_{gs} and V_{ds}, for a F20 process GEC-Marconi chip MESFET transistor and figures 12 and 13 compare the extracted and measured scattering parameters.

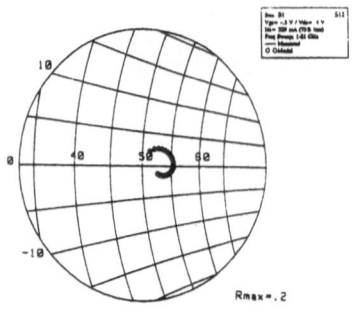

Fig.13a - S12 parameters   **Fig.13b - S21 parameters**

I.6) - Large signal model extraction

In figure 14 we can see the equivalent circuit of a non-linear model of a GaAs-MESFET transistor. We consider four typical non-linearities: the schottky junction (I_{gs}, C_{gs}), the channel current (I_{ds}) and the breakdown effect (I_{dg}).

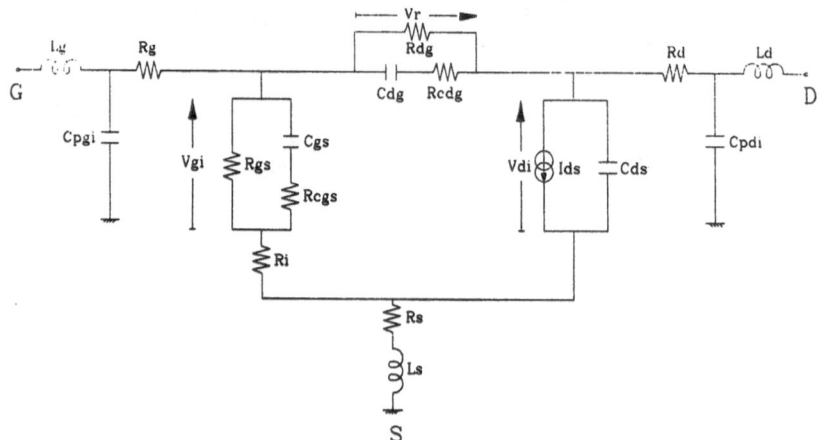

Fig. 14 - Non-Linear GaAs-MESFET Model

a) Schottky Junction: As we can see in figure 14, the Schottky junction is represented by a I_{gs} non-linear current source in parallel with a non-linear capacitor C_{gs}. These non-linearities along with the resistors R_g, R_i and R_s represent a real Schottky junction. The authors [1] [17] use different equations to represent I_{gs} and C_{gs} but the most common form of modelling these functions is using the Schottky diode equations given by (1) and (2):

$$
C_{gs} = \begin{cases} C_{gso}\left(1 - \dfrac{V_{cgs}}{V_{bi}}\right)^{-\gamma} + C_{gse} & V_{cgs} < K V_{bi} \\[4mm] C_{gso}\,(1-K)^{-\gamma} + C_{gse} & V_{cgs} \geq K V_{bi} \end{cases}
\tag{1}
$$

$$
I_{gs} = \begin{cases} I_{ns}\left(e^{\alpha_s V_{gi}} - 1\right) & \alpha_s V_{gi} \leq M_s \\[4mm] I_{ns}\left(e^{M_s}\left(1 + (\alpha_s V_{gi} - M_s) + \dfrac{(\alpha_s V_{gi} - M_s)^2}{2}\right) - 1\right) & \alpha_s V_{gi} > M_s \end{cases}
\tag{2}
$$

Where Vgi is the internal gate voltage.

Figure 4 shows the experimental I_{gs} curves. In this case I_{ns}= 4,023 pA and α_s= 32,17. In figures 15 and 16, the linear extraction from scattering parameters and the fitting using equation (2) respectively versus V_{gs}, V_{ds} external voltages, can be observed. Both kinds of measurement have been done for a F20 GEC-Marconi chip MESFET transistor.

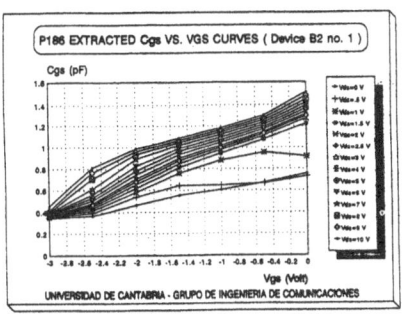 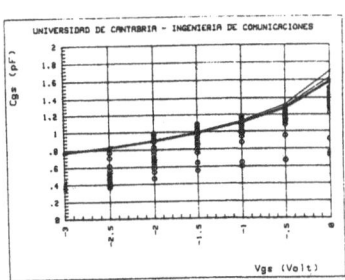

Fig. 15 - Linear extraction of C_{gs} Fig.16 - Fitting curves of C_{gs} using (2)

b) Breakdown Effect: In the same way as the Schottky junction, there are several ways of modelling this effect [18]. However, a very useful equation for modelling the I_{dg} Breakdown current is (1). In this case, the parameters of the equation are not typical of a Schottky diode and the variable is the drain gate reverse voltage, function of V_{gi} and V_{di}. Figure 17 shows the experimental and fitting curves of the Breakdown current of a MGF-1802 packaged transistor.

c) Non-linear modelling for the Ids Channel Current:

A lot of authors have developed very interesting models for the channel current source I_{ds} [17] [19]. A simple way of obtaining a non-linear model is to fit the I_{ds} equation to DC characteristics along with a resistor (R_{lf}) in parallel with the I_{ds} current source. R_{lf} is used to fit the output impedance, in this case the S_{22} parameter at different operating points.

Therefore R_{lf} is a bias dependent non-linear function given by equation (3). With this philosophy a non-linear model for a 14*170 μm GEC-Marconi MESFET chip, using the modified Materka approach, has been obtained. Figure 18 represents the experimental and simulated DC characteristics and in figure 19 we can observe comparisons between MDS simulations using the total extracted non-linear model and the experimental scattering parameters at the operating point V_{gs}=-2.4 V, V_{ds}=4 V.

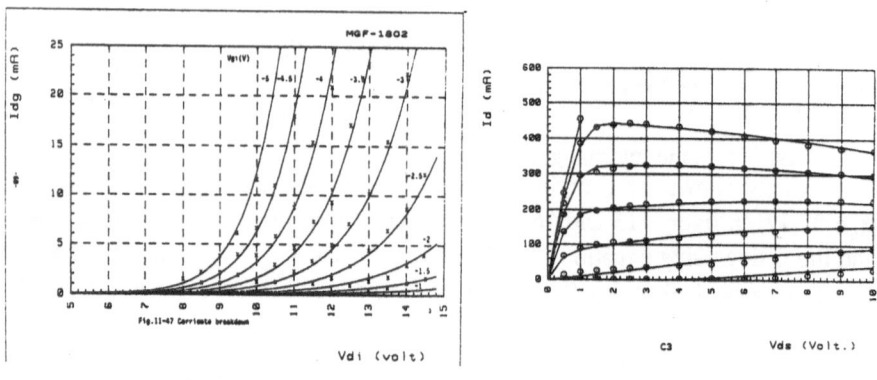

Fig.17 - Breakdown current	Fig.18 - Exp and fitted DC
(MGF-1802)	characteristics

Other way of modelling the current source can be to obtain the parameters of I_{ds} from pulsed I/V characteristics at a given operating point using, for instance, the same Materka equation. This single source model is only valid at the quiescent point where the characteristics have been measured. In this case it is not necessary to implement R_{lf} as the before case. Figure 20 shows the experimental and fitted characteristics at V_{gs}=-1 V and V_{ds}= 4 V for a 2*250 Siemens MESFET chip and figure 21 represents simulations and experimental results at the same operating point using the total non-linear model.

However, a simple non-linear equation for modelling this current source does not take into account phenomena such as the low frequency dispersion. The prediction of the correct working mode of a transistor requires a model capable of reproducing AC, low frequency and RF behaviour. The difference between the slopes of the DC curves and the transconductance and output conductance, extracted from Scattering parameter measurements, is very important with respect to the large signal behaviour of the transistor. To solve these problems, different options have been developed [2] [10] [11] [20].

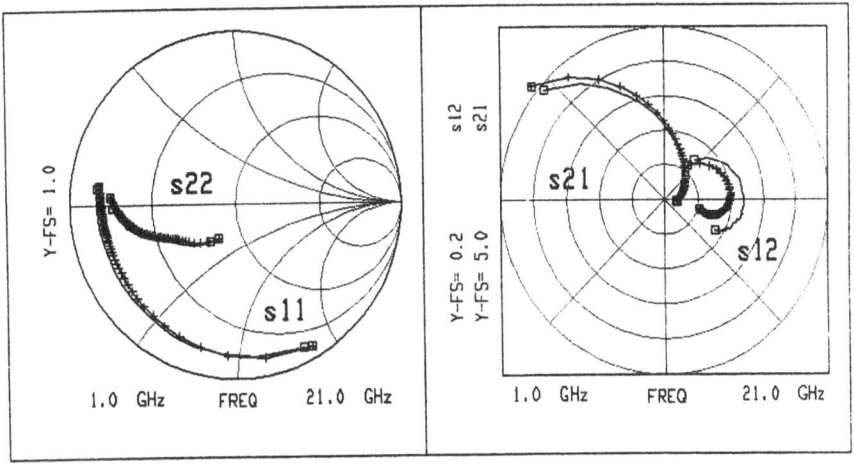

—— : Experimental
++++ : Model

Fig.19 - Experimental and simulated S parameters.

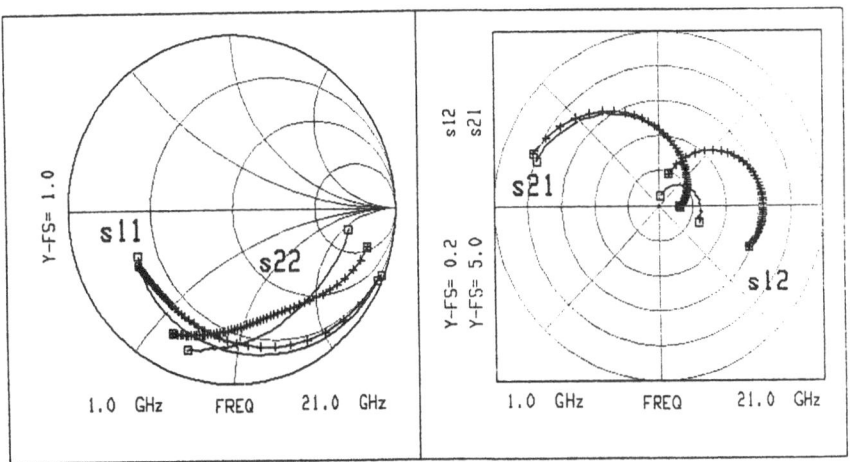

—— : Experimental
++++ : Model

Fig.20 - Experimental and simulated S parameters.

The most simple form of representing the low frequency dispersion is to introduce a series resistor-capacitor path parallel to I_{ds} current [2] (Fig.22). The capacitor function is to eliminate the resistor influence at DC operation, in this case I_{ds} are the static characteristics. This capacitor has a big value and it has no influence at high frequencies (above a few KHz). This RC branch can reproduce the variation versus the frequency of the output conductancebut it does not allow us to follow the frequency dispersion of the transconductance.

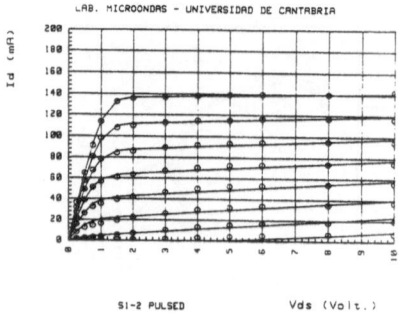

LAB. MICROONDAS - UNIVERSIDAD DE CANTABRIA

SI-2 PULSED   Vds (Volt.)

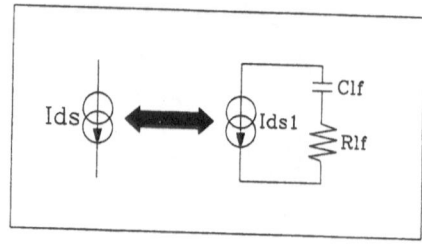

Fig.21 - Experimental and fitted pulsed characteristics

Fig.22 - I_{ds} current modelled with Output Resistance

To reproduce the frequency dispersion in both transconductance and output conductance it is necessary to develop more sophisticated models such as the double I_{ds} current source model [10] [11]. The idea is to change the resistor for another nonlinear current source (Fig.23). In this case, the source I_{ds1} represents the low frequency behaviour and I_{ds2} applies a correction to source I_{ds1} in order to simulate the RF operation. The total slope of the two sources g_m and G_{ds} is obtained from scattering parameter measurements.

A very interesting solution of double source model I_{ds} has been presented by [11]. In this case, I_{ds1} represents the DC characteristics in parallel with a non-linear source I_{ds2}. The characteristic of this second source is that the slopes of the curves are the difference between the slope of the DC characteristics and the RF slope at a particular operating point. The total I_{ds} can be written by equation (3).

$$I_{ds}(V_{gs}, V_{ds}) = I_{ds_{DC}}(V_{gs}, V_{ds}) + \left(g_{m_{RF}}(V_{gs_{DC}}, V_{ds_{DC}}) - g_{m_{DC}}(V_{gs_{DC}}, V_{ds_{DC}}) \right) V_{gs_{RF}} + \tag{4}$$
$$+ \left(G_{ds_{RF}}(V_{gs_{DC}}, V_{ds_{DC}}) - G_{ds_{DC}}(V_{gs_{DC}}, V_{ds_{DC}}) \right) V_{ds_{RF}}$$

$$I_{ds}(V_{gs}, V_{ds}) = I_{ds_{DC}}(V_{gs_{DC}}, V_{ds_{DC}}) + \tag{5}$$
$$+ \int_{V_{gs_{DC}}}^{V_{gs_{DC}}+V_{gs_{RF}}} g_{m_{RF}}(v, V_{ds_{DC}}) \, dv + \int_{V_{ds_{DC}}}^{V_{ds_{DC}}+V_{ds_{RF}}} G_{ds_{RF}}(V_{gs_{DC}}, v) \, dv$$

A second possibility is presented by [10]. This model is also a double nonlinear current source but it has two important differences: The most important is that the slopes of I_{ds2} are the difference between the slope of DC characteristics and the RF slopes at every point of the curves. These slopes are not constant as we have seen in the above case. The second difference is that the I_{ds} non-linear equation is derived for good fitting near pinchoff operating points. The mathematical form of the equation is given in (4). On the other hand, [21] presents the same idea for modelling I_{ds} but extracting the equation by integration methods.

d) Operating point dependent non-linear GaAs-MESFET model:
Comparisons between MDS simulations and experimental results of a packaged
GaAs-MESFET NE72084 transistor, charged by 50 Ohms at the input and output
ports in an operation point, using a non-linear model of simple I_{ds} current source
[4] have demonstrated better results when the model has been extracted from pulsed
measurements than using the quasi-static approximation (model extracted from S
parameters in the interest band). Furthermore, a non-linear current source model of
I_{ds} extracted from dc and pulsed measurements is capable of reproducing the low
frequency dispersion phenomena. However these models are only valid at one
operating point. Therefore, it is not possible to use these techniques to obtain a
universal model for reproducing the transistor behaviour in DC, low frequency
dispersion, S parameters and RF large signal at any operating point. That is a big
handicap for the designers and the foundries.

To solve this problem, we have developed a non-linear double current source
model of I_{ds} which, knowing its DC and pulsed characteristics at a few operating
points, reproduces the I_{ds} current behaviour in any operation point by only one
fitting. Therefore, the complete model of the transistor will be capable of
responding under DC, low frequency dispersion, S parameters and RF large signal
conditions, at any operation point. In figure 23 we can see the total I_{ds} current
source where:

I_{ds1} simulates the DC characteristic curves as a function of the internal
independent variables (V_{gi}, V_{di}). In this case, to model I_{ds1}, the Quadratic Curtice
equation has been used but the correct way of representing this non-linear current is
to use the equation of the literature with the best fitting.

I_{ds2} is a function of the independent voltages (V_{gi}, V_{di}) and the quiescent
operating point given by:

$$I_{ds2}(V_{gi}, V_{di}, V_{gcci}, V_{dcci}) = I_{ds}(V_{gi}, V_{di}, V_{gcci}, V_{dcci}) - I_{ds1}(V_{gi}, V_{di}) \qquad (6)$$

To represent the I_{ds} current, we have developed a non-linear equation not only
dependent on the internal variables (v_{gi}, v_{di}) but on the operation point V_{gcci} and
V_{dcci}. This allows the reproduction of the transistor behaviour at any operating
point. The total equation is given by equation (7).

C_{lf} is a linear capacitor that contributes to simulate the low frequency
dispersion.

$$I_{ds} = \frac{\beta \, [V_{gi} - V_t]^2 \ [1 + (\lambda_0 + \lambda_1 V_{dcci}) \, V_{di}]}{|V_{gcci} - 1|^P \ [|a| + V_{dcci}]^q} \ \tanh \left[\frac{\alpha \, V_{di}}{[|b| + V_{dcc}]^r} \right] \qquad (7)$$

Figures 24 and 25 show the experimental and fitted curves of DC and pulsed
characteristics of the a 6*50 μm GEC-Marconi GaAs_MESFET chip transistor.

In figure 26 comparisons between the simulated curves using the total model
and the experimental pulsed characteristics of the transistor at different quiescent
operating points can be seen. Taking into account that the model has been extracted
measuring characteristics at the quiescent bias point V_{gcc}=0 Volts, V_{dcc}=5 Volts,
it can be observed that our equation reproduces the curves of the transistor at any

one bias point showing good agreement with the experimental measurements. Furthermore, the total non-linear model has been implemented in the MDS nonlinear simulator.

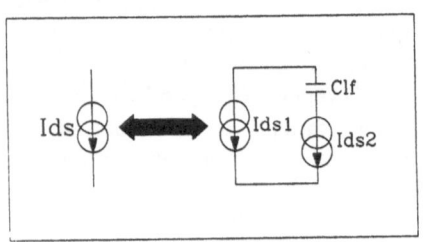

Fig.23 - I_{ds} current modelled by two non-linear sources

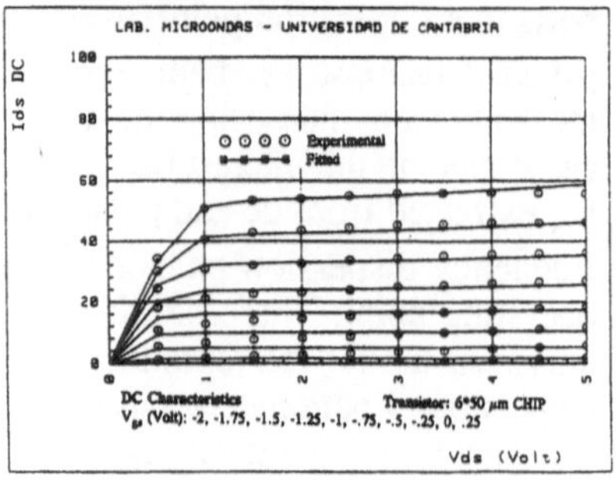

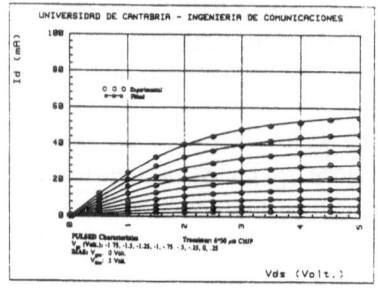

Fig.25 - Experimental and fitted pulsed curves

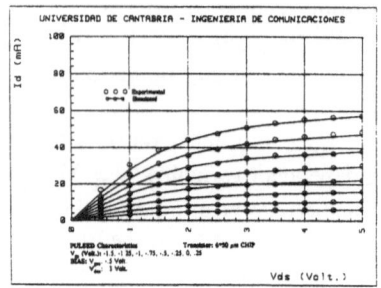

Fig.24 - Experimental and fitted DC characteristics

Fig.26 - Exp. and simulated pulsed curves

Figure 27 shows simulations of output power as a function of input power of the device loaded by 50 Ohms at the input and output ports. In figure 28 the experimental results which correspond to the same bias, frequency and power conditions have been represented showing good agreement.

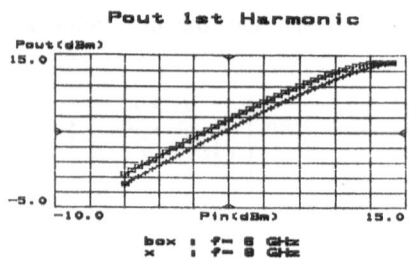

Fig.27 - Simulated Pout/Pin curves

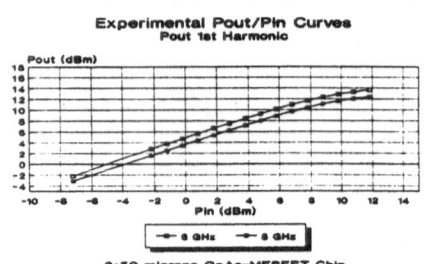

Fig.28 - Experimental Pout/Pin curves

273

Finally, a complete large signal MESFET equivalent circuit for a 6000 μm
D07A PML process chip power transistor, using the Materka model [6] has been
obtained. The model has been extracted from pulsed characteristics at the operating
point V_{gs}=-2.5 V, V_{ds}=3 V (figure 7). Figure 29 shows the simulations and
experimental results of the output power behaviour versus large signal analysis at
the interest operating point.

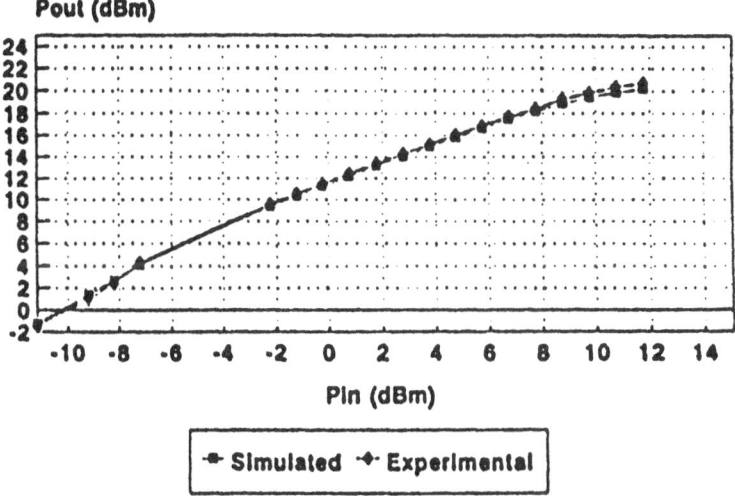

Fig.29 - Experimental Pout/Pin curves.

II - LOAD/SOURCE-PULL MEASUREMENTS

II.1) - Introduction

The load/source-pull is a measurement technique for the experimental
determination of the 'best' output and/or input load of a transistor. It is described
here for power applications, but it is used also e.g. for noise characterisation, being
then a source-pull technique. The simple underlying principle is to measure the
transistor in conditions as close as possible to the actual operating ones, i.e. with
the actual bias, input and output loads and power level as in the final circuit, and to
optimise them on the measurement bench by systematically scanning them or
iteratively searching the 'best' condition (e.g. maximum output power, maximum
efficiency, etc.)[23,24]. The set-up must replicate as accurately as possible the
actual circuit environment of the transistor and at the same time allow a precise
measurement of the quantities of interest, i.e. input and output power, impedance,
intermodulation, harmonic content, etc.

A general scheme for a load/source-pull system is given in figure 30 ; actual
implementations may differ for several characteristic features. They can have active
or passive loads, be manual or automatic, have an 'in-situ' evaluation of the
impedances or require switching systems or pre-characterisation, allow or allow not
control over impedances at harmonic frequencies, allow the measurement of only
output power and gain or also intermodulation, etc. In general they all require a
calibration procedure, that may or may not include fixture calibration. We will
examine these aspects in the following.

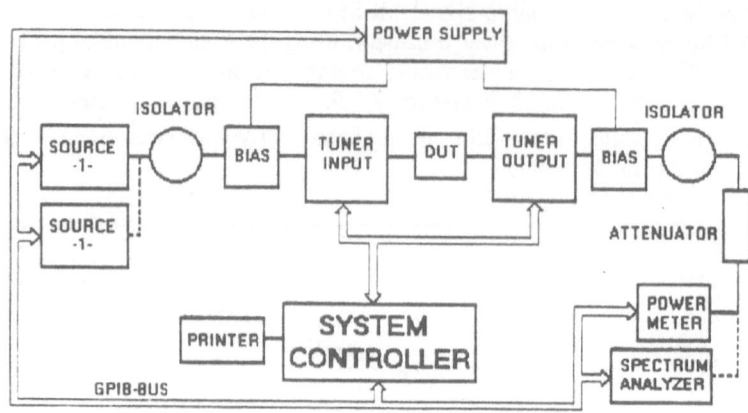

Fig.30 - A general load/source-pull set-up

II.2) - Passive and active loads

Passive tuners are two-port components that present a variable impedance at one port when loaded with 50 Ω at the other one. They are usually realised with double slugs in a coaxial cable, or with double stubs in waveguides (figure 31). The mechanical tuning elements are driven by motors controlled by a computer for automatic measurements. Passive tuners are simple, but their intrinsic losses do not allow the realisation of highly reflective loads, close to the edge of the Smith chart (figure 32). The losses vary with the impedance, requiring careful tuner characterisation if the transmitted power must be evaluated. Moreover, measurement uncertainty greatly increases when the impedance transformation ratio from 50 Ω increases. If the tuners are electronic, i.e. realised with PIN diodes, care must also be taken not to exceed their linearity region.

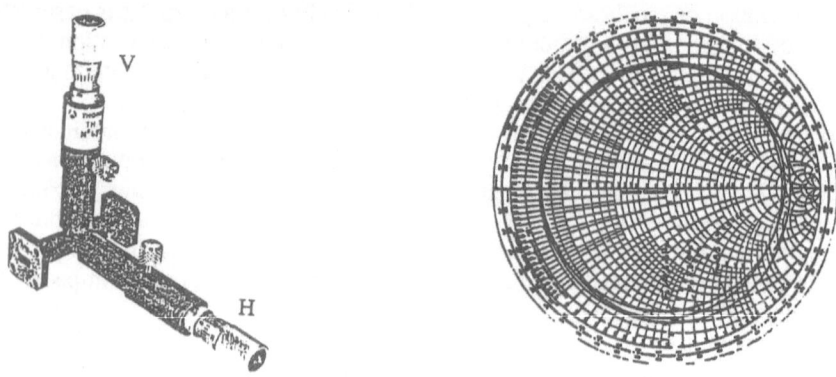

Fig.31 - A passive tuner   Fig.32 - Limit region of a passive tuner

Active loads are 'artificial' loads, obtained by injecting into the output of the transistor a wave with controlled amplitude and phase relations with the wave coming out of the transistor itself, so to 'synthesise' the desired reflection coefficient [25]. This is usually realised by splitting the signal of the microwave power source, and feeding one part to the input of the transistor and the other one to the output through a variable attenuator and phase shifter (figure 33). This system allows the realisation of reflection coefficients with unity amplitude; its practical implementation is however critical, especially at high frequencies, because all the components of the set-up must be kept soundly in place by the mechanical arrangement to avoid phase and amplitude fluctuations. Automatic measurements are easily obtained using controlled attenuators and phase shifters.

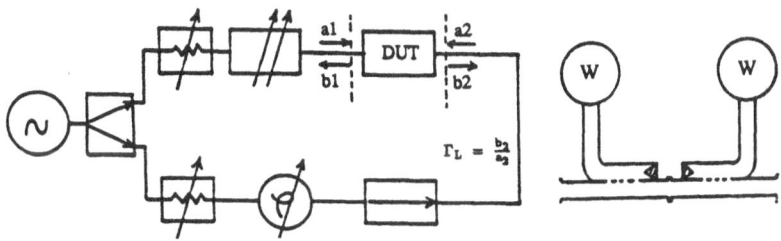

Fig.33 - The active load scheme   **Fig.34 - Incident- and reflected-wave power meters**

II.3) - Power and impedance measurements

The arrangement of the meters for the quantities of interest vary with the set-up. Incident/reflected input/output powers are usually evaluated by means of power meters on the coupled arm of directional couplers inserted in the main transmission path (figure 34). The power losses of the elements between the meters and the transistor are usually compensated via a calibration procedure; output power, power gain and efficiency are then easily computed.

If intermodulation is to be measured, two microwave power sources at neighbouring frequencies are combined at the input, and a Spectrum Analyser is placed, e.g., after the output tuner, at the end of the main transmission path (figure 30); Carrier-to-Intermodulation level can thus be evaluated.

There are three main ways of evaluating the load/source reflection coefficient: in the first and simplest one the passive tuners are pre-characterised with the aid of a Network Analyser for as many positions (i.e. points on the Smith chart) as possible, assuming perfect reproducibility and mechanical (or elecronic) control under measurement, and interpolating for positions not actually previously measured. The reflection coefficient is therefore 'a priori' known when setting the tuner. The measurements are very fast and reasonably accurate.

A second method is to place a switch at the output of the tuner, commutating between the transistor for the power measurement and a Reflectometer for impedance evaluation. The advantage of this arrangement is simplicity, but non-perfect reproducibility of the switch lowers the accuracy of the system; switching-off of the transistor is also required when the tuner is disconnected for safety of operation. The procedure turns out to be rather complicated and inaccurate.

276

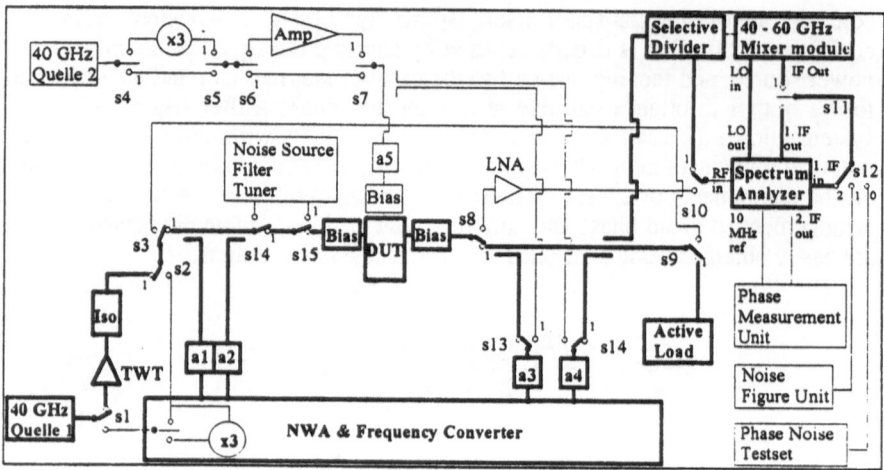

Fig.35 - A combined measurement test set-up

A third method is the 'in-situ' sampling of incident/reflected waves by means of directional couplers inserted between the transistor and the tuner; a Network Analyser (equipped with a four-port converter) is connected to the coupled arms (figure 35, a combined test set-up from IMST). This is also the only possible method of impedance evaluation with an active load system. No switching or pre-characterisation is required, and reproducibility is intrinsically ensured; however, since the couplers do not work in a 50-Ω system, a complete two-port calibration and error-correction procedure is required for the correct evaluation of the impedances (figure 36).

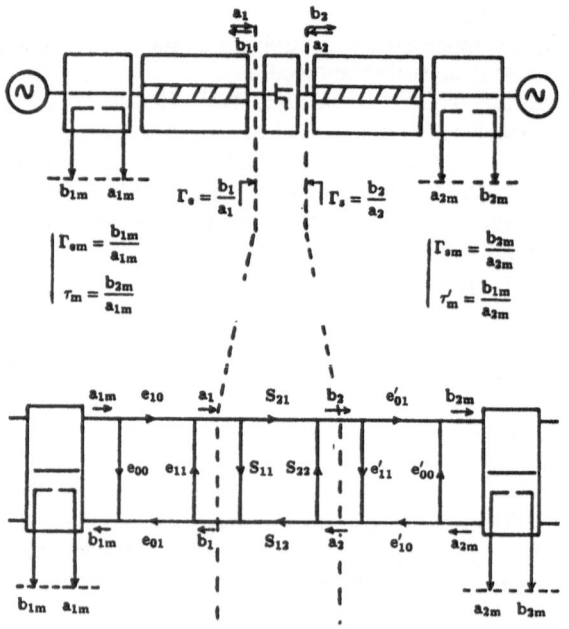

II.4) - Harmonic loading

The harmonic impedances are controlled by using multiplexers at the output (and input, if required) of the transistor; individual tuners, or individual active-load paths equipped with frequency multipliers for reflected-wave generation, must be used for each harmonic frequency. In case no harmonic control is envisaged, $50\text{-}\Omega$ loading is usually provided to the harmonics by the isolators or circulators in an active load system, if their bandwidth is large enough; unpredictable impedances are on the other hand usually presented by passive tuners. It must be remembered that harmonic control can affect the results for a 10+15%.

II.5) - Calibration

Calibration of the set-up is performed with either microstrip or coaxial (or waveguide) standards. In the first case the standards are usually less accurate and more difficult to realise, but include the correction of fixture non-idealities within the calibration; for on-wafer measurements coplanar standards are easy to fabricate, but the high probe losses can affect the accuracy of the calibration. In the case of coaxial or waveguide standards calibration is performed at the plane of the fixture, that must be separately characterised and de-embedded. A TRL-type of calibration proves to be suitable for error correction [26].

III - PHYSICAL MODELLING

Device simulation based on physical model is an extremely useful tool for linking device design and technology to device behaviour and performance, especially for MMIC's. A number of physical models based on several simulation approaches are already established since many years [27]; however, their application to circuit performance prediction and optimisation is still very limited. The main obstacle to the integration of device modelling and circuit analysis lies in the very high CPU-time requirements of the coupled simulation; this is especially true for large-signal simulation.

Physical models are based on the semi-classical Boltzmann transport equation [28]. A detailed description at collision level of the transport phenomena can be obtained from its solution with statistical Monte Carlo techniques [29]. This approach gives accurate results but involves high computational costs; instead, macroscopic transport equations based on a phenomenological description of collisions are usually solved with deterministic approaches.

Most models of this group are derived by taking moments of the Boltzmann equation to obtain a set of conservation equations, and then making additional assumptions [30]. They are usually divided into energy transport (or hydrodynamic) models and drift-diffusion models, depending on the inclusion or exclusion respectively of the energy conservation equations. Energy transport models [31,32] are less efficient from a computational point of view, but accurately describe overshoot and energy-dependent phenomena; drift-diffusion models on the other hand still show good agreement with experiments for short devices down to 0.5 μm, while being comparatively faster [33]. In both cases equations can be considered in two dimensions (i.e. on a section of the device) [32] or in only one

278

dimension [34,35,36], assuming that the transport phenomena are essentially one-dimensional from source to drain, and assuming a simplified scheme for depletion depth calculation (the so-called quasi-2D model). Further simplifications can eventually lead to an analytical solution [37]. Further modelling issues are the description of phenomena such as surface effects and buffer traps, and thermal behaviour.

The inclusion of any such model into a non-linear circuit analysis algorithm can lead to severe problems. Non-stationary large-signal physical simulation is usually performed in the time domain (e.g. SPICE) or Harmonic Balance (e.g. LIBRA, MDS, Microwave Harmonica, etc.) non-linear circuit analysis algorithm. However, the device phenomena have much smaller time constants than the embedding microwave circuit and consequently a much shorter time step for the numerical integration; an extremely high number of iterations of the CPU-time consuming physical model is therefore required for the analysis of a microwave period.

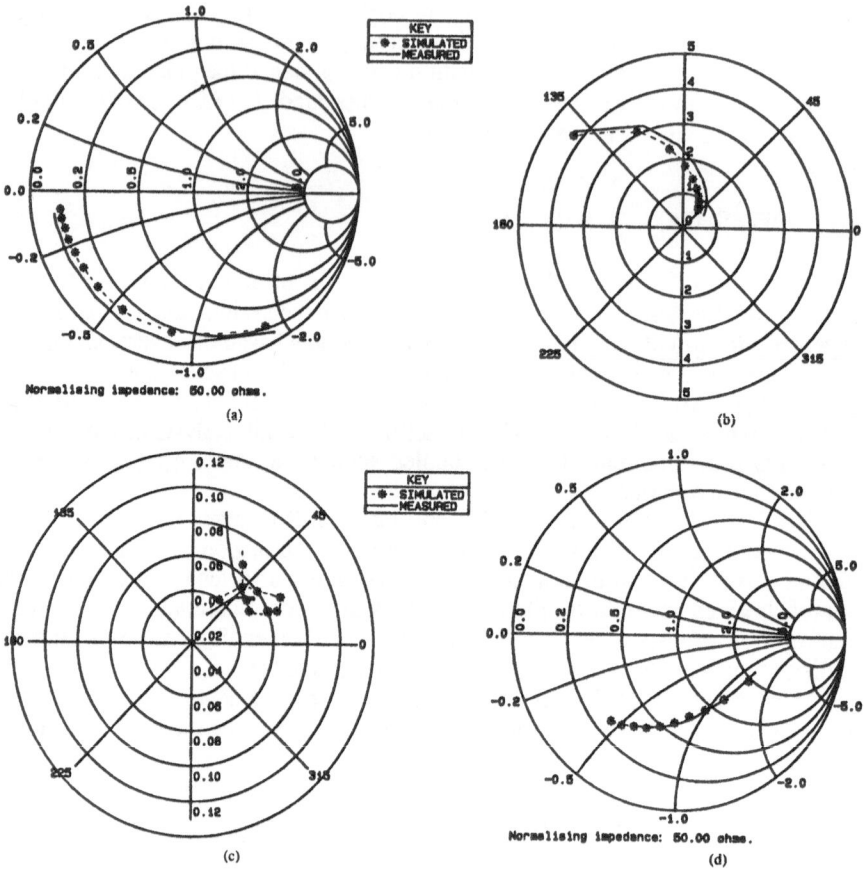

Fig.37 - Simulated and measured S-parameters of a power
MESFET

Two solutions have been so far proposed: in the first case, a linearised small-signal analysis of the device is performed at many bias points, together with the simulation of the DC I-V characteristics, and a large-signal equivalent-circuit model is extracted very much as from measured data. This procedure allows a very fast circuit analysis as well as an accurate device characterisation, and an easy interfacing between physical models and existing CAD tools. An advanced two-dimensional drift-diffusion model [33] and a quasi-two-dimensional hydrodynamic model [36] have so far been used for the preliminary 'measurement-like' intrinsic device simulation, with satisfactory results. The small-signal parameters of the device are in both cases close to measurement results, and the large-signal simulations show good agreement with experiments (figure 37 and figure 38, from [36]).

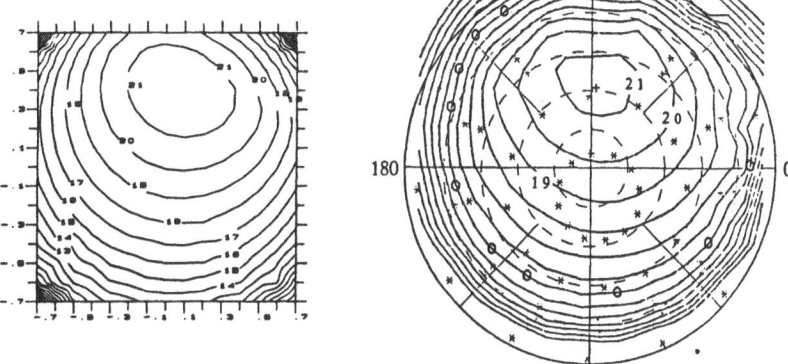

Fig.38 - Simulated (left) and measured (right) constant output power contours

In the second approach a quasi-two-dimensional hydrodynamic model in the time domain is directly linked to a harmonic-balance algorithm. The simplifying assumption is made that the intrinsic transitory phenomena of the device can be neglected, since the (microwave) driving voltages and currents have much longer time constants; the same time step for the analysis of the circuit is therefore used also for the intrinsic device[38,39]. Circuit analysis is much slower than in the previous case because of the greater computational burden of the physical model with respect to an equivalent circuit, but variables like temperature and traps occupancy which depend on the large-signal working conditions of the circuit can be consistently adjusted during the non-linear analysis (figure 39, from [39]). Results are in good agreement with experiments (figure 40 and figure 41, from [39]).

Thermal modelling has also been investigated intensively [40]. The models belong to two main types: the thermal resistance models, where the source of heat is taken as known from the simulation of transport phenomena and only the heat equation is solved [41], and the thermal and transport coupled models, where the transport and heat diffusion equations are consistently and simultaneously solved [42]. While the second group exhibits greater accuracy and physical consistency, models belonging to the second group are much faster, and can be used for a coupled non-linear thermal and circuit analysis [43] in cases when the temperature cannot be assumed to be constant in time (e.g. pulsed operating conditions).

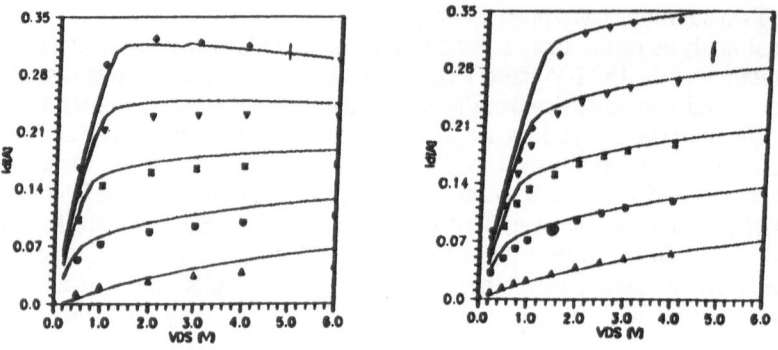

Fig.39 - Simulated and measured DC (left) and pulsed (right) I-V curves

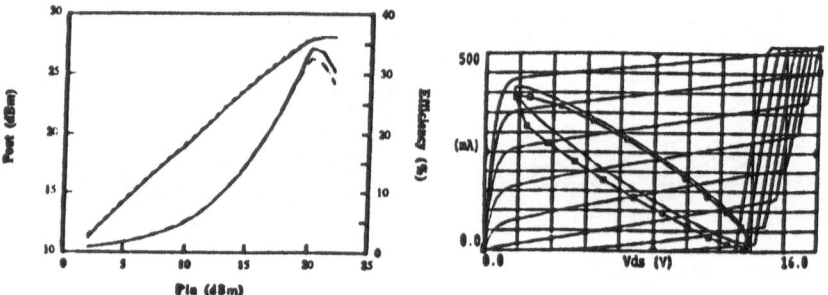

Fig.40 - Simulated and measured output power and efficiency

Fig.41 - Simulated and measured load curves

REFERENCES

[1] C. Rauscher, H.A. Willing, "Simulation of Nonlinear Microwave FET Performance Using a Quasi-Static Model", IEEE Trans. MTT-27, No.10, October 1979.

[2] C. Camacho-Peñalosa, C. S. Aitchinson, "Modelling Frequency Dependence of Output Impedance of a Microwave MESFET at Low Frequencies", Elect. Letts., Vol 21, N° 12, 1985, pp. 538-539.

[3] E.Allamando, Y.Bonnaire, "Nonlinearities, of GaAs Submicrometer FET: New Model of Characterization and Modelization", 18th European Microwave Conf., pp.243-248, Stockholm (Sweden), Sept. 1988.

[4]   T. Fernández, A. Mediavilla, A. Tazón, J.L. García "Low Frequency Dispersion Measurements for Nonlinear Microwave MESFET Modelling", GaAs'92 European Gallium Arsenide Symposium. ESTEC, Noordwijk, The Netherlands, 27-29 April 1992.

[5]   T.Fernández, Y.Newport, A.Tazón A.Mediavilla, "Extracting Advanced Large Signal MESFET Models from DC, AC and Pulsed I/V Measurements", MIOP'93, Sindelfingen, Germany, May 1993.

[6]   HSPICE User's Manual, Version H8907, Meta-Software, Campbell, CA, 1989 (Materka Model).

[7]   J.M.O'Callaghan, J.B.Beyer," A Large Signal Nonlinear Model from Small Signal S- Parameters", IEEE MTT-S Digest, 1989, pp. 347-350.

[8]   M. Paggi, P. H. Williams and J. M. Borrego, "Nonlinear GaAs MESFET Modelling Using Pulsed Gate Measurements", IEEE MTT Symp. Digest, 1988, pp. 229-231.

[9]   A. Platzker, A. Palevsky, S. Nash, W. Struble and Y. Tajima, "Characterization of GaAs Devices by a Versatile Pulsed I-V Measurement System", IEEE MTT-S Digest, 1990, pp.1137-1140.

[10] Thomas J. Brazil, "A Universal Large-Signal Equivalent Circuit Model for the GaAs MESFET", 21st European Microwave Conf., pp. 921-926, Stuttgart (Germany), Sep. 1991.

[11] H. Sledzik and I. Wolff, "Large Signal Modeling and Simulation of GaAs MESFETs and HFETs", Int. Journal of Mic. M.-W. Comp.-Aided Eng., Vol.2, Nº1, pp.49-60, 1992.

[12] H. Fukui,"Determination of the Basic Parameters of a GaAs MESFET", Bell Syst. Tech. J., 58(3),1879,pp.772-797

[13] T.Fernández, Y.Newport, J.M.Zamanillo, A.Mediavilla, A.Tazón, "High Speed Automated Pulsed I/V Measurement System", 23rd European Microwave Conf., Madrid (Spain), September 1993.

[14] G. Dambrine, A. Cappy, F. Heliodore, E. Playez, "A New Method for Determining The FET Small-Signal Equivalent Circuit", IEEE Trans. on MTT, Vol. 36, Nº 7, 1988.

[15] E. Arnold, Richard Golio, Monte Miller and Bill Beckwith, "Direct Extraction of GaAs MESFET Intrinsic Elements and Parasitic Inductance Values", MTT-S Digest, 1990, pp. 354-362.

[16] M. Berroth and R. Bosch, "High Frequency Equivalent Circuit of GaAs FETs for Large- Signal Applications", IEEE Trans. MTT, Vol. 39, Nº 2, Febr. 1991, pp. 224-229.

[17] T.Kacprzak, a.Materka, "Compact DC Model of GaAs FET's for Large-Signal Computer Calculation", IEEE J. Solid-State Circuits, Vol.SC-18, April 1983, pp.211-213.

[18] T.A. Winslow, D.Fan, R.J.Trew, "Gate-Drain Breakdown Effects Upon the Large Signal Performance of GaAs MESFETs", MTT-S Digest 1990, pp.315-318.

[19] W. R. Curtice and M. Ettemberg, "A Nonlinear GaAs Fet Model for Use in the Design of Output Circuits for Power Amplifiers", IEEE MTT-33, Dec 1985, pp.1383-94.

[20] T.Fernández, Y.Newport, J.M.Zamanillo, A.Tazón A.Mediavilla, "Modelling of Operating Point Non Linear Dependence of Ids Characteristics from Pulsed

Measurements in MESFET Transistors", 23rd European Microwave Conf., Madrid (Spain), September 1993.

[21] V.Rizzoli, A.Constanzo, A.Neri, "An Advanced Empirical MESFET Model for Use in Nonlinear Simulation", 22nd European Microwave Conf., pp.1103-1108, Sept. 1993.

[22] F.Filicori, G.Vanini, A.Mediavilla, A.Tazón, "Modelling of Deviations Between Static and Dynamic Drain Characteristics in GaAs FETs", 23rd European Microwave Conference, Madrid (Spain), September 1993.

[23] F.N.Sechi, 'Design procedure for high-efficiency linear microwave power amplifiers', IEEE Trans. on Microwave Theory Tech., vol. 28, n. 11, pp. 1157-1163, Nov. 1989

[24] J.M.Cusack, S.M.Perlow, B.S.Perlman, "Automatic load contour mapping for microwave power amplifier", IEEE Trans. Microwave Theory Tech., vol. MTT-22, n. 12, pp. 1146-1151, Dec. 1974

[25] Takayama, "A new load-pull characterisation method for microwave power transistors", IEEE 1976 MTT-S Digest, pp.218-220, June 1976

[26] A.M.Khilla, "Accurate measurement of high-power GaAs FET terminating impedance improves device characterisation", Microwave J., May 1985, pp.255-263

[27] S.Selberherr, *Analysis and Simulation of Semiconductor Devices*, Vienna, Springer Verlag, 1984

[28] K.Bløtekjær, "Transport equations for electron in two-valley semiconductors", IEEE Trans. Electron Devices, vol. ED-17, pp.38-47, Jan. 1970

[29] C.Jacoboni, L.Reggiani, "The Monte Carlo method for the solution of charge transport in semiconductors with applications to covalent materials", Rev. Mod. Phys., vol. 55, n. 3, pp.645-705, July 1983

[30] P.A.Sandborn, A.Rao, P.A.Blakey, "An assessment of approximate nonstationary charge transport models used for GaAs device modelling", IEEE Trans. Electron Devices, vol. ED-36, n. 7, pp. 1244-1253, July 1989

[31] W.R.Curtice, Y.H.Yun, "A temterature model for the GaAs MESFET", IEEE Trans. Electron Devices, vol. ED-28, pp.954-962, Aug. 1981

[32] C.M.Snowden, D.Loret, "Two-dimensional hot-electron models for short-gate-length GaAs MESFET's", IEEE Trans. Electron Devices, vol. ED-34, pp.212-223, Feb. 1987

[33] G.Ghione, C.U.Naldi, F.Filicori, "Physical modelling of GaAs mesfet's in an integrated CAD environment: from device technology to microwave circuit performance", IEEE Trans. Microwave Theory Tech., vol. MTT-37, n. 3, pp.457-468, March 1989

[34] B.Carnez, A.Cappy, A.Kaszynski, E.Constant, G.Salmer, "Modelling of a submicron gate field-effect transistor including effects of nonstationary electron dynamics", J.Appl. Phys., vol. 51, pp.784-790, 1980

[35] P.A.Sandborn, J.R.East, G.I.Haddad, "Quasi-two-dimensional modelling of GaAs MESFET's", IEEE Trans. Elctron Devices, vol. ED-34, pp. 985-991, 1987

[36] R.R.Pantoja, M.J.Howes, J.R.Richardson, C.M.Snowden, "A large-signal physical MESFET model for computer-aided design and its applications", IEEE Trans. Microwave Theory Tech., vol. MTT-37, n. 12, pp.2039-2045, Dec. 1989

[37] R.J.Trew, J.B.Yan, D.E.Stoneking, "GaAs power MESFET performance sensitivity to profile and process parameter variations", IEEE Trans. Microwave Theory Tech., vol.MTT-36, pp.1873-1876, Dec. 1988

[38] G.Halkias, H.Gérard, Y.Crosnier, G.Salmer, "A new approach to the RF power operation of MESFET's", IEEE Trans. Microwave Theory Tech., vol. MTT-33, pp. 817- 825, 1985

[39] F.Giannini, G.Leuzzi, M.Kopanski, G.Salmer, "Large-signal analysis of quasi-2D physical model of MESFET's", Electron. Lett., to be published

[40] A.Benvenuti, F.Bonani, G.Ghione, C.U.Naldi, "Thermal models for III-V devices and FET thermal simulations", ESPRIT Project 6050 'MANPOWER', Report ED09/ED10, June 1993

[41] G.N.Ellison, "Methodologies for thermal analysis of electronic components and systems", in *Advances in thermal modelling of electronic components and systems*, A.Bar-Cohen and A.D.Kraus ed., IEEE Press, New York, 1993

[42] L.L.Liou, J.L.Ebel, C.I.Huang, "Thermal effects on the characteristics of AlGaAs/GaAs heterojunction bipolar transistors using one-dimensional simulation", Solid-State Electron., vol. 35, pp.579-585, 1992

[43] V.Rizzoli, A.Lipparini, V.Degli Esposti, F.Mastri, C.Cecchetti, "Simultaneous thermal and electrical analysis of nonlinear microwave circuits", IEEE Trans. Microwave Theory Tech., vol. MTT-40, n. 7, pp.1446-1455, July 1992

4.7   Coplanar Waveguide Modelling

R. Kulke and T. Sporkmann

Institut für Mobil- und Satellitenfunktechnik, D-47475 Kamp-Lintfort, Germany

The theoretical analysis of coplanar waveguide structures is almost as mature as the analysis of conventional planar microstrip structures /4.5.1-4.5.4/. However, even today only very few companies utilise CPW technology for their circuit design. On a hybrid coplanar circuit design it has been shown for instance, that coplanar circuits allow up to 30% of space reduction compared to microstrip circuits. On the other hand, the successful design of complex coplanar MMIC's has also been demonstrated. Despite these success stories and despite some of the natural advantages of CPW structures this has not lead to a broad acceptance of this technology. One part of the problem is, that most of the numeric simulators, capable to predict complex planar structures, are only 2D or 2.5D simulators. In the CPW case however, especially with finite ground width, it is absolutely necessary to support real 3D simulation since a multi-layer airbridge technology or bond wires between the ground are mandatary for high frequency circuit design. Despite all this experience, no commercial CAD program offers a complete range of coplanar discontinuities up to now.

Aim of the project. A small portion of the ESPRIT-CLASSIC project is, to establish coplanar waveguide elements within a European CAD environment. The 3D finite difference algorithms described in /4.5.3/ and /4.5.4/ are utilised as basis of this work. For the application of numerical techniques within a CAD environment, and especially for circuit optimisation or synthesis it is clear, that a parametric description of any discontinuity is compulsory.

For this reason, the generic numerical procedure has been adapted to pre-defined parameterised structures for coplanar discontinuities such as opens, a short, corner, step, gap, tee and a cross. In addition, three different airbridge types are already included in the parameteric description of the coplanar discontinuities. Figure 4.5.1 depicts the various types of airbridges. All three types are available for the elements bend, tee junction and cross. Airbridge and underpass (types 1 and 2 of figure 4.5.1) may be connected at any position in a circuit to compress the coplanar odd mode. In addition to the discontinuities and the airbridges some

---

[1] "Institut für Mobil- und Satellitenfunktechnik", Moerser Straße 316, D-4132 Kamp-Lintfort

[2] PROJECT 6016: "Components for Large Signal Sixty GHz GaAs Integrated Circuits"

coplanar lumped elements like the interdigital and MIM capacitor, the spiral inductor, the transformer, the thin film resistor and the capacity line complete the catalogue of coplanar elements. A model for the simulation of coplanar coupled lines is in preparation. Table 4.5.1 (at the end of the chapter) gives an overview of the available models, the proper equivalent circuits, the schematic symbols and an example of the layout.

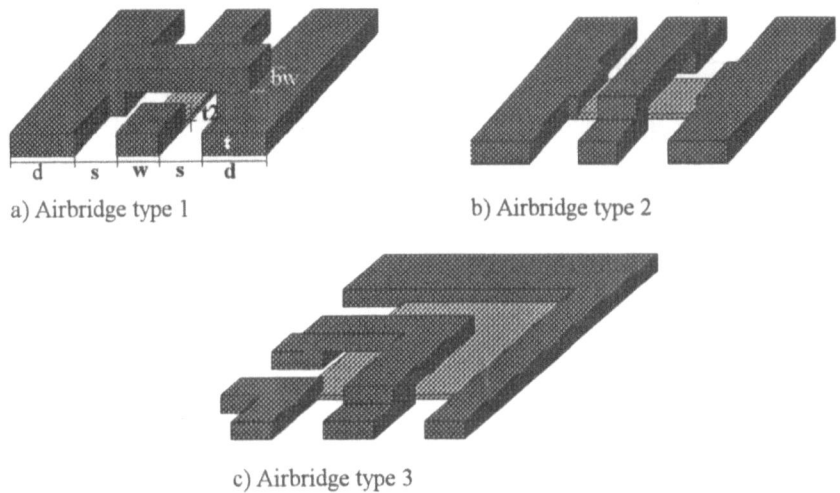

a) Airbridge type 1   b) Airbridge type 2

c) Airbridge type 3

Fig. 4.5.1. Three different types of airbridges available:
    a) underpass of center line,
    b) underpass of ground connection,
    c) metallization under discontinuity

Equivalent circuits constituted out of lumped capacitors, inductors or resistors are then calculated from these parameterically described discontinuities. The values of these equivalent circuits are then stored and maintained within an n-dimensional data bank. The correlation between the geometry's and the equivalent circuits is kept also as information within this data bank. Thus, efficient circuit analysis as well as optimisation even based on accurate interpolations becomes possible. Figure 4.5.2 gives a global overview of the projects aim.

The individual C-routines for the CPW elements are represented by the blocks on the left hand side of figure 4.5.2. The library block is then followed by a sophisticated interface to a European and finally also to US-CAD tools. The interface includes the numerical adaptation to the mentioned programs, the n-dimensional data bank management for quick analysis and interpolation as well as layout and schematic definitions for the individual CPW elements. The justification of this work is placed on the right hand side of figure 4.5.2.

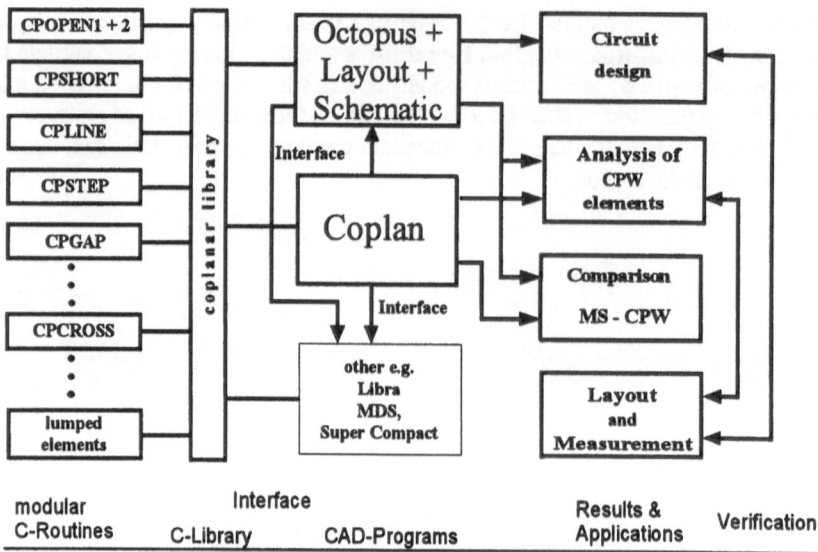

Fig. 4.5.2. Block diagram showing the implementation of the CPW elements into a CAD environment

It can be seen, that parametric studies of the CPW components will take place as well as a circuit design for a 30GHz to 60GHz up-converter. In addition, the microstrip and coplanar technologies will be compared side by side on a numerical and measurement basis. To do so, the IMST developed in co-operation with Daimler Benz and DASA a GaAs wafer with passive coplanar components and simple circuits to verify the above described models by measurement. Figure 4.5.3 depicts one half of the wafer with zoomed sections of the 4 main groups for the measurements. In addition to these groups, there exist some more 'special' structures to determine for example a coplanar to slotline transition, the influence of the coplanar odd mode or a typical broadband bias network. The wafer design has been completed, thus, the measurements can start immediately after the wafer fabrication.

Justification. It is clear that CPW circuits feature several advantages compared to conventional planar microstrip circuits. However, the hybrid case must be compared independently from the foundry oriented circuit design. For this summary the focus is on MMIC design where multi layer technology is standard and thus, airbridges or under-paths can be applied without any expenses. In fact the thinning of the wafer and the complete backside processing including the expensive via hole etching can be avoided by using CPW circuits. Therefore, not only the size of the circuit can be reduced, but also the expenses for the technology runs can be minimised. In addition, the introduction of many via holes within even a simple circuit decreases the yield of most circuits by a large amount.

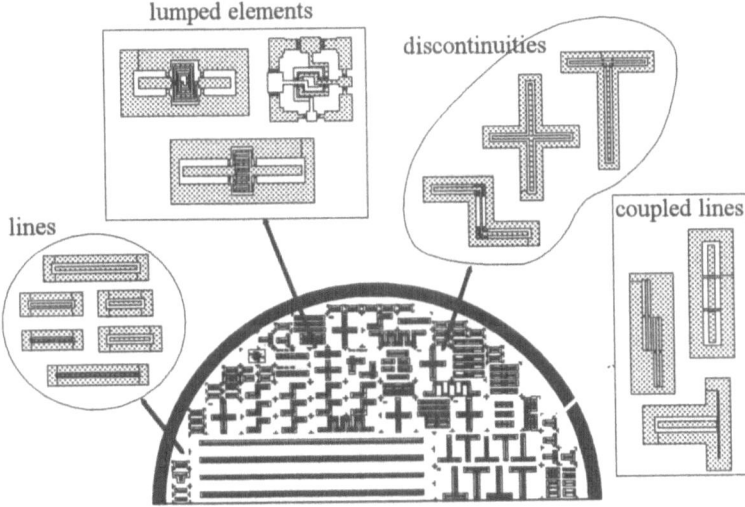

Fig. 4.5.3. GaAs test wafer with passive coplanar circuits - the zoomed sections indicate the main groups for the measurement

Other advantages such as reduced line-to-line coupling, small shielding sensitivity and efficient ground contact to the active devices favour the application of the CPW technology. This numerically verified behaviour is in contrast with the common understanding of the CPW's behaviour. Up to know there is a broad acceptance, that the field of CPW lines are less confined than those of microstrip lines. This would result in an increase of the environmental sensitivity (shielding for instance). Our work as well as the work in /4.5.5/ shows, that the opposite is the case. It is true however, that CPW lines have slightly higher loss than microstrip lines, but with a reduced circuit size this deficiency can be compensated for. The only main disadvantage of circuit designs utilising coplanar lines is the fact, that many CPW discontinuities excite odd mode waves which then travel on the adjacent CPW lines almost without absorption. This requires special care for the circuit design phase.

Results. An intensive parametric analysis of all coplanar discontinuities has been made. To do so, the IMST uses in addition to the above described software package, which includes the finite difference formulation, three further programs. One of them is the program "Stingray", which solves 2 dimensional problems of multiple coupled lines in arbitrary layers. The basic theory in thid program is the method of finite elements. Figure 4.5.8 depicts the coupling capacitance of two coupled lines each in microstrip and coplanar technique. Full wave analysis has been made by utilising Spectral Domain as well as Finite Difference Time Domain formulation. The curves in figure 4.5.7, 4.5.9 and 4.5.10 show the frequency dependent behaviour of the excitation of higher modes and the matrix coefficient S_{41} of coupled lines. More results of Spectral Domain and FDTD were published in /4.5.5/ and /4.5.6/. The following diagrams are an extraction of the analysis, which have been made in the past. We wish to exhibit, that the software packages are able to describe the full behaviour of all coplanar discontinuities.

Figure 4.5.4 depicts the parametric calculation of the characteristic impedance's of the coplanar waveguide as a function of line width and the spacing to ground. It becomes evident, that there is a great range of characteristic impedance's (30Ω-90Ω) and plenty line configurations for a single impedance value.

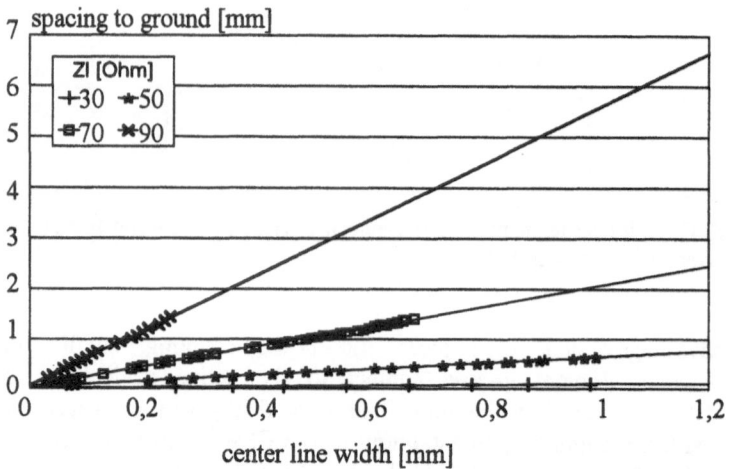

Fig. 4.5.4. Parametric analysis of a coplanar waveguide: Characteristic impedance ε_r=12.9, h1=600, h2=1500, h3=1000μm, t=3μm, d=1000μm

Figure 4.5.5 depicts the change in the characteristic impedance for microstrip and coplanar lines as function of the shielding distance h_2. It can be seen, that for the microstrip case the shielding must be 10 times the substrate thickness away from the line in order not to affect the characteristic impedance anymore. In the CPW case, a factor of only 2 is needed. Thus, the distance to the shielding in the coplanar case can be much closer.

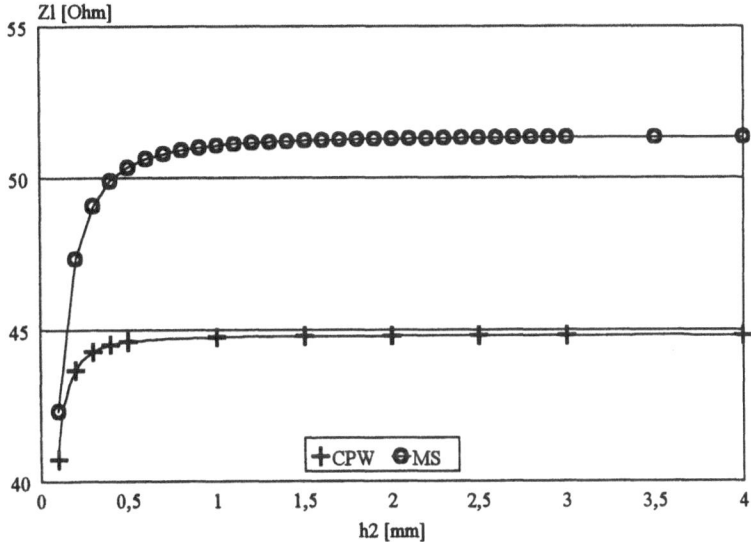

Fig. 4.5.5. Comparison of CPW with microstrip lines as a function of shielding (h_2=distance to cover plate), Substrate parameter:
MS-line: h=150μm; ε_r=12.9; w=100μm
CPW-line: h=600μm; ε_r=12.9; w=100μm; s=50μm; d=3000μm

One goal of our work is the comparison of coplanar and microstrip structures. Such a calculation is given in figure 4.5.6. Two structures have been analysed. The CPW case in figure 4.5.6.a has about half the size of the microstrip structure in 4.5.6.b. It can be see, that such a structure has only little dispersion even at 60GHz while the microstrip TEE clearly shows strong dispersion effects. Selecting a CPW-50Ω line with dimensions as in the micrstrip case would lead to a similar frequency response. As a result of this investigation, we can say, that due to the additional degree of freedom, the dispersion in the CPW lines can be shifted towards higher frequencies. This investigation was made with an airbridge of type 3 (metallization area under the discontinuity) to suppress higher wave modes, although the Finite Difference formulation gives no information about the excitation of the odd mode. To compare various airbridges, a Spectral Domain simulation of a coplanar bend with three different type of airbridges was made.

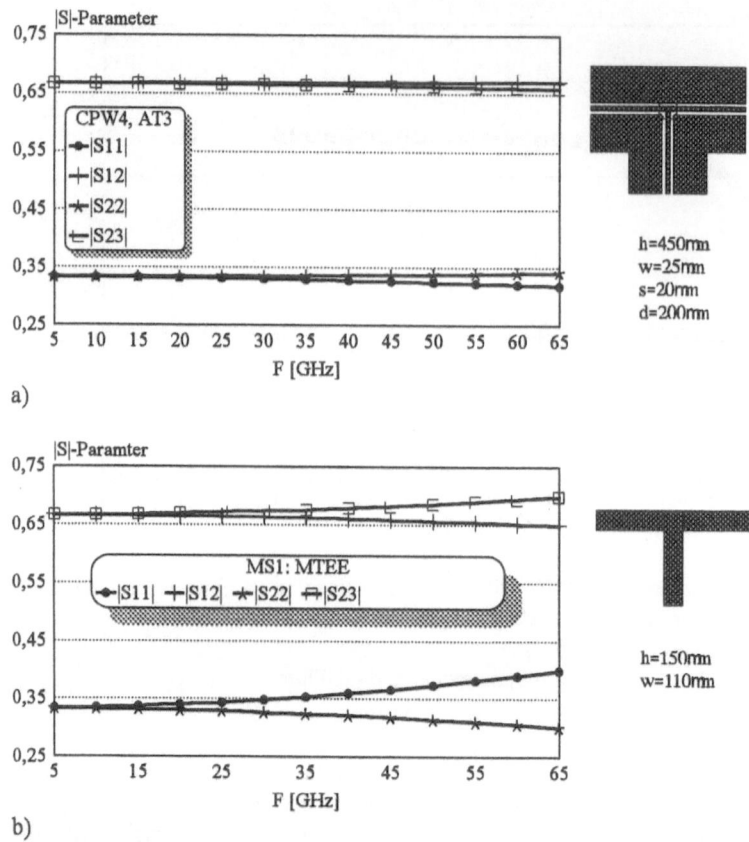

Fig. 4.5.6. CPW-Tee in comparison with a MS-Tee (50Ω configurations)
a) CPW-Tee with airbridge of type 3
b) MS-Tee

291

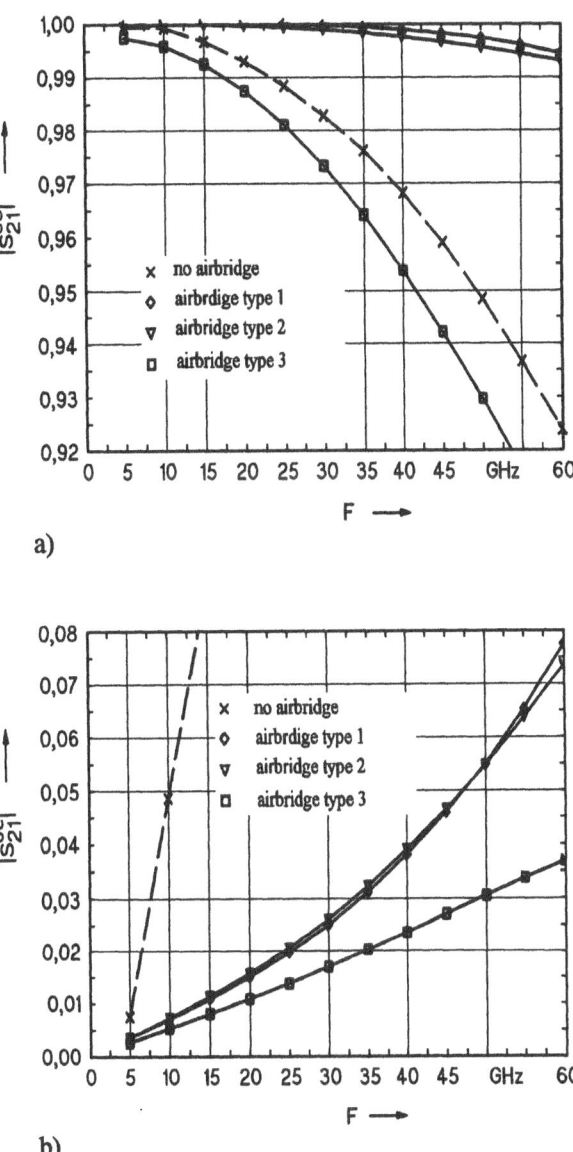

a)

b)

Fig. 4.5.7. Spectral domain analysis of the coplanar bend under the consideration of even/odd mode of excitation/transmission and different types of airbridges (line parameters: ε_r=12.9, substrate height h=200μm, line width at port 1 and 2 w=75μm, distance to ground s=50μm and ground width d=200μm
a) excitation even - transmission even
b) excitation even - transmission odd

The results are presented in figure 4.5.7.a and b. We consider the transmission behaviour of the bend for two cases (even and odd) of excitation and transmission. Figure 4.5.7.a (even excitation at port 1 - even transmission to port 2) shows, that the airbridges of types 1 and 2 have the best transmission behaviour. Using no airbridge or airbridge type 3, the even mode will be attenuated by the bend. The next survey was made with an excitation of the even mode at port 1. At port 2 we consider the odd mode. When using the coplanar bend without airbridge, the odd mode is excited at this discontinuity even at lower frequencies. Here, the best transmission behaviour is achieved with the airbridge type 3.

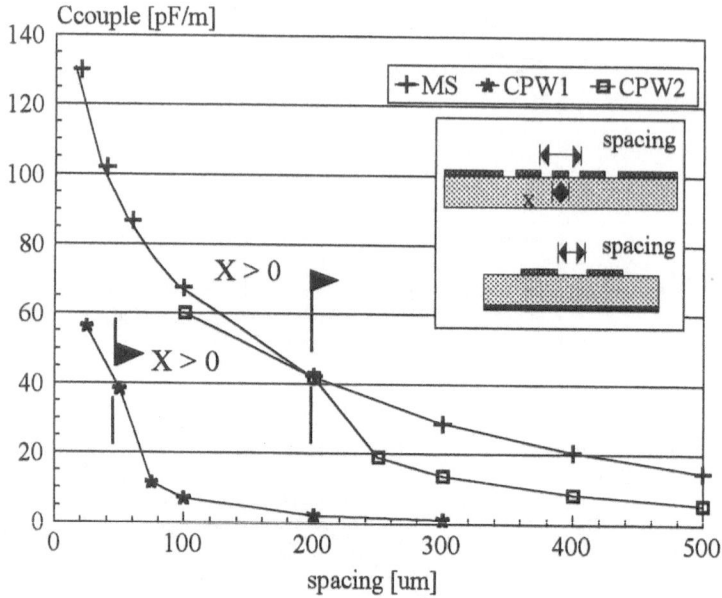

Fig. 4.5.8. Comparison of CPW lines with microstrip lines: line-to-line coupling
Substrate parameter: h=450μm, ε_r=12.9
MS-line: w=330μm; CPW1: w=30μm, s=25μm; CPW2: w=140μm s=100μm

Another important part of our work is the comparison of CPW and MS coupled lines. Such an example is given in figure 4.5.8. The 1/distance behaviour occurs in the microstrip as well as in the CPW case. However, it can be seen, that the line-to-line coupling is reduced by a factor of two to one hundred in the case CPW lines are utilised. This leads to the conclusion, that proximity effects can drastically be reduced and therefore, the package density can be increased. These calculation were made with the finite elements formulation. Regarding the figures 4.5.9 and 4.5.10, we get similar results, when executing the spectral domain analysis. Figure 4.5.9 depicts the frequency and line-to-line distance dependent investigation of coplanar coupled lines. The unwanted coupling S_{41} is always better than -40dB. In the case of microstrip coupled lines (see figure 4.5.10) the worst case comes to the values of -10dB for S_{41}. Also, there is a greater dispersive behaviour than in the CPW case.

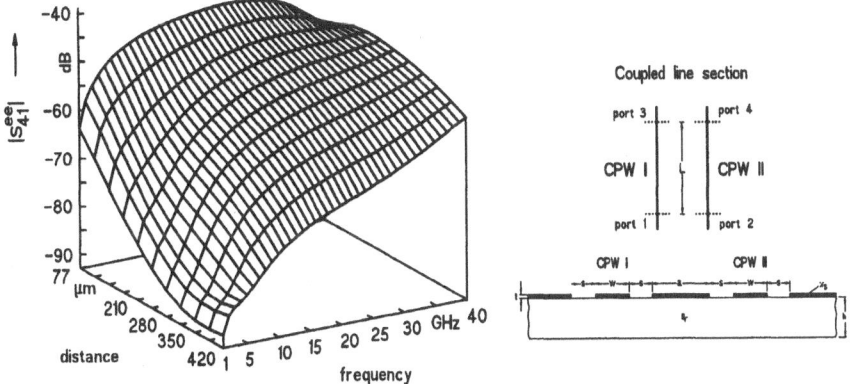

Fig. 4.5.9. Spectral Domain analysis of two coupled coplanar lines, h=410μm

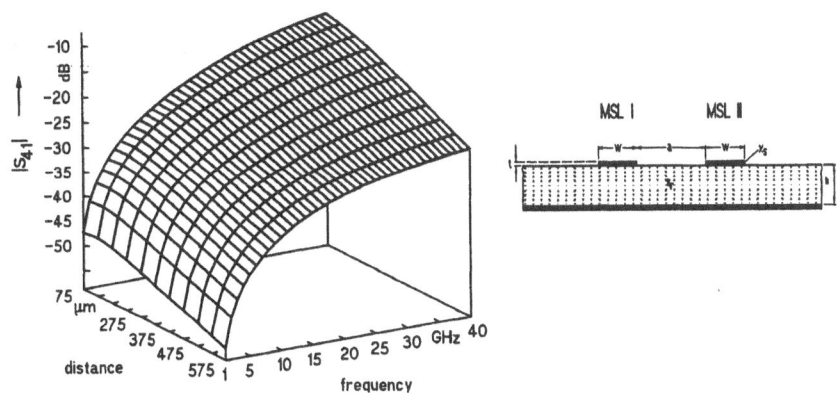

Fig. 4.5.10. Spectral Domain analysis of two coupled microstrip lines, h=150μm

One of the last aspects to be considered, is the interface between CPW and microstrip lines. In the first case, the microstrip chip is mounted on the coplanar line, so that the grounds of both structures are directly contacted. The two lines are connected with a small bond wire. Such a structure is depicted in figure 4.5.11 together with the simulation results. Figure 4.5.11.a shows the characteristic impedance's of the single CPW and MS line. The different dispersive behaviour becomes evident. The results of the transition calculation (S_{11}: circle markers) are presented in figure 4.5.11.b. The curve with the triangle markers is the ideal reflection coefficient of the transition resulting from the different impedance's. The difference between the two curves in figure 4.4.11.b is due to the non ideal transition.

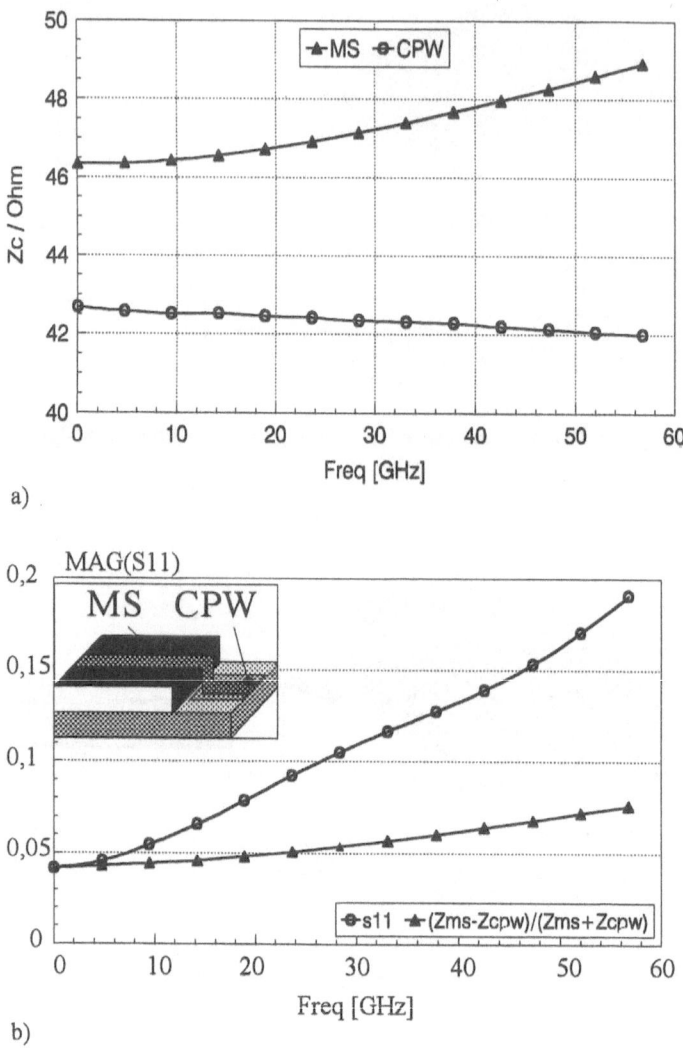

a)

b)

Fig. 4.5.11. Coplanar to microstrip transition - simulated with Finite Difference Time Domain (FDTD) software
a) characteristic impedance's of the single CPW and MS line
c) reflection coefficient of the ideal and real CPW to MS transition

The other CPW to MS transition, we want to discuss, is the so called Flip-Chip configuration. A coplanar line on a GaAs substrate (h=400µm, s=75µm, w=75µm) is mounted headlong on a circuit in microstrip technique, so that the center line of the CPW configuration and MS line are in the same layer. The MS circuit has the substrate parameter of Alumina: ε_r=9.8, h=125µm and w=125µm. An opening is in the Alumina substrate below the CPW block. The ground areas of the two line types are connected with metallization strips at the front sides of this opening. The results of the simulation and the geometry are depicted in figure 4.5.12. The curves of $|S_{11}|$ and $|S_{12}|$ are calculated for one half of the circuit (one CPW to MS transition).

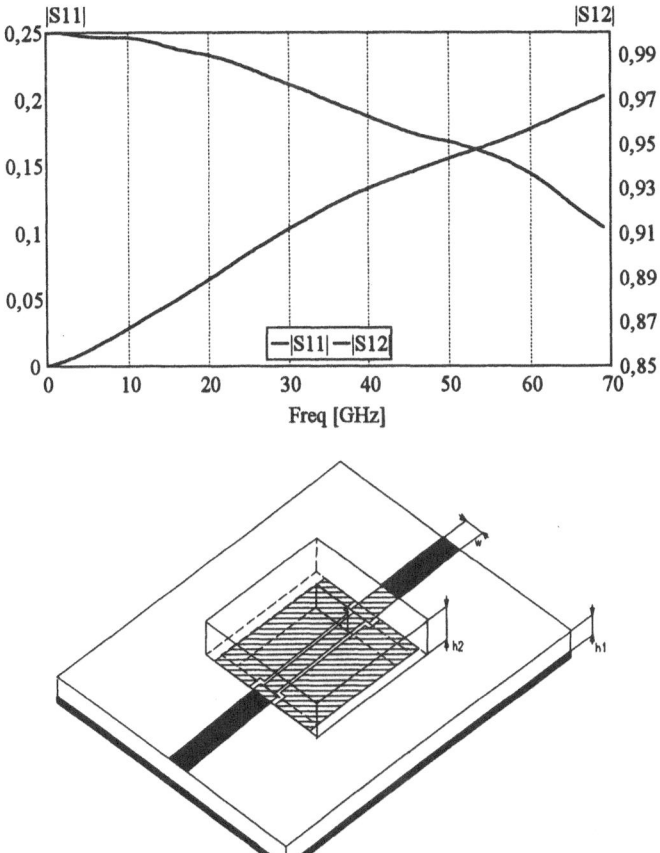

Fig. 4.5.12. Coplanar to microstrip transition - Flip-Chip - simulated with the Finite Difference Time Domain program
CPW-block: ε_r=12.9, h_2=400µm, s=75µm, w=75µm
MS-circuit: ε_r=9.8, h_1=125µm, w=125µm

Summary. The CPW oriented activities of the IMST within and beyond ESPRIT-CLASSIC have been shown. The CAD oriented application of 3D finite difference algorithms including three different airbridge types has been described. From this theoretical background results a catalogue of at least 18 parametric coplanar elements. The use of a lookup table and an interpolation algorithm are an enhancement of the models within this circuit design module. These features are linked together in a C-library, which can be built into any microwave CAD software package. Two additional aids for the circuit designer are the schematic element catalogue and the tool for the actual circuit layout, which can be applied to different foundry processes. The justification of CPW as an alternative to MS has been discussed. In the future we will verify the coplanar models with measurements. Especially for this task a complete GaAs wafer with passive coplanar circuits has been developed. To demonstrate the influence of the excitation of odd wave modes a spectral domain simulation of a coplanar bend with different airbridges have been shown. At last, two coplanar to microstrip transitions were discussed.

References

/4.5.1/ J.S. McLean, T. Itoh, "Fullwave Analysis of the Radiative Properties of Short-Circuit Discontinuities in Modified Coplanar Stripline", IEEE MTT-S Digest, 1992, pp. 203-206.

/4.5.2/ S. G. Pintzos, "Full-Wave Spectral-Domain Analysis of Coplanar Strips", IEEE Trans. Microwave Theory Tech., vol. 39, No. 2 February 1991, pp 239-246.

/4.5.3/ M. Naghed; I. Wolff, "A Three-Dimensional Finite-Difference Calculation of Equivalent Capacitances of Coplanar Waveguide Discontinuities", IEEE MTT-Digest, 1990, pp.1143-1146.

/4.5.4/ M. Naghed; M. Rittweger; I.Wolff, "A New Method for the Calculation of the Equivalent Inductances of Coplanar Waveguide Discontinuities", IEEE MTT-S Digest, 1991, pp.747-750.

/4.5.5/ T. Becks; I. Wolff, "Investigations on the Reduction of Package Densities in Coplanar Circuits", paper accepted for IEEE MTT-S, 1993.

/4.5.6/ M. Rittweger; M. Werthen; I. Wolff, "FDTD simulation for microwave packages and interconnects", IEEE MTT-S Workshop on EM Modelling of Microwave Packages and Interconnections, Atlanta, June 1993

IMST Institut für Mobil- und Satellitenfunktechnik	Coplanar Circuit Elements Part I		
Physical	Equivalent Circuit	Schematic Element	Layout

Tab. 4.5.1. Catalogue of coplanar elements - Part 1

	Coplanar Circuit Elements Part II		
Physical	Equivalent Circuit	Schematic Element	Layout

Tab. 4.5.1. Catalogue of coplanar elements - Part 2

4.8 Planetary MOVPE Reactor Modelling

C Waucquez and J Marchal
Polyflow s.a., Louvain-la-Neuve, Belgium

1 Introduction

Our objective is to develop a mathematical model to predict momentum, energy and mass transfers in a horizontal planetary CVD reactor. The planetary reactor is a III-V multiwafer device which combines a rotation of the substrate with a rotation of the satellites. This design guarantees a good homogeneity in the batch of wafers as well as an excellent uniformity on each wafer.

Several models have been proposed in the literature to simulate CVD reactors. Although early models were based on the assumption that the flow is nearly one-dimensional [1-2], there is actually an agreement in the literature that the derivatives in all spatial directions must be taken into account [3]. Neglecting second order derivatives (diffusion) in one of the spatial dimensions indeed allows the use of a parabolic solver in one direction and thus greatly reduces the computational effort, but important effects such as the presence of vortices or the flow stability cannot be predicted, because the model is based on the hypothesis that the fluid is not recirculating and that the flow remains evolutionary *in space*.

Due to complexity of the geometry and boundary conditions in the Planet reactor, we have selected a model based on the finite element technique. Our technique is similar to the method described in [3]. A global heat exchange model which takes into account radiation in the reactor chamber as well as heat conduction in all solid parts is included. We use a kinetic model based on the assumption that decomposition of trimethylgallium (TMGa) limits the reaction mechanism. The reaction is mass transfer limited in the bulk flow, and the reaction on the substrate is described by a first order kinetic mechanism. Steady-state and implicit time-dependent (precision limited adjustable time step) versions of the code have been developed, the latter in order to check the time stability of the steady-state flow.

The paper is organized as follows. Section 2 briefly describes the planetary reactor. We then recall in Section 3 the field equations and boundary conditions. Numerical technique is addressed in Section 4. Finally we report typical simulations results: in Section 5 we discuss two-dimensional solutions while in Section 6, three-dimensional solutions are presented.

2 Planet Reactor

A complete description of the Planet reactor has been published by Frijlink in [1]. A schematic view of the reactor chamber is presented in Fig. 1. It includes two gas inlets and one rotating substrate, on which seven 2" wafers or five 3" wafers are placed. The carrier gas is hydrogen, the group III precursors and dopants entering by the outer entrance, whereas the group V precursors enter by the inner entrance. The wafers are located on satellites, which are levitated and rotated by a secondary gas flow. A complete description of the gas foil rotation system can also be found in [1]. The rotation of the main platform is typically 30 [rpm] while the rotation of the satellites is about 180 [rpm].

The substrate is heated from below with the radial distribution of the electrical power of the heating lamps under control of the operator. Although the substrate is made of carbon (which is an excellent heat conductor), conduction in the substrate must be modeled, because the temperature cannot be measured accurately on the upper surface of the substrate, and because minor temperature differences can have a major influence on the growth mechanism on the wafers.

The design of the ceiling of the reactor chamber provides an additional control over the temperature distribution in the reactor. The top of the chamber is made of a quartz plate, and a gap exists between the quartz and the steel part. A mixture of argon and hydrogen is located between the quartz plate and the reactor chamber. The layer is so thin that only conduction across the gap must be taken into account, and convection in Ar/H_2 mixture can be neglected. The conductivity of argon and hydrogen being different, this mechanism provides the operator with a control on the conductivity in the top ceiling plate, by means of controlling the Ar/H_2 ratio.

Modeling the inlet geometry is important, because the temperature distribution of the gas along the distributor, which is close to the edge of the satellite, is influenced by the design of the inlet region. A bad design of the entry section can also induce a secondary vortex *before* the distributor, due to the flow resistance in the radial direction. This vortex, if present, has only little influence upon the uniformity of the epitaxial layer, but can lead to difficulties in operating the reactor in unsteady conditions.

The multiwafer Planet system is built by Aixtron GmbH (Aachen, Germany) and Laboratoires d'Electronique Philips (Limeil-Brévannes, France) as a part of an EEC Esprit project (5003 PLANET).

3 Basic Equations

Let Ω_f be the fluid part of the domain, the solid part being designated by Ω_s (the conductivity is dependent on position to take the variety of solid materials into account).

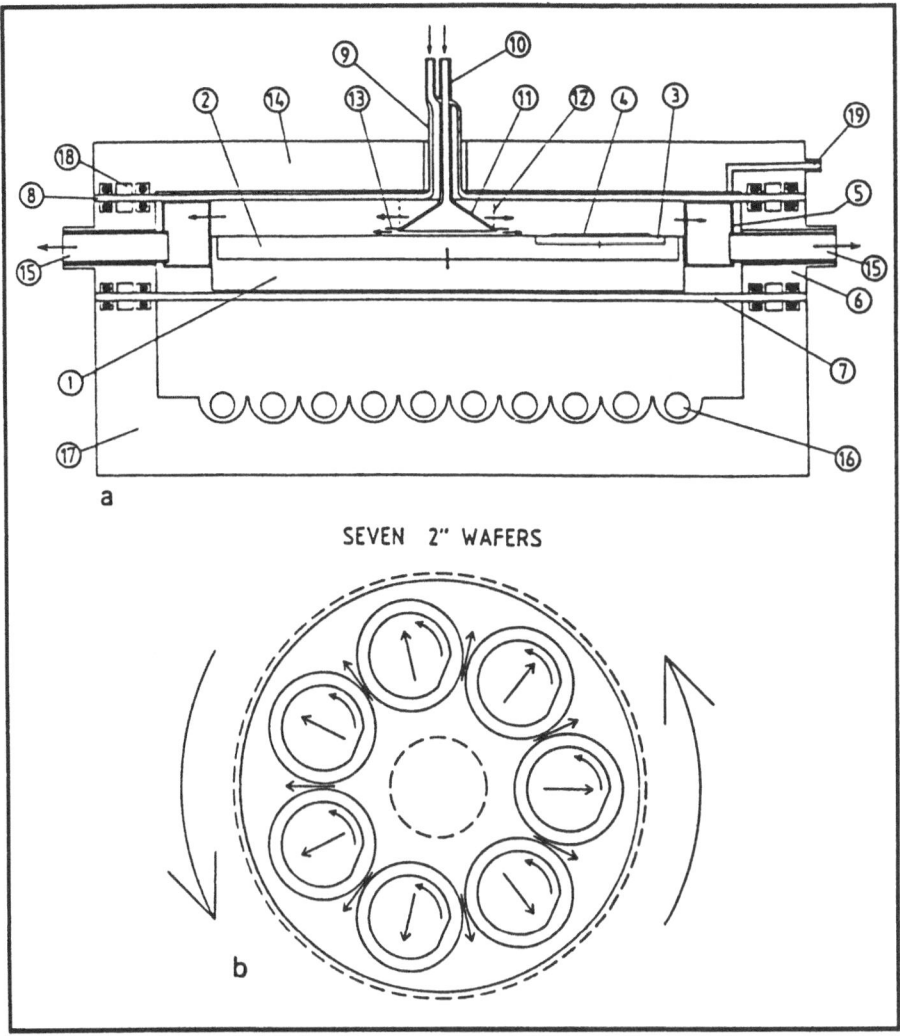

SEVEN 2" WAFERS

Figure 1. (a) Cross-section of the reaction chamber through the main axis. The reactive gas flow is indicated by arrows: (1) stationary part of graphite substrate holder; (2) main rotating disk; (3) satellite disk; (4) 2 inch wafer; (5) molybdenum exhaust gas collector; (6) water cooled stainless steel ring; (7) bottom quartz disk; (8) ceiling quartz disk; (9) outer entrance for main H_2 carrier flow; group III precursors and dopants; (10) inner entrance for second H_2 carrier flow and group V precursors; (11) cone; (12) cylindrical entrance grating; (13) deflector ring; (14) water cooled aluminium top plate; (15) gas exits; (16) 2 kW infrared lamp; (17) water cooled elliptic mirror; (18) vacuum ring; (19) inlet Ar/H_2 mixture for ceiling temperature control. (b) Top view of the substrate holder. The directions of rotation and the flow of the reactor atmosphere are indicated.

3.1 Momentum transfer

A momentum balance over an elementary gas volume leads to the following equation:

$$\rho[\frac{\partial v}{\partial t} + v.\nabla v] - \rho g - \nabla.\sigma = 0, \text{ on } \Omega_f \tag{3.1}$$

where ρ, v and σ stand for the gas density, velocity and Cauchy stress tensor respectively, whereas g is the gravity vector. ∇ stands for the gradient operator. The stress tensor in equation (3.1) is based on the following constitutive equation:

$$\sigma = -pI + \lambda I\nabla.v + \mu[\nabla v + \nabla vT] \tag{3.2}$$

where μ is the shear viscosity, and λ stands for the 'viscosity' in an expanding flow. As no data is available for λ, we have assumed that λ obeys the so-called Stokes' relationship $(3\lambda + 2\mu = 0)$. This hypothesis is compatible with the model used in [3]. In equation (3.2), p is the *variation* of the pressure with respect to a reference pressure P_0 which is also the pressure level at which the reactor operates. As the Mach number in the reactor is very low, compressibility effects can be neglected. This means that all parameters are *algebraic* functions of the absolute temperature T and pressure level P_0, but not of pressure variation p.

A mass balance over an elementary gas volume leads to the following mass conservation equation:

$$\frac{\partial \rho}{\partial t} + \nabla.[\rho v] = 0, \text{ on } \Omega_f \tag{3.3}$$

It must be noted that although $\nabla.v$ is not zero, the model is not compressible because there is no dependence of the density upon the pressure variation.

The ideal gas law is used to compute the density as a function of temperature:

$$\rho = \frac{P_0 M_w}{R_g T} \tag{3.4}$$

where M_w is the molecular weight of the carrier gas, R_g is the ideal gas constant and T is the temperature.

3.2 Energy transfer

The energy equation reads:

$$\rho c_p[\frac{\partial T}{\partial t} + v.\nabla T] - \nabla.[k\nabla T] = 0, \text{ on } \Omega_f \tag{3.5}$$

where c_p and k are respectively the volumetric heat capacity and conductivity of the gas. In the solid domain Ω_s, equation (3.5) is also taken into account.

However, the advection term vanishes and k now stands for the conductivity of the solid.

The temperature dependence of all parameters strongly couples the momentum, mass conservation and energy equations.

3.3 Mass transfer

The concentrations of III-V precursors in the carrier flow are so low that the mass transfer equations can be decoupled from the flow problem. This means that first we solve the problem for momentum and energy transfers, the concentration of TMGa being computed afterwards on the basis of a known velocity field.

In the case of GaAs systems, it appears from physical considerations that a single reaction limits the kinetic of the system. This chemical equation is written as:

$$Ga(CH_3)_3 + AsH_3 \rightarrow GaAs + 3\ CH_4 \tag{3.6}$$

whereas the kinetic mechanism in the bulk flow is limited by the transport of TMGa. Let x be the partial pressure of TMGa in the flow. The transport equation for this species is written in the form:

$$c[\frac{\partial x}{\partial t} + v.\nabla x] - \nabla.[cD[\nabla x + \alpha_t x \nabla[\ln T]]] = 0, \text{ on } \Omega_f \tag{3.7}$$

where the flux of TMGa takes both diffusion and thermodiffusion into account. In equation (3.7), c is the total concentration of the carrier gas ($c = \rho/M_w$), D is the diffusion coefficient of TMGa in the carrier gas and α_t is the thermal diffusion factor.

3.4 Boundary conditions

Momentum equation

The following boundary conditions have been used:

- on solid walls: $v = 0$
- along $\partial\Omega_{inlet}$:   fully developed velocity profiles
- along $\partial\Omega_{outlet}$:   outflow conditions
- along $\partial\Omega_{susceptor}$: no slip conditions

Along the distributor, the direction of the flow is supposed to be aligned with the grid. Such a constraint comes naturally in the discretized system by means of a restriction on the interpolation/weighting functions.

Energy equation

The energy equation is solved in the full domain. Along all interfaces between solid parts and fluid parts, we impose the continuity of the temperature as well as continuity of the heat flux.

Along the top ceiling plate, conduction in the Ar/H$_2$ layer is modeled as follows:

$$Q_{conduction} = k_{mix}(T - T_{steel}), \text{ on } \partial\Omega_{top} \tag{3.8}$$

where $Q_{conduction}$ is the heat flux density, k_{mix} is the transfer coefficient (which depends on the conductivity of the Ar/H$_2$ mixture), T is the temperature along $\partial\Omega_{top}$ and T_{steel} is the (imposed) temperature of the steel part.

Radiation between the susceptor and the ceiling plate as well as radiation between the susceptor edge and the ambient are modeled as a jump in the heat flux density:

$$Q_{radiation} = \varepsilon\sigma(T^4 - T_{susceptor}^4), \text{ on } \partial\Omega_{ceiling plate} \tag{3.9a}$$
$$Q_{radiation} = \varepsilon\sigma(T^4 - T_{ambient}^4), \text{ on } \partial\Omega_{susceptor edge} \tag{3.9b}$$

where $Q_{radiation}$ is the heat flux density, ε is the emissivity, σ is the Boltzman constant, $T_{susceptor}$ is the temperature of the susceptor, and $T_{ambient}$ is the temperature of the ambient.

In all inlet regions, room temperature is assumed, whereas the conductive heat flux is zero at the outlet. At the bottom of the susceptor, we impose a heat flux density.

Mass transfer

In the inner inlet region, the flux of TMGa is zero, whereas the relative concentration is equal to 10-4 in the outer part of the inlet region. All walls except the susceptor are assumed to be non-reacting. This means that the global flux of concentration:

$$\mathbf{J}.\mathbf{n} = - cD[\nabla x + \alpha_t x\nabla[\ln T]].\mathbf{n} \tag{3.10}$$

is zero along those walls. Along the susceptor and the satellites, a first order kinetic model is used to predict the mass flux as a function of the partial concentration:

$$\mathbf{J}.\mathbf{n} = k_{surf}cx, \text{ on } \partial\Omega_{susceptor} \tag{3.11}$$

where

$$k_{surf} = \gamma\sqrt{\frac{R_gT}{2\pi M}} \tag{3.12}$$

In equation (3.12), γ is the reaction probability, and M is the molecular weight of TMGa. In the case of a high reaction probability, the reaction at the surface of the wafer is limited by mass transfer, and condition (3.11) is equivalent to imposing a vanishing concentration along the boundary in the limit $k_{surf} = \infty$. Therefore we use the value $\gamma = 1$, and we recover the TMGa flux density by equation (3.11) once the concentration field has been obtained.

3.5   Physical parameters

Standard power law expressions were fitted to material data of H_2 carrier gas. All dependencies are summarized in Table 1. We also list values and correlation for the transport coefficients.

Table 1. Hydrogen Properties and TMGa Transport Properties

- Viscosities, heat capacity and thermal conductivity of H_2:

$$\lambda, \mu, c_p, k = f(T) \sim a(T/T_0)^b$$

coefficients	units	a	b	T_0
λ	$[N \cdot s \cdot m^{-2}]$	$-.5973\ 10^{-5}$	$.6780$	300
μ	$[N \cdot s \cdot m^{-2}]$	$.8960\ 10^{-5}$	$.6780$	300
c_p	$[J \cdot kg^{-1} \cdot K^{-1}]$	$.1431\ 10^{+5}$	$.3100\ 10^{-1}$	300
k	$[W \cdot m^{-1} \cdot K^{-1}]$	$.1830$	$.7100$	300

- Diffusion coefficient of TMGa in H_2:

$$D = 2.32\ 10^{-9}\ (T)^{1.71}\ (1/P_0),\ T\ in\ [K]\ and\ P_0\ in\ [atm]$$

- Thermal diffusion factor of TMGA in H_2:

$$\alpha_t = 0.50361 + 0.144231\ 10^{-2}\ T - 0.68947\ 10^{-6}\ T^2,\ T\ in\ [K]$$

---

4   Numerical Technique

Our method is based on a standard finite element procedure. The Green's theorem is applied to equations (3.1), (3.3), (3.5) and (3.7), and approximation subspaces are selected for the velocity, pressure, temperature and concentration fields. Let us denote those subspaces V^h, P^h, T^h and X^h. The discrete problem is formulated as:

find $(v^h, p^h, T^h, x^h) \in V^h \times P^h \times T^h \times X^h$
such that $\forall\ (\underline{v}^h, \underline{p}^h, \underline{T}^h, \underline{x}^h) \in V^h \times P^h \times T^h \times X^h$,

$$< \rho[\frac{\partial vh}{\partial t} + vh.\nabla vh] - \rho g \; ; \; \underline{v}h > + < \sigma \; ; \; \nabla \underline{v}h >$$

$$- << \sigma.\mathbf{n} \; ; \; \underline{v}h >> = 0 \quad (4.1)$$

$$< \frac{\partial \rho}{\partial t} + \nabla.[\rho vh] \; ; \; \underline{p}h >$$

$$= 0 \quad (4.2)$$

$$< \rho c_p[\frac{\partial Th}{\partial t} + vh.\nabla Th] \; ; \; \underline{T}h > + < k\nabla Th \; ; \; \nabla \underline{T}h >$$

$$- << k\nabla Th.\mathbf{n} \; ; \; \underline{T}h >> = 0 \quad (4.3)$$

$$< c[\frac{\partial xh}{\partial t} + vh.\nabla xh] \; ; \; \underline{x}h > + < cD[\nabla xh + \alpha_t xh\nabla[\ln Th]] \; ; \; \nabla \underline{x}h >$$

$$- << cD[\nabla xh + \alpha_t xh\nabla[\ln Th]].\mathbf{n} \; ; \; \underline{x}h >> = 0 \quad (4.4)$$

In those expressions $<;>$ is the standard L^2 scalar product on the domain, whereas $<<;>>$ represents the integral along the boundary. All functions of Vh, Th and Xh are assumed to satisfy the essential (Dirichlet) boundary conditions.

Equations (4.1) to (4.3) are highly non-linear, due to the temperature dependence of all parameters. In the case of a steady-state problem, solutions of the algebraic system cannot be obtained directly for any value of the flow rate. We solve the system by increasing the flow rate progressively within an evolution procedure (continuation technique). At each iteration, a full tangent matrix (Newton-Raphson technique) is evaluated and the resulting linear system is then solved by Gaussian elimination with the use of a frontal method.

Equation (4.4) is decoupled from previous equations and is linear in the concentration field.

Once the concentration field has been obtained, the flux density of TMGa is evaluated along the wafer by (3.11). As the boundary conditions for the concentration field do not take rotation of the wafer into account, the flux density is then averaged on the wafer by the following formula:

$$\tau^*(r^*) = \lim_{n \to \infty} \frac{1}{n} \sum_{i=1}^{n} \tau(\sqrt{x^2 + y^2}), \text{ on } \partial\Omega_{\text{wafer}} \quad (4.5)$$

where

$$(x, y) = (x_0 + r^* \cos \frac{2\pi i}{n}, y_0 + r^* \sin \frac{2\pi i}{n}) \quad (4.6)$$

In equation (4.6), (x_0, y_0) are coordinates of the centre location of a satellite. Averaging the flux on the wafer afterwards is equivalent to supposing that the wafer rotation has no influence on the depletion of TMGa. This hypothesis is justified by the high reaction probability on the wafer.

5 Two-dimensional Results

In this model, we analyse the flow in a plane passing through the axis of the substrate and the centre of one satellite. It has been found essential to include all details of the reactor design in the model, i.e. the complete entry section, the top ceiling plate, the inlet grating, the deflector and the carbon substrate, in order to predict the energy transfer accurately. Our present model is 2-D 1/2. The velocity field has three components, and we assume that all derivatives in the azimuthal direction can be neglected. The motion is described in a cylindrical system of coordinates where (e_r, e_z, e_θ) are respectively the unit vectors aligned in the radial, vertical and azimuthal directions. The components of the velocity vector are designated by u, v and w. The radial dependence of w along the susceptor comes in the system as a boundary condition. All scalar fields (p, T and x) are two-dimensional.

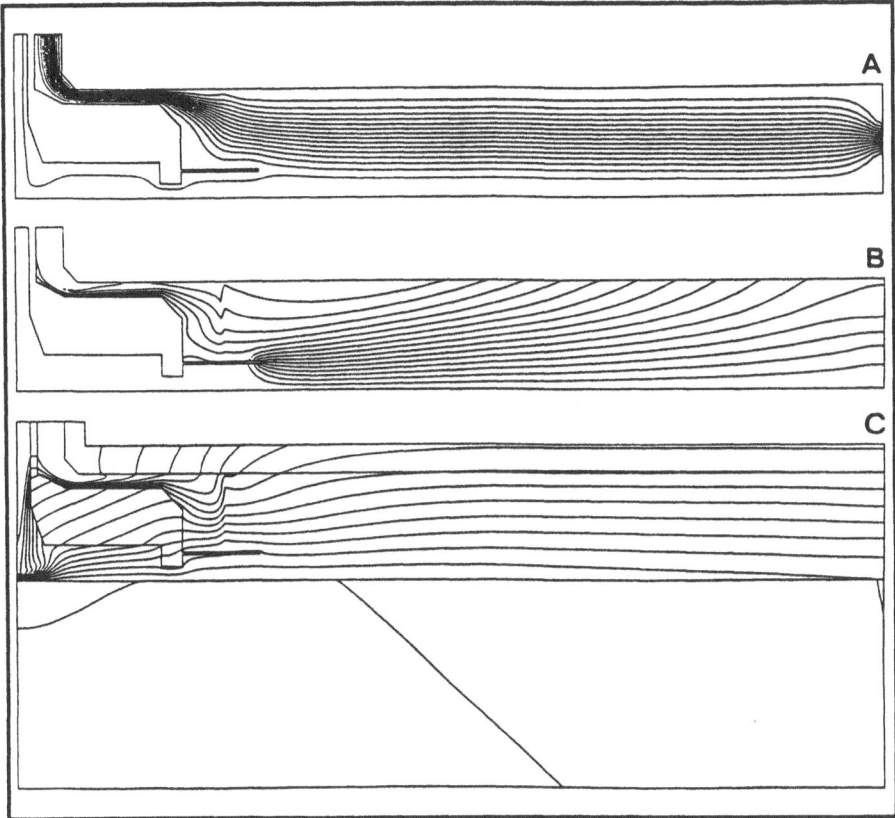

Figure 2. Effect of 50% of nominal flow rate on streamlines (A), concentrations (B) and isotherms (C).

The first set of results has been obtained at atmospheric pressure, i.e. $P_0 = 10^5$ [Pa]. The nominal flow conditions are the following:

$$Q_{III} = 39/43\ Q,\ Q_V = 4/43\ Q,\ Q_{nom} = 43\ [SLM]$$

Fig. 2 is relative to 50% of nominal flow rate. We present the streamlines, the contour lines of the TMGa concentration in the fluid part and the isotherms in the complete domain.

At very low flow rates, one can observe the presence of a vortex located above the satellite. This vortex is caused by the centrifugal force and occurs when the azimuthal velocity component w on the edge of the satellite is important with respect to the horizontal velocity component u. However this recirculating zone completely disappears at about 10% of nominal flow rate. This means that in all practical conditions, the reactor does not present any recirculations in the main chamber.

Fig. 3 shows the flux density of TMGa as a function of the radial coordinate for various flow rates, for 2" (left hand side) and 3" (right hand side) applications respectively. The boundaries of the substrate and the satellite have been indicated by single and double vertical lines respectively. One can observe the presence of a peak in the flux density, corresponding to the diffusion of TMGa from the upper entrance (grating) into the substrate region. For 2", this peak is located outside of the wafer region, so it does not influence the uniformity of the layer. For 3" however, the peak is located within the wafer region and it will strongly degrade the uniformity.

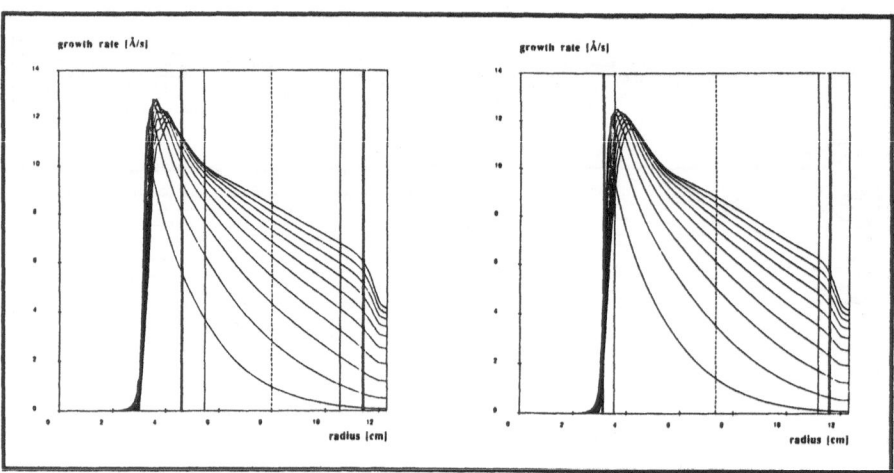

Figure 3. Growth rate profiles of the deposit for various sets of carrier flow: 2" (left) and 3" (right) applications.

309

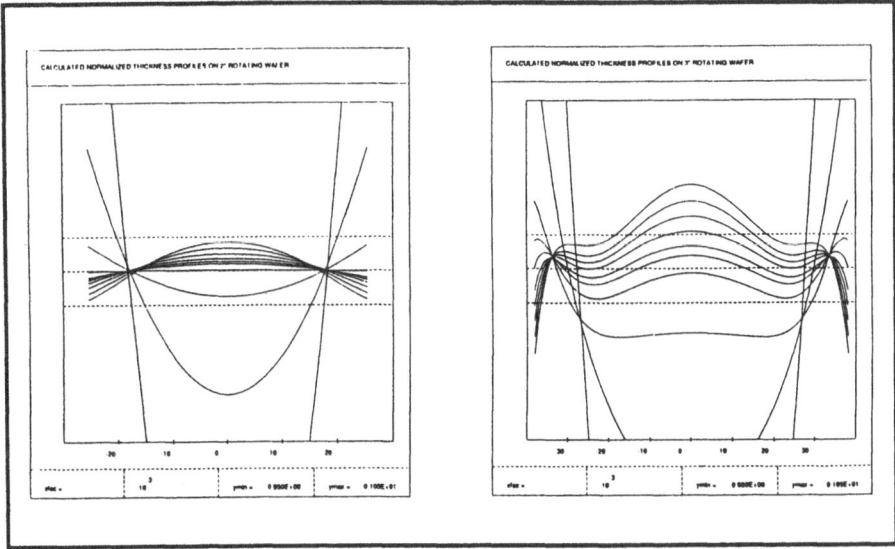

Figure 4. Normalized thickness profiles (deviation with respect to an arbitrary reference value) on rotating wafer for various sets of carrier flow: 2" (left) and 3" (right) applications.

Once the rotation of the satellite has been taken into account, as described in Section 4, one obtains the flux density on the rotating wafer, which is shown in Fig. 4 for various flow rates, for 2" (left hand side) and 3" (right hand side) applications respectively. These profiles have been normalized with respect to the integral of the flux density over the wafer. They are also an image of the uniformity of the layer, because the number of molecules of TMGa reacting on the substrate on an elementary section will be proportional to the flux density, the operating time (in steady-state condition) and the elementary area.

One can observe that the uniformity is very good for a large range of operating conditions. This can also be seen in Fig. 5, where we have plotted the maximum deviation (on the wafer) of the normalized flux density as a function of the total flow rate, for 2" (left hand side) and 3" (right hand side) applications respectively. One can observe that the best uniformity is obtained at 40% of nominal flow rate for 2" wafers, and at 50% of nominal flow rate for 3" wafers.

In Fig. 5, we also report the efficiency in the reactor. The efficiency is defined as:

$$\frac{\text{\#mol reacting on a selected area (wafer, susceptor, ...)}}{\text{\#mol entering the reactor}}$$

where #mol stands for the number of molecules of TMGa.

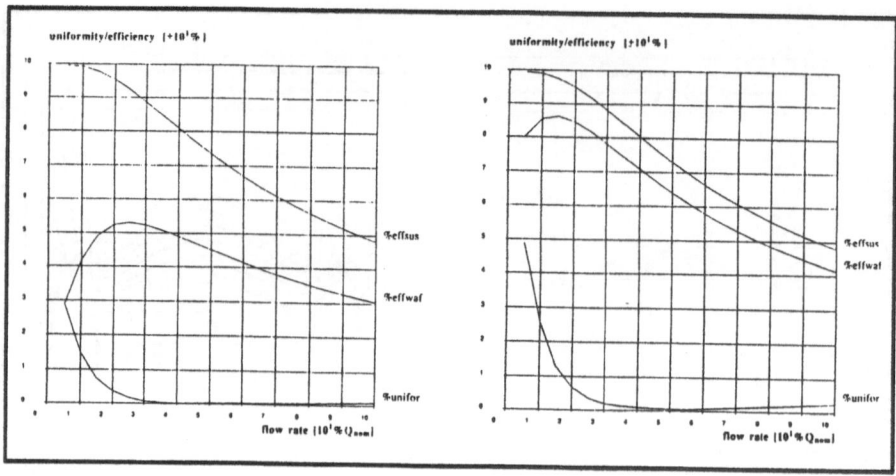

Figure 5. Uniformity and efficiencies versus flow rate: 2" (left) and 3" (right) applications.

Sensitivity analysis

We have analyzed the sensitivity of the uniformity of the epitaxial layer with respect to various parameters, namely:

- total flow rate
- operating pressure
- rotation speeds
- inner/outer flow rate ratio
- conductivity of the Ar/H_2 layer
- global dimensions
- heating profile

Sensitivity analysis has been reported in full in [4]. From all the above in the tasklist, it appeared that the uniformity was most sensitive to the total flow rate.

Comparison with experimental results

In Fig. 6, numerical results have been compared to experimental measurements of the relative uniformity. The squares indicate experimental results, whereas solid lines correspond to numerical simulation. Although this comparison is somewhat difficult, because experimental measurements in the range of 1...2 percents of uniformity are difficult, this comparison is essential to validate the mathematical model.

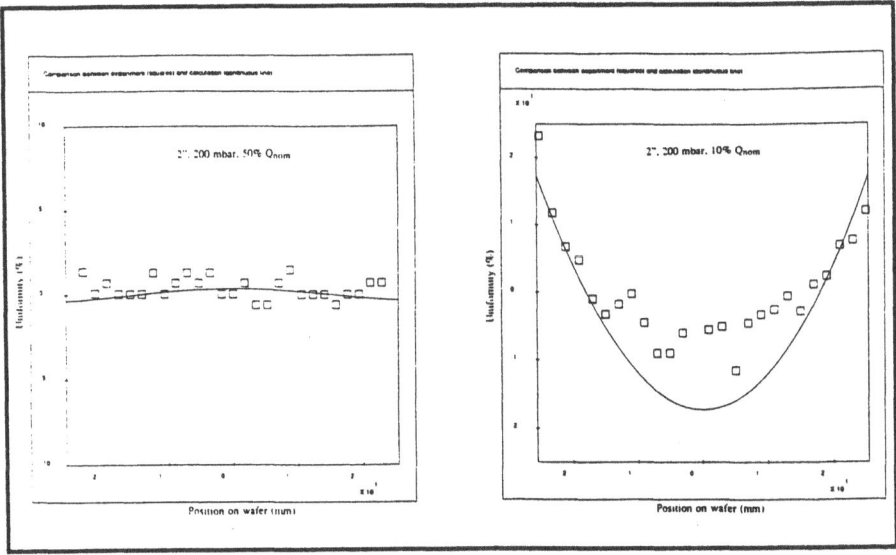

Figure 6. Comparison between measured (squares) and calculated (continuous line) 2" (left) and 3" (right) relative uniformities. Non standard flow conditions.

Two uniformity results are available for comparison [4-5-6]. They have both been obtained by Mr Ambrosius from Philips Optoelectronics Centre (Eindhoven, The Netherlands) on 2" wafers at $P_0 = 200$ [mbar].

Left hand side result is relative to 50% of nominal flow rate. Such a flow rate has been recommended in view of an optimum in relative uniformity with respect to the flow rate. Obviously, the order of magnitude of the uniformity agrees with the measurements, but the dispersion of the experimental results does not allow a fine comparison.

Right hand side result is relative to 10% of nominal flow rate, i.e. a flow regime intentionally very far from the optimum. In this case, the deviation in uniformity is within the range of about ±20%. This large value of the non-uniformity makes the measurement easier. It must be noted that the sign of the curvature predicted by the model corresponds to the experiment and that the comparison can even be better if the numerical or experimental curve is shifted vertically (different normalizations).

6 Three-dimensional Results

The planetary rotation of the wafers makes the problem inherently three-dimensional. In view of the important computer time and memory requirements, it has been decided to spend variables on velocity predictions in the reactor chamber rather than on temperature predictions in reactor enclosure. This allows to capture flow patterns with reasonable precision while keeping the computer effort within a tolerable limit. The computer-time is a few days on a fast RISC workstation (5 Mflops).

In order to handle boundary conditions that are steady-state for *all* rotating pieces, decision was made to solve the flow equations in a reference frame attached to the main plateau, i.e. a mobile reference frame, rather than in an inertial system. Thus, solutions are the ones an observer sees when moving along with the susceptor. This is but easy to convert solutions, once obtained, into a fixed reference frame.

In what follows, calculations have been carried out for 2" applications at atmospheric pressure.

6.1 Flow behaviour in a symmetric sector

As long as a 'n'-fold symmetry for a platform bearing 'n' wafers holds, the computational domain restricts to 1/n of the whole piece (Fig. 7). Identical kinematic behaviour from sector to sector is thus assumed. By 'identical', it is meant that the temperature and velocity fields periodically recover same values along the circumferential path.

Without injection area

First three-dimensional calculations have been undertaken on computational domains restricted to sectors downstream of the injection area. The direction of

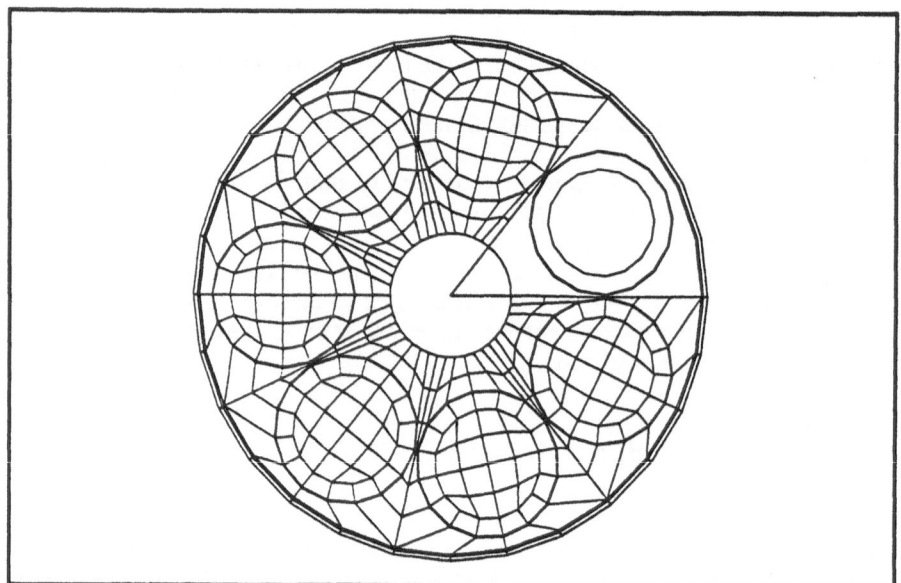

Figure 7. Computational domain of a symmetric sector, with main chamber and injection area. Top view.

the upper inflow is supposed to be aligned with the direction of the slits of the distributor. Three-dimensional flows for a range of flow rate have been analyzed. Main results sum up in what follows.

Although relatively coarse grids are used, the three-dimensional model compares quite well to the two-dimensional one: flow patterns from 3-D cross-sections fairly match 2-D solutions.

The existence of vertical vortices over the satellites has been revealed. Contrary to two-dimensional models where the assumption of vanishing second order derivatives in the azimuthal direction is made, important effects such as vertical vortices can now be predicted because the model allows the fluid for recirculating in *any* direction. The main rotation causes the centre of the vortex to move. This phenomenon is well depicted in Fig. 8: for increasing height of successive horizontal planes, the location of the cyclone 'eye' changes.

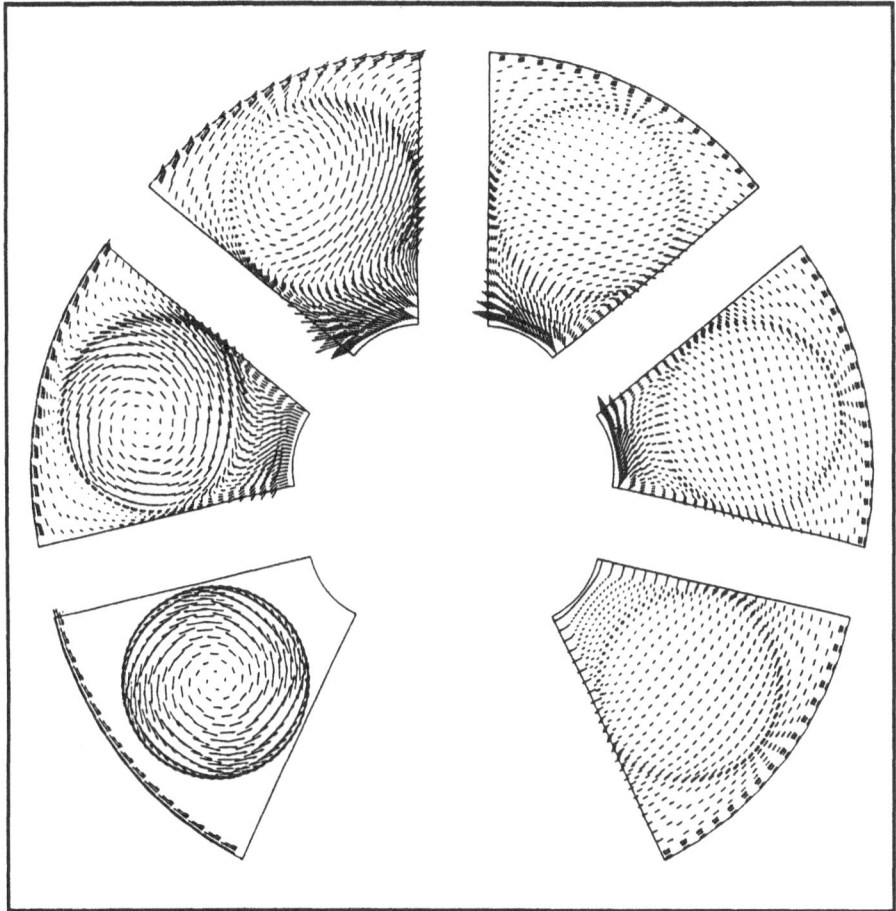

Figure 8. Vector plots at various heights of horizontal cutting planes, from susceptor level to ceiling level, starting lower left (clockwise direction).

314

There is a pumping action by spinning satellites. The rotation action also stabilises the flow reducing the possibility for nonaxisymmetric flows. Actually, the rotation dampens out the effect of slightly nonaxisymmetrically placed inlets and outlets.

The residence time may be quite long: at 50% of nominal flow rate, a particle, when close to the susceptor, crosses successive wafer areas, prior to exiting somewhere roughly at the opposite of the injection point. The traces of fluid particles are complex.

The assumption that the satellite rotation has no influence on the depletion of TMGa is almost verified, as can be seen in Fig. 9, but still, a small glide effect is clearly visible.

Present three-dimensional simulations show better uniformity when increasing the total flow rate by a large amount. A flow rate of about 50% of nominal flow rate in view of the optimal choice in relative uniformity as suggested by the sensitivity analysis in 2-D is not yet proved in 3-D.

Including injection area

In this model, the computational domain covers one entire symmetric sector. This allows to investigate the effect of various obstacles on the flow field. In particular, the specific design of the entry section (the deflector forces the lower flow, i.e. the group V products, to move almost radially whereas the distributor forces the upper flow, i.e. the group III products, to move almost tangentially) may result in unforeseenable changes in flow patterns in the neighbourhood of the satellite edges, leading to large deposit variations over wafer.

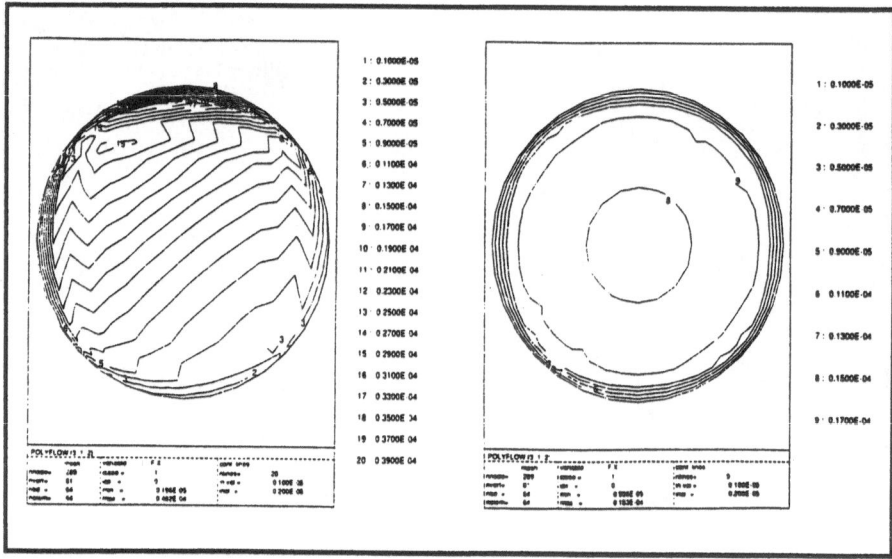

Figure 9. Pre- (left) and post- (right) normalized growth rate profiles on 2" wafer at nominal conditions.

315

Effects of various flow rates have been studied. As long as 30% of nominal flow rate is concerned, the distributor does not influence the direction of the flow as strongly (about 25° with respect to the radial direction) as assumed (70° with respect to the radial direction). It must be noted that the distributor has been discretized into a *single* layer of finite elements. As the distributor represents a serious obstacle to the motion of particles at high speeds, it also comes out that intensive mesh refinements are required. However this is a formidable computational task.

6.2 Flow behaviour in full reactor

An ultimate step in the simulation has been the prediction of the stability of the three-dimensional flow in the reactor. For this analysis, only the momentum and energy equations have been considered (the transfer reactants do not affect the flow

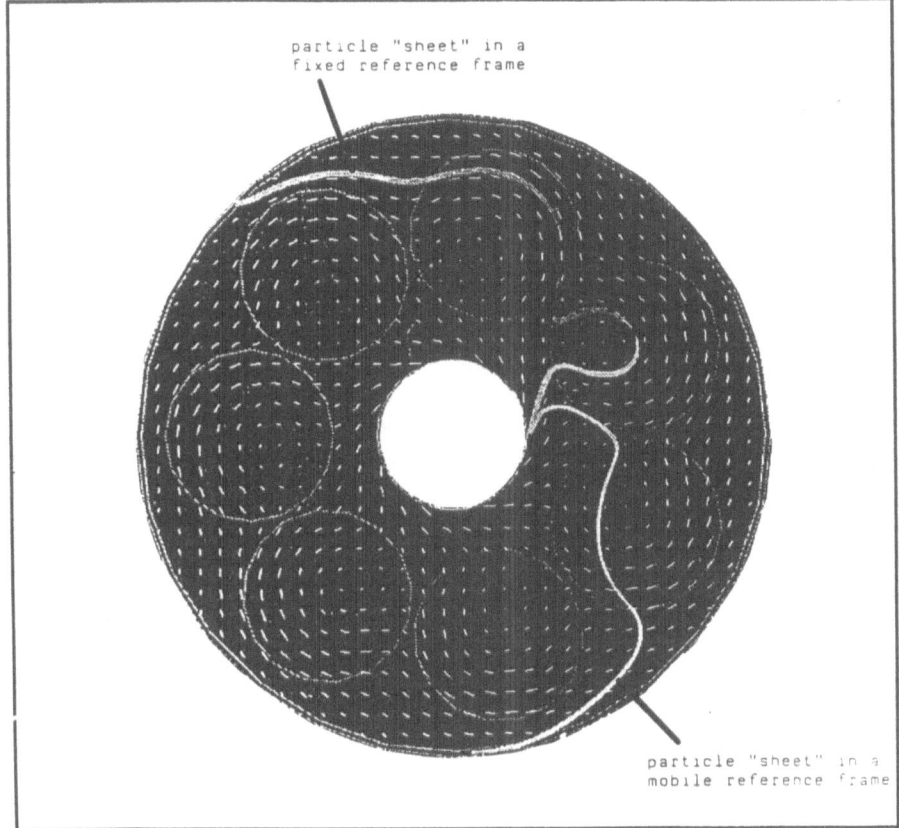

Figure 10. Particle 'sheet' trajectories close to susceptor in mobile and in fixed reference frames at 50% of nominal flow rate.

field). The mesh covers the entire reaction chamber, and time-dependent simulation has been used to allow for development of instabilities.

Starting from scratch, the symmetric behaviour of the flow with respect to symmetric sectors has been confirmed. No instabilities have been detected. Actually, symmetry breaking can only occur under conditions where multiple stable flows exist. Which of the flows are realized depends on the start-up procedure and parameter variations. If the inlet is symmetric, if thermal boundary conditions on wall and susceptor are symmetric, then the flow field will be symmetric. In case of an off-axis placed inlet tube or temperature variations across the susceptor, it is non-symmetric patterns that are likely to develop.

Although equations are *solved* in a mobile reference frame, solutions can be *expressed* in a fixed system of coordinates. The vortex induced by satellite rotation is clearly evidenced in Fig. 10, where velocity vectors in a plane orthogonal to the main axis are shown. For sake of comparison, particle sheets in both reference frames are presented.

7 Conclusions

The heat transfer within MOVPE reactors is complex, involving radiation heat transfer between solid surfaces, convection in the gas phase, as well as conduction in the gas, the reactor wall and the susceptor assembly. Since the chemical reactions are temperature activated, the temperature distribution is critical in achieving uniformity. A complete model with *all* solid components (ceiling plate, injection cone, deflector, distributor and three-parts susceptor) has been used to explore how heat transfer characteristics and changes in operating parameters affect flow fields and deposition performance in the Planet reactor.

Our results have shown that the uniformity of the epitaxial layers was very good in all practical operating conditions. Numerical results have been compared to experimental measurements of the relative uniformity. Qualitative and quantitative agreement has been obtained and has allowed to use the model for improving the original design of the reactor. In particular, the nominal flow rate has been adapted to 50% of the one initially indicated in the project. The optimal flow rate, in terms of uniformity, is thus one of the output of the simulation. A sensitivity analysis has also been performed, but is not reported in the present paper.

Three-dimensional simulations of the Planet reactor have finally been carried out. Important effects such as the presence of vertical vortices, long residence time, complex particle paths and recirculation zones have come to light. The results have demonstrated that the side effects at the wafer edge can be reduced and thus the uniformity can be improved by increasing the distance between the distributor and the satellite.

8 Acknowledgement

This work was partly supported by the European Community under project PLANET 5003.

References

1. Frijlink, P.M. (1988) J. Crystal Growth, 93, 207-215.

2. Coltrin, M.E., Kee, R.J., and Miller, J.A. (1986) J. Electrochem. Soc., 133, 1206-1213.

3. Fotiadis, D.I., Boekholt, M., Jensen, K.F., and Richter, W. (1990) J. Crystal Growth, 100, 577-599.

4. Esprit PLANET 5003 Multiwafer Planet MOVPE Reactor (1993) Final Report.

5. Ambrosius, H.P.M.M., Linders, R.W.M., Waucquez, C., and Marchal, J.M. (1991) APCT'91 (Advanced Processing and Characterization Technologies), Clearwater Beach.

6. Frijlink, P.M., Nicolas, J.L., Ambrosius, H.P.M.M., Linders, R.W.M., Waucquez, C., and Marchal, J.M. (1991) 7th International Conference on Vapour Growth and Epitaxy, Nagoya.

5.1   Manufacturing Issues

C L Sansom, J A Turner and D J Warner
GEC-Marconi Materials Technology Ltd., Caswell, Northants NN12 8EQ, UK

1.   Manufacturing Strategy

In comparison with the Silicon IC industry, the strategic management of GaAs manufacturing is only now beginning to assume importance. Led by the GaAs digital businesses, manufacturing strategy has moved GaAs production from the R & D laboratory through job shop and batch processing into either group (cell) manufacture or full line. The positioning of a particular activity is dependent on the maturity of the product or process plus the product volumes involved. Typically within the lower volume, high added value analogue MMIC markets small batch manufacture is the norm (Figure 5.1).

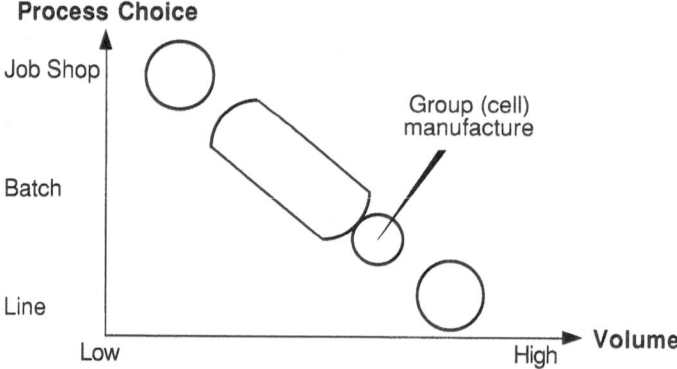

Figure 5.1   Process Choice and Volume

Product/Process Focus   Focusing in the strategic sense is concerned with the need to operate each plant with a limited, concise, manageable set of products, technologies, volumes and markets. Focus analysis provides one of the essential inputs into the corporate strategy development by highlighting the degree of match between the marketing and manufacturing strategies. If the operation is product focused, facilities are designed on a general purpose basis and therefore flexible. If process focused, the operation meets the needs of a narrow range of products, normally with high volumes and/or similar process requirements.

A representation of focus within the GaAs industry is shown in Figure 5.2.

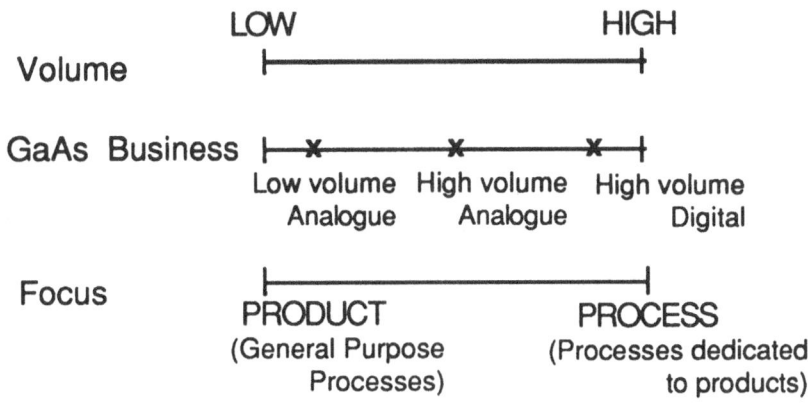

Figure 5.2   GaAs Product/Process Focus

The acceptability of such a focus position depends on the maturity of the particular product involved, i.e. the position on the product life cycle at the time, as shown in Figure 5.3.

For example, if most of a company's products are in the growth/maturity stage of their life cycle, a strategic shift from product to process focus is advisable, requiring dedicated processes to meet the volume demand.

Process Span : A Strategic Choice   The breadth and direction of a company's process span is a measure of its links and relationships at either end of the process spectrum (Suppliers at one end, Customers at the other). Within the electronics industry the most famous example of successful widening of process span concerns Texas Instruments, who successfully integrated forward from chips to calculators, displacing Bowmar as market leader. World recessionary pressures and over capacity have tended to force the GaAs businesses to concentrate resources on core activities, and into a tendency to reduce the operational span. This is particularly true for labour intensive activities such as assembly, and capital intensive niche activities such as III-V heterostructure epitaxial growth. The growth in the European sub-contract market sector in recent years suggests that this trend will continue.

Quality Function Deployment (QFD)   QFD presents in a formal matrix way the requirements of the customer at each stage in a product's development. The needs of the customer (whether internal or external) are transformed into operational requirements and personnel implications. Used effectively the technique can be cascaded down from the strategic requirements to ensure that the customers' needs are satisfied at all stages in the design-to-manufacture cycle.

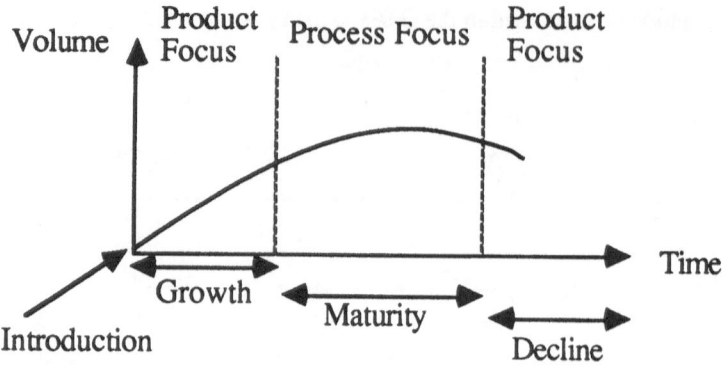

Figure 5.3   Focus Emphasis and the Product Life Cycle

Used extensively in Japan, the methods are achieving rapid growth in European manufacturing circles. The example shown here in Figure 5.4 translates a GaAs MMIC requirement into operational requirements for a fictitious GaAs Foundry.

The operating effects and human factors are rated and analysed in a formal way to ensure a good match to the customer's requirement. The chart looks daunting at first sight, but always has the same basic layout. The customer's requirements are listed vertically on the left side of the plan (low prices, short cycle time...) with an importance weighting on a scale of 1-10 (10 being most important). These features translate to operational and human factors listed across the top of the chart (yield, packing density....). The relationship between a requirement and a factor is rated as strong (score 9), medium (score 3) or weak (score 1). These scores are written into the main matrix. The importance of a particular factor is scored by summing the individual (factor score x requirement importance) products for a particular factor. For example, the absolute technical importance score for 'yield' is $(9 \times 8) + (6 \times 3) = 90$.

There is additional optional information that may be displayed. We may require a factor to be either maximised, minimised or to hit a certain target value. This is shown at the top of the chart, together with a triangular matrix that reflects any inter-factor correlations. The customer can also rate our current performance against his requirements, producing customer rating indices. As a Foundry supplier we may have internal target values for our individual factors, and compare these with our current capability in an 'Engineering competitive assessment'. Finally, we may wish to check that all factors are under control and monitored in a suitable manner, as shown near the base of the chart.

A chart of this complexity requires much study, as there are many interactions and trade-offs to consider. For a full explanation of QFD techniques see for example references 1-4.

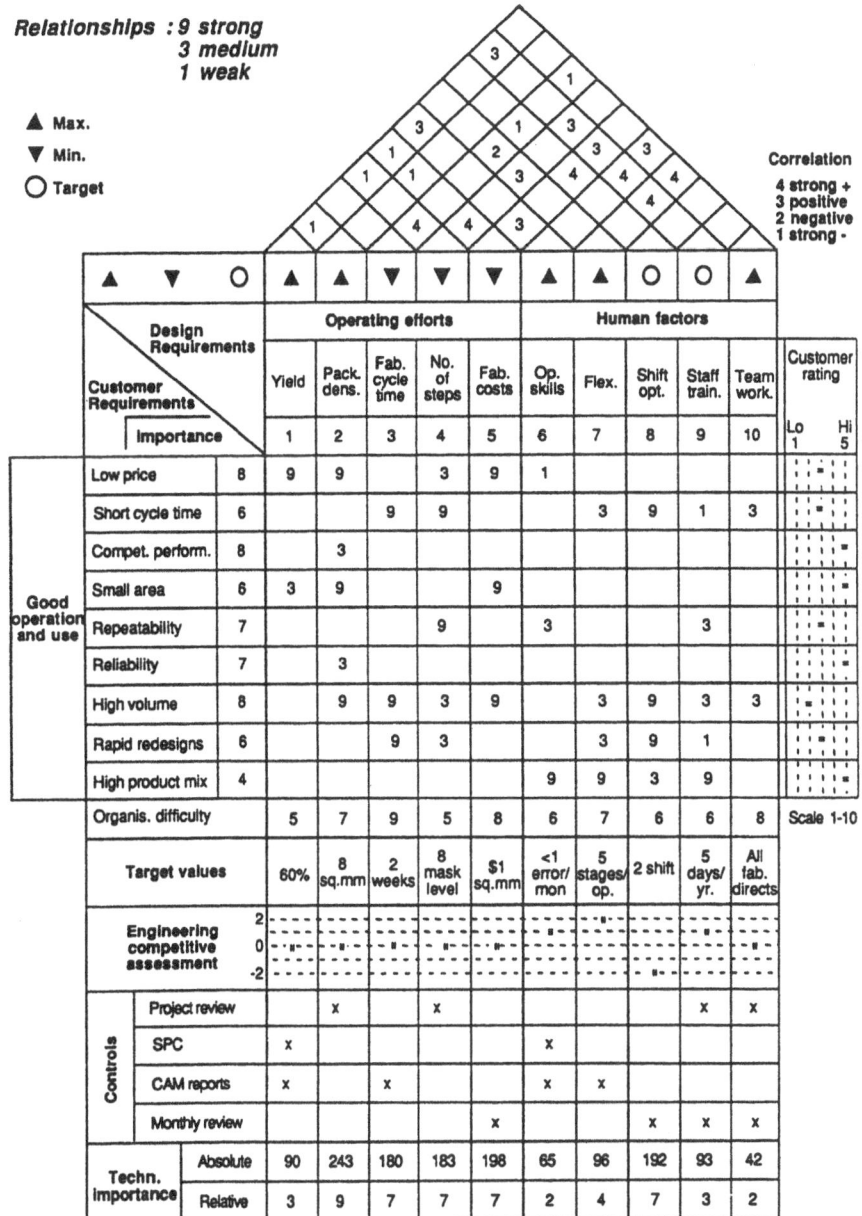

Figure 5.4 Quality Function Deployment

2 Wafer Fabrication : Cost and Yields

This section concentrates on analogue GaAs MMICs, although many of the arguments and formulae are equally applicable to digital GaAs circuits.

The yield of functioning GaAs MMICs can be described in terms of the following individual yield figures of merit:

a) Throughput yield   This refers to the loss of whole wafers through, for example, wafer breakage. This yield increases as the process maturity increases and as techniques such as automated wafer handling are adopted.

b) Component yield   This refers to the loss of individual circuits owing to the catastrophic failure of one or more components.

c) Probe yield   Failure of a functioning circuit to meet specification accounts for this yield loss.

d) Visual yield   Failure to pass visual inspection criteria results in yield loss.

e) Other   For example, burn-in failure.

Increasing circuit complexity has tended to maintain component yields at <70%, and it is this yield metric that is receiving the most attention.

Component Yields   It is over 30 years since Hofstein and Heiman (5) published the Poisson yield model for a silicon insulated gate FET,

$$\text{yield } Y = e^{-NA_G D} \qquad\qquad \text{equation 5.1}$$

where A_G is gate area of each transistor
       D is the average defect density (defects/unit area)
       N is the total number of transistors

Defect densities in real semiconductor manufacturing processes appear to vary from wafer to wafer and lot to lot. Therefore a constant defect density rarely exists. The modelling of such varying defect densities was first proposed by Murphy (6), producing the well known formula for the mixed Poisson yield model,

$$\text{yield } Y = \int_0^\infty = e^{-AD} f(D)dD \qquad\qquad \text{equation 5.2}$$

where A = defect sensitive area
             D = defect density
             $f(D)$ = defect density distribution

However, care is needed when applying equation 5.2. The formula does not hold when severe defect clustering occurs. This can arise within new or highly complex processes, such as those used to fabricate GaAs transistors. Nevertheless, the formula is freely used in GaAs yield modelling.

Norris and Barratt (7) found no strong evidence of such defect clustering in their analysis of GaAs MMICs. As a result they appropriately used Poisson statistics to derive the following yield for MMICs produced at Lockheed Sanders:

$$\text{yield } Y \approx A \exp\left\{ -[D_F.F + D_C.C] \right\} \qquad \text{equation 5.3}$$

where

	A	fixed (process dependent) constant
	F	FET gate periphery
	C	MIM capacitor area
	D_F, D_C	corresponding defect densities per unit area

If Murphy's model does hold, it is possible to plot a family of probe yield curves relating % die yield to die size for a range of defect densities (per unit area), as shown in Figure 5.5. Superimposed on the plot are corresponding stages in the development of GaAs MMIC manufacturing, from research to volume production (8).

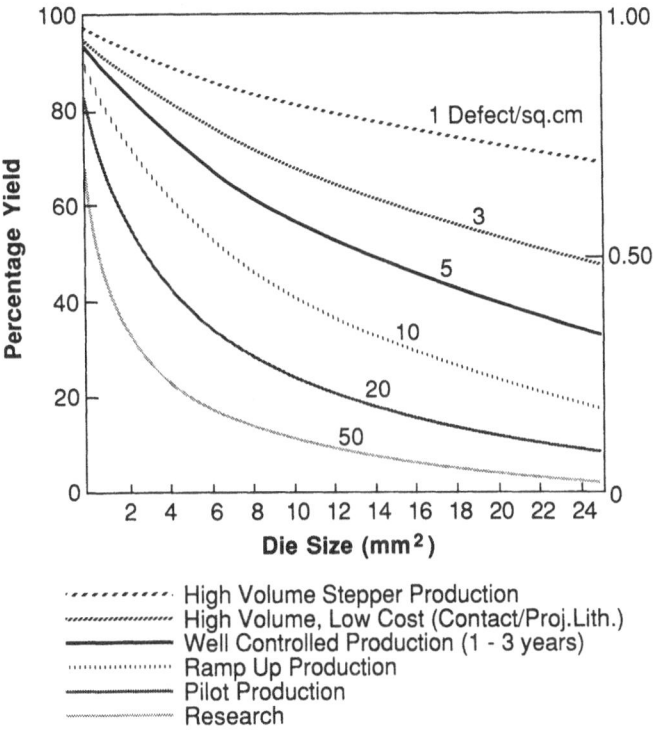

High Volume Stepper Production
High Volume, Low Cost (Contact/Proj.Lith.)
Well Controlled Production (1 - 3 years)
Ramp Up Production
Pilot Production
Research

Figure 5.5   Die Yield Against Die Size

324

Experience Curves The experience curve is a long term strategic concept. It combines the effects of numerous factors relying on a competent and effective management team to systematically exploit the experience cost opportunities which exist (9). The curves shown in Figure 5.6 demonstrate the relationship between cumulative chip deliveries and relative cost for curves of 70 - 90%. By comparison, a 70% experience curve described the cost of random access memory (RAM) components in the period 1976-84. The curves are particularly useful during production ramp-ups to aid decisions on future pricing strategies. Clearly it is important to develop an accurate cost model in order to achieve meaningful results. Inaccurate apportioning of overhead costs, for example, can lead to misleading conclusions, a fact that has led to the increased use of activity-based costing techniques.

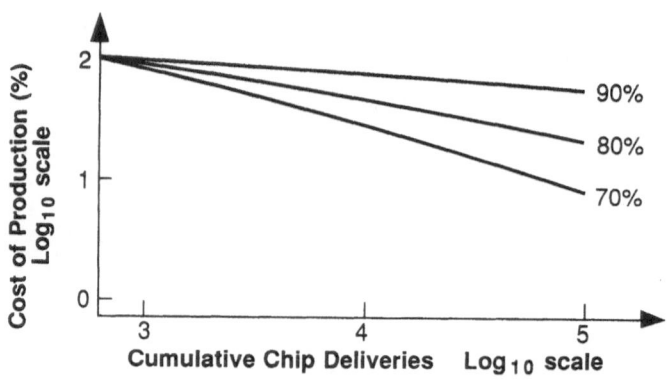

Figure 5.6   Experience Curves

3   Process Control

The increasing use of GaAs MMICs in high volume, low cost commercial systems imposes new demands on MMIC manufacturing. The importance of low costs per chip requires great emphasis to be placed on the quality of the MMIC product, to ensure high yields of chips that meet specification from each wafer started. This in turn means that the processes that are used to fabricate the MMICs are well-controlled, are individually of intrinsically high yield, and that the equipments on which the processes are run are well-understood and well-maintained. Long established techniques are available to manufacturers of GaAs MMICs that can assist in reaching the goal of high quality production while simultaneously reducing the levels of waste, and thereby reducing chip costs. Throughout industry in Europe, North America and Japan, statistical process control (SPC) is the primary tool for monitoring and controlling production processes.

Statistical process control was first mooted in the USA in the 1930s, with Walter A Shewhart and W Edwards Deming at the forefront of its application. In the years following the Second World War, the techniques of SPC were embraced by the Japanese, under Deming's tutelage, and contributed greatly to the economic revival of Japan (10). The Japanese (silicon) semiconductor industry was one of the many to benefit. MMIC manufacturers have been quick to appreciate the cost advantages that SPC can bring, and many GaAs companies have written about the introduction of SPC as GaAs leaves Research and Development to become an established manufacturing technology (11-14).

Background to SPC Statistical Process Control uses the historical day-to-day variation in results from a process for comparison with the latest measurements as a means of assessing whether the process is continuing to produce items of the desired quality (15). If "abnormal" measurements occur (that is, measurements that do not conform to the historical pattern of data), then immediate preventive action can be taken to ensure that the process is restored to its required state before any further processing is carried out. This emphasis on immediate feedback of measurements to adjust the process is the key difference between the SPC philosophy and the "traditional" method of fault detection. The latter's emphasis on in-process and post-processing inspection as a means of detecting and discarding unacceptable items is notoriously fallible and relies on repeated inspections that delay processing, add no value to the product and still do not guarantee that all faulty items are prevented from reaching the customer. SPC implementation can result in the near-elimination of in-process inspections and massive reductions in testing, while helping to ensure greatly improved product quality. It gives operators confidence that processes are stable and initiates production stoppages only when processes are demonstrably producing inferior items.

In the context of SPC a "process" can be any activity from the combination of materials, people and techniques that go to produce a finished MMIC to the analysis of despatch records and customer complaints. Anything that can be measured, such as electrical test data or numbers of defective items, can be subjected to the controls of SPC. The "statistical" aspect links present data to the data patterns of the recent past, thereby providing the feedback that allows faltering processes to be controlled.

The key to SPC is the comparison of measured data collected from a process with the normal, or Gaussian, distribution. For a fixed, stable process, the data measured from run to run, or from wafer to wafer, will not be identical, but will form a distribution about some central (mean) value. This reflects the natural variation inherent in any production process: for example the small differences occurring in chamber pressure, gas flow or power level during nominally identical dielectric deposition runs, in addition to measurement errors, will result in slightly different thickness measurements from run to run (Figure 5.7). The data should approximate to a normal distribution provided no special causes are present that can be shown to distort the pattern of measured results. Such special, "assignable", causes might be inconsistent placement of test pieces from which measurements are taken, or the use of more than one, non-identical, equipments for the processing, or two operators performing the task in subtly different ways.

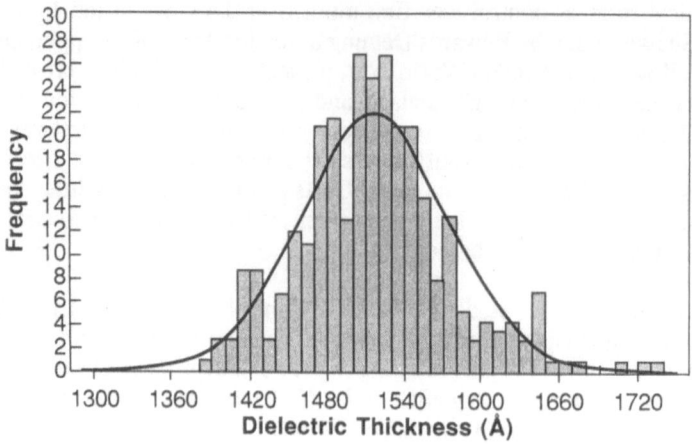

Figure 5.7   Typical MMIC Process Data

It is usual for measurements in production to be taken on a sample basis rather than measuring every possible site on every wafer to gain knowledge of the entire population of items in the distribution. This approach is reflected in the data of Figure 5.7, which are of one measurement per process run. However, provided all assignable causes have been eliminated and the data truly represents the output from a stable process, the Central Limit Theorem can be invoked to infer information about the entire population knowing only the distribution of the sample data. The normal distribution is of such a shape that the probability of a

measurement value falling between the mean ($\overline{X}$) minus one standard deviation (s) and the mean plus one standard deviation is 68.26%. The probability of a

measurement falling between $\overline{X}$ - 2s and $\overline{X}$ + 2s is 95.46%, and for $\overline{X}$ ± 3s is 99.73%.

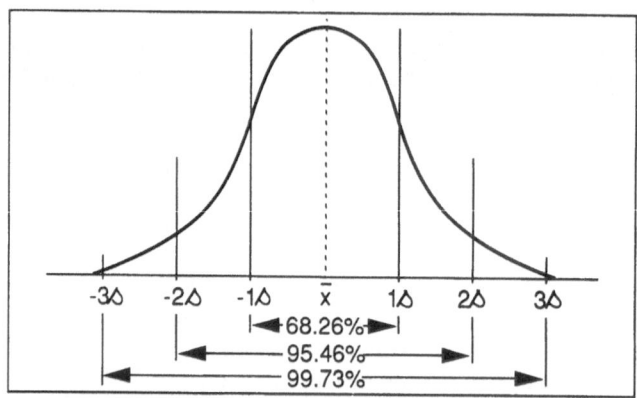

Figure 5.8   Probability and the Normal Distribution

327

It can be seen from Figure 5.8 that the probability of a measurement falling outside the bulk of the distribution, beyond $\overline{X}$ - 3s and $\overline{X}$ + 3s, is 0.27%. That is, 0.135% of the data for normal distribution should fall below the value $\overline{X}$ - 3s and 0.135% above the value $\overline{X}$ + 3s ; i.e. about 1 in 1000 measurements should each fall above or below the $\overline{X} \pm$ 3s range. These $\overline{X} \pm$ 3s values provide reference lines (known as control lines) against which new data can be compared and also allow the "quality" of the process to be assessed, with reference to the specification assigned to the measurement.

Measurements from a process are gathered on a sample basis (for example, three measurements every 100 runs) and the data plotted sequentially. The mean value and the difference between the highest and lowest value (the range) are usually plotted. After at least 20 sets of measurements from this process (for statistical validity) the distribution of data can be assessed by calculating the mean of the mean values $\overline{\overline{X}}$) and the mean of the range values ($\overline{R}$) over the 20+ runs (Figure 5.9).

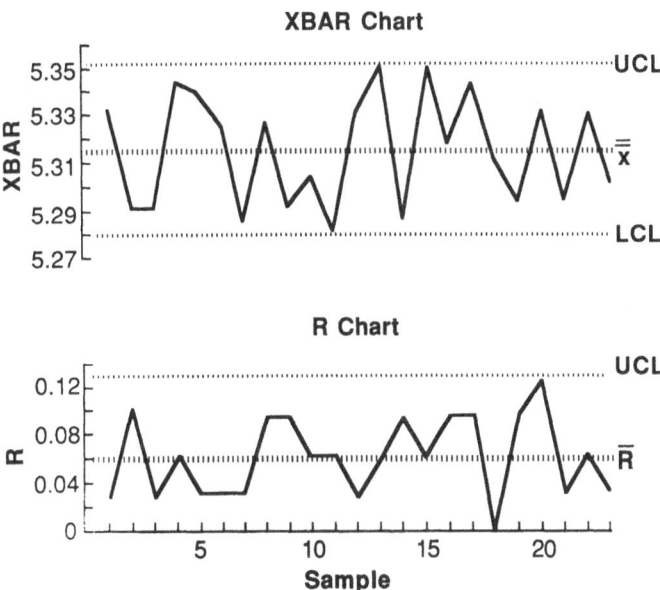

Figure 5.9   Mean and Range Charts for Occasional Samples

328

Control limits for the process are calculated using the formulae:

$$UCL_{\overline{X}} = \overline{\overline{X}} + A_2\overline{R}$$

$$LCL_{\overline{X}} = \overline{\overline{X}} - A_2\overline{R}$$

$$UCL_R = D_4\overline{R}$$

$$LCL_R = D_3\overline{R}$$

where UCL, LCL are respectively the upper and lower control limits for each

plot, and A_2, D_3 and D_4 are tabulated factors converting $\overline{R}$ into an estimate of standard deviation, taking into account the number of measurements forming the sample (3 in the case of the example in Figure 5.9). These limit lines are also shown in Figure 5.9.

The historical data, in the form of a "typical" mean value $(\overline{\overline{X}})$ and a "typical"

range $(\overline{R})$, have thus been used to provide reference lines against which further data points can be compared. Provided future data points from the sampled measurements fall between these control limits, the process has not changed. Therefore if the process was originally producing items within specification, the sampled data confirm that it is still doing so, and thus the need for extensive testing and inspection is removed. Note the proviso concerning the specification : a process can be in statistical control yet still not produce items that meet specification.

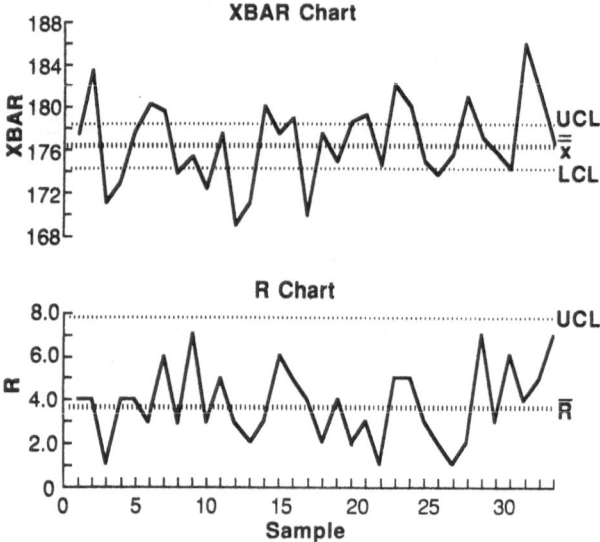

Figure 5.10 MMIC Process Data Analysed by the Mean and Range Method

Continuous versus Batch Processing The example presented above of sampling every 100 process runs is not typical of MMIC processing, especially in the early days of manufacture following transfer from a Research and Development environment. In these circumstances it is far more usual to take measurements at several sites on each wafer, or on one wafer in a batch of wafers. This method of sampling the total potential data pool is quite different from the sampling procedure described above, and different charting techniques are

required as a consequence (14 - 16). If the data are presented in the $\overline{X}/R$ form they will generally appear completely out of control (Figure 5.10), since there are now two principal sources of variation in the data (assuming assignable causes have been eliminated from the process): not only the within-sample variation, the range, but also the run-to-run variability (Figure 5.11).

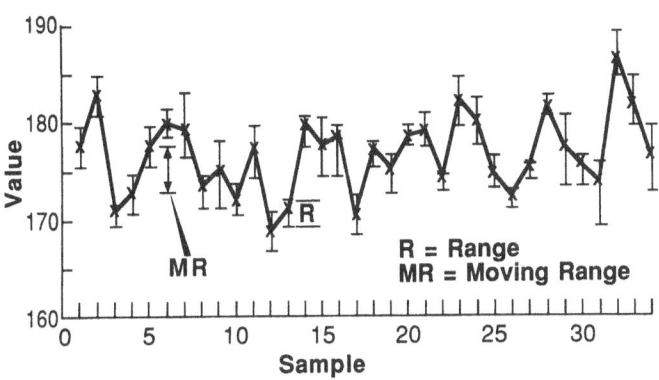

Figure 5.11   Comparison of Within-Batch and Batch-to-Batch Variability

In order to take this into account, it is usual to calculate the successive differences in the mean value from batch-to-batch (the moving range) and plot the mean and moving range as a chart pair (Figure 5.12). The average of the moving range values now provides the estimate of standard deviation necessary for the calculation of control limits, where the sample size for selection of tabulated factors is now 2. An assessment of the control of within-batch variation may be gained in exactly the same way by charting the successive range values and their associated moving ranges.

Application of SPC to MMIC Fabrication Computer aided manufacturing systems, such as Workstream (formerly known as COMETS) and PROMIS (17,18), are essential in modern MMIC facilities for the storage of measurement data from standardised test structures and its immediate, or subsequent, analysis.

Control charts for SPC can be readily set up to provide instant feedback to technician and engineering staff of the state of each process making up the overall fabrication route. Charts can be presented on the terminal at the time of data entry and any violations of the control limits or statistically unusual trends of data points will cause the process to be stopped, pending investigation of the reason for the violation. By the statistical rules employed, about 1 in 1000 measurements might be expected to cause a chart violation through natural variation, but assignable causes of violation should be rigorously pursued and eliminated.

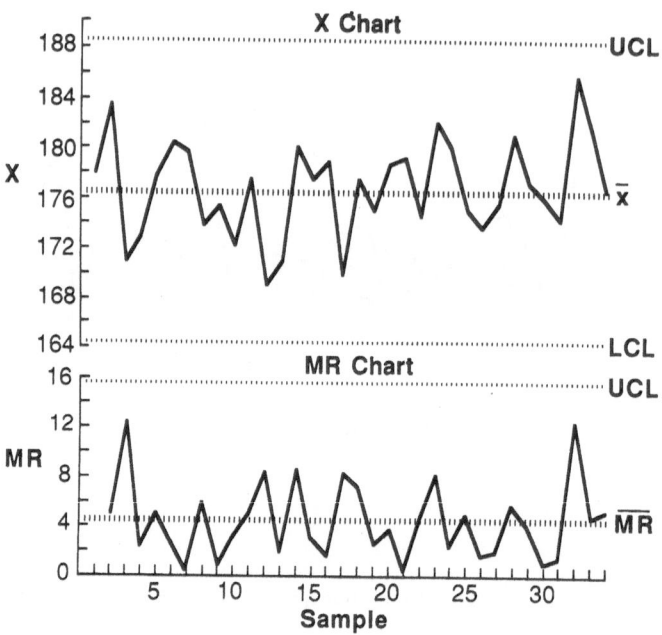

Figure 5.12   Data Analysis by the Moving Range Technique

As mentioned earlier, a process in statistical control does not necessarily guarantee products within specification. The calculations to generate control limits provide a means of measuring the capability of a process to produce items within specification by the simple division of the tolerance allowed by 6 times the standard deviation of data from the process. The centring of the data within the specification is measured by dividing the difference between the process mean and the nearer control limit by three times the standard deviation for the process.

These parameters, C_p and C_{pk} respectively, relate directly to process yield, since they record the likelihood of the data distribution for the process producing items outside the specification   limits.   The C_p index is related to process yield according to the curve in Figure 5.13. A figure of 1.33 for both C_p and C_{pk} (the two indices are equal when the process data are centred within the specification) is reckoned to be the minimum necessary to guarantee less than 1 in 1000 items out of specification.

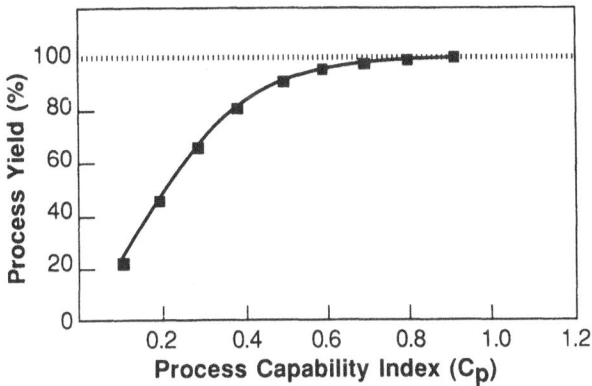

Figure 5.13   Relationship Between Cp Index and Process Yield

The philosophy of continuous improvement requires that process capabilities are regularly re-assessed and, where possible, process modifications set in train to improve yields still further. The use of statistical design of experiments can be of great benefit here, in reducing the number of trials required to gather statistically meaningful data on the process and thus leading to process enhancement. Software engineering tools such as RS/Discover (19) can assist in experimental design as well as data analysis, leading to efficient deployment of resources.

4   Continuous Quality Improvement

ISO9000 ISO9000 sets out standards for quality systems, informing suppliers and manufacturers of what is required of a quality oriented system. The standards identify the basic disciplines and specify the procedures and criteria to ensure that products or services meet the customers' requirements. Many companies now insist that they will only purchase from suppliers that are certificated to ISO9000 standards.

The following key conditions must be met in order to achieve certification:

1   One manager must co-ordinate and monitor the quality system, and see that prompt and effective action is taken to ensure that the requirements of ISO9000 are met.

2.   Organisation, structure, resources, responsibilities, procedures and processes must be documented and understood by the appropriate personnel.

3.   Establish a documentation control system.

4.   Ensure product identification and traceability of materials.

5.   Ensure appropriate control mechanisms, especially in manufacturing operations.

6.   Perform inspection and test as required, with appropriate controls.

7.   Establish a system to identify the inspection status of product during all stages of manufacture. Written control procedures are necessary.

8.   Non-conforming product should be clearly identified. Document all non-conforming details for the record, and document all corrective actions.

9.   Corrective action procedures must be in evidence.

10.   An effective records storage and retrieval system must be in operation.

11.   Carry out internal quality audits to check effectiveness of the quality system.

12.   Carry out appropriate training of personnel, and keep training records.

13.   Use statistical techniques where appropriate.

Quality Costs   The traditional model divides quality costs into 4 segments.

1. **Prevention costs**   These costs arise from any action taken to investigate, prevent or reduce defects and failures.

2. **Appraisal costs**   These costs rise from an assessment of the quality achieved.

3. **Internal failure costs**   These are costs arising within the manufacturing organisation of the failure to achieve the quality specified.

4. External failure costs These arise outside the manufacturing organisation of the failure to achieve the specified quality.

Within GaAs manufacturing the following are examples of the above:

Prevention costs Quality and process control, planning, training, audits, design reviews, calibration, supplier assessment.

Appraisal costs Inspection, test.

Internal failure costs Scrap, rework, repair, failure analysis, concessions.

External failure costs Customer complaints, liability, repair, warranty.

Traditionally the GaAs manufacturing activity has exhibited high appraisal and internal failure costs (through in-process inspection, 100% test and high scrap and rework rates). In common with many other industries the emphasis is now on prevention, moving away from extensive inspection and test as the yields increase.

Total Quality Despite being eager to take up some of the key tools of Total Quality (or Continuous Improvement as it is perhaps more appropriately known), the GaAs semiconductor industry has not been in the forefront of practitioners. However there are signs that as the business matures the benefits that other industries have reported from adopting a CI strategy may also attract GaAs executives and engineers.

5   Shop-Floor Scheduling

Semiconductor batch processing in general, and small batch GaAs wafer fabrication in particular, present difficult problems for the Production Controller. A typical GaAs MMIC front face process is shown schematically in Figure 5.14. Each concentric ring represents a process stage (or mask level). A batch completes a revolution of the innermost circle, proceeding then to the larger adjacent circle, and so on. Since batches are routed through the same equipment many times, scheduling soon becomes an art rather than a science.

There are two practical methods for developing schedules of work to be done by production resources. They are forward scheduling and backward scheduling. Both methods may be used manually or by computer. Forward scheduling is the way a Production Controller will plan his/her work in the short term. 'Do one job, when you've finished it - do the next job which should be ready for you by then'. It is the only practical method for finite capacity scheduling, since by planning jobs one after another the capacity limit is respected. The finish date is the result of a schedule. The problem with this method is that it is difficult and requires high levels of data accuracy if schedules for more than the very short term are to be developed. Finite capacity scheduling software can be costly.

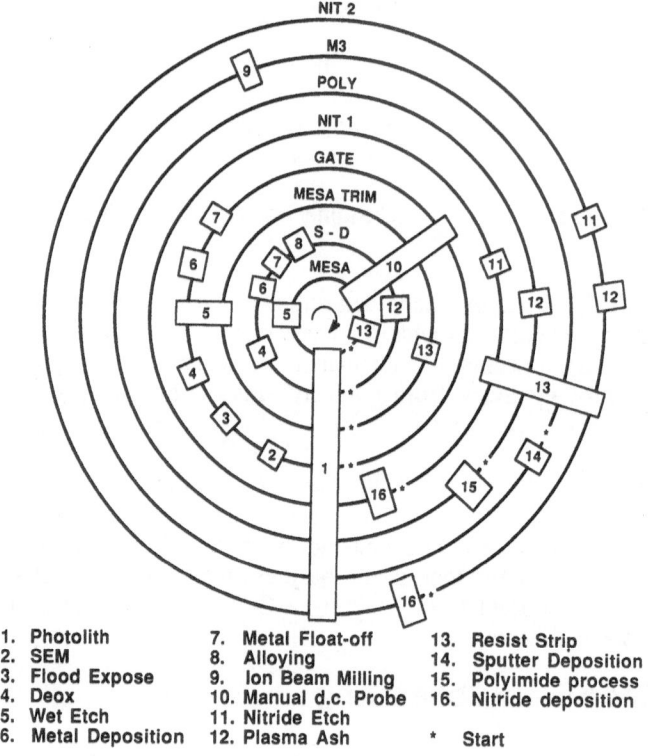

1. Photolith
2. SEM
3. Flood Expose
4. Deox
5. Wet Etch
6. Metal Deposition
7. Metal Float-off
8. Alloying
9. Ion Beam Milling
10. Manual d.c. Probe
11. Nitride Etch
12. Plasma Ash
13. Resist Strip
14. Sputter Deposition
15. Polyimide process
16. Nitride deposition
* Start

Fig 5.14   MMIC Process Representation

In backward scheduling the plan for the work is developed by starting from the desired completion date and working backwards. The last operation is planned and the time required is calculated. Usually some slack time is inserted then the preceding operation is planned. For each step a start and finish time is determined. Backward scheduling is cheap and easy, but no account of capacity is taken whilst the schedule is being calculated (although this can be done afterwards). Conventional MRP is backward schedule based.

A third alternative, bottleneck scheduling, is less widely used. Often referred to as OPT (Optimised Production Technology) this computer-based package accepts that it is unrealistic to attempt to balance the line. That is, bottlenecks will always exist. In the OPT approach the schedule is constructed around the needs of the bottleneck resource. Backward scheduling ensures that the bottleneck is never starved. An hour lost at a bottleneck is an hour lost for the total system. Forward scheduling downstream will create an accurate, practical schedule thereafter.

Finally, it is worth considering the merits of Just-in-time (JIT) scheduling, which has revolutionised shop floor scheduling in many manufacturing environments and has indeed been piloted in photolithography work cells. There are many misconceptions regarding JIT, the two principal ones being that JIT is only applicable to volume manufacture; and that demand must vary little over time. Neither of these is actually true. In addition the wider definition of JIT - 'Reduction of Waste'- can be a useful lever to reduce all non value-adding activities, not just queuing time and inventory levels in batch manufacture.

If a software scheduling system is adopted care must be taken to ensure it interfaces with existing wafer fabrication databases. Most modern MMIC Foundries utilise CAM (Computer-Aided-Manufacture) systems for WIP tracking, equipment and plant monitoring, materials management and yield statistic gathering. Modular CAM systems such as PROMIS (17) and WORKSTREAM (18) have their own shop-floor scheduling packages, and it can be difficult to integrate the CAM system with other bought-in scheduling software packages.

6   Equipment Issues

The transfer of MMIC processes from development to full production can entail significant changes in equipment requirements. The drive towards high chip yields at minimal cost often conflicts with the very high capital costs involved in equipping clean rooms in which to produce those chips. Semiconductor processing equipment originally intended for silicon wafer fabrication can sometimes be modified to handle the heavier and more fragile gallium arsenide wafers. Often, however, purpose-built equipment is required and the relatively small sales volumes to be expected within the GaAs industry render individual equipments disproportionately expensive. A common compromise is to modify the equipments used in the MMIC development phase, for example, by adding cassette-to-cassette handling mechanisms, and place increased emphasis on preventive maintenance schemes to keep equipment functioning, rather than buying new. CAM systems are again of great benefit here. The wafer tracking functions of the CAM system automatically register equipment usage, allowing routine maintenance to be scheduled after a chosen number of runs. The CAM also allows a complete maintenance record to be kept, and provides valuable data on equipment uptimes and reliability.

As manufacturing volumes are increased, equipment and process techniques require review. For example, the lithography methods employed may be dependent on wafer numbers for maximum processing efficiency. A first move might be to set up a "lithography cell", where the group of lithography equipments (photoresist and developer tracks, aligners) are used full time by a limited number of personnel to ensure that this potential bottleneck is controlled effectively. As volumes increase further, the added expense and complexity of wafer stepper technology might be invoked to guarantee very high yields as well as good throughput levels.

Lithography Issues   One of the most important aspects of GaAs processing is the definition of the small geometries necessary for the high speed operation of the devices. Many lithographic techniques are now available and in this section we will examine the options that are open to the process technologists for the definition of 0.5 micron and smaller metal line widths.

In the mid 1960s when optical lithography first became a commercial reality, contact and proximity printing was used. This approach was abandoned some 15 years ago in the silicon industry but is still one of the major methods of lithography used in GaAs fabrication processes. Mask wear and contamination through its closeness to photoresist-covered wafers were two of the main reasons for its abandonment by the silicon industry and certainly when volumes exceeding 2000 wafers per week are encountered this is a serious consideration. In GaAs however, where production numbers are more modest such problems are not so prevalent and new processing techniques have solved most of the contamination problems.

An additional problem of this approach at the 0.5 micron line resolution level is the procurement of the contact mask. It is only recently that mask makers have developed the necessary technology to allow reproducible geometries at these small dimensions to be achieved. However, as wafer sizes increase beyond the typical 3" diameter this type of lithography will become more difficult as contact over larger areas is almost certain to be non-uniform and alignment tolerances will become unacceptable.

Already GaAs fabrication lines are moving towards a stepper technology which removes many of the problems of contact lithography. Steppers operate on a projection principle and do not rely on mask contact with the resist. They are also capable of reducing the mask image so that 5x or 10x reticules can be used, reducing the problems of producing masks with superfine geometries. In the past this technique has been limited to resolving dimensions of 0.6 - 0.8 microns. However in the past two years or so several major breakthroughs in stepper design have been made which will mean that this approach will be technically and economically viable for the manufacture of integrated circuits with feature sizes down to 0.2 microns.

The resolution limit of this technique (the Rayleigh criterion) is defined as

$$R = \frac{K\lambda}{NA}$$

where NA is the numerical aperture, λ the wavelength of the incident radiation and K a constant, classically 0.7 - 0.8.

In order to improve the resolution limit, R has to be minimised. Currently systems have an NA equal to 0.55 - 0.6, though new equipments have values closer to 0.7. For practical reasons I-line illumination is used, where $\lambda = 356nm$ (20). If 0.2 micron resolution is required, K is the remaining factor that can be optimised and recent developments, such as phase shift masks and off-axis illumination (21 - 23), have shown that K values of 0.4 may be achievable. These equipment and operational tricks together with improvements in processing techniques, such as more sensitive resists and the use of electron cyclotron resonant etching systems, gives real hope for the resolution of 0.2 micron geometries in the near future.

The alternative solutions to sub 0.5 micron geometries   There are four additional approaches to lithography that offer promise for the production of <0.5µm geometry device structures: electron-beam direct-write, ion-beam direct-write, X-ray steppers, and laser-based deep-UV steppers. If we examine each in detail, we are able to judge the viability of the various techniques in terms of ultimate resolution and cost.

Electron-Beam Lithography   This technique is by far the most mature of the four approaches; although it is not yet an accepted silicon processing tool, the GaAs industry is using its high-resolution capability in a fine line mixed lithographic process. In this technique, only the fine lines are directly written with the electron beam, the remaining coarser features are defined by conventional optical lithography. GaAs MESFET-based ICs with 0.3µm gate lengths have been produced, albeit only at SSI complexity, using this approach.

Throughput is a major disadvantage of this approach to achieving small dimensions. The beam diameter must be small with a correspondingly small beam current, which proportionally increases the resist exposure time. A further disadvantage is its inability to write closely spaced lines due to the proximity effect. Upon entering the resist, the beam is scattered outside the initially finely defined "strike" area, and is also backscattered from the substrate, causing the resist to be exposed over a larger area than intended. However, this problem can be alleviated by using a multilevel resist technology. The scattering in the electron-beam-sensitive resist is minimised by having only a very thin layer of resist on top of a thicker layer of electron-beam-insensitive material. The underlying thicker resist layer is preferentially removed by subsequently exposing it through the developed pattern in the top electron beam exposed layer.

Some reporters suggest (24) that the need for the electron beam process to supersede optical lithography has disappeared and hence the real applications are:
1.   Reticle making;
2.   Semi-customised circuits;
3.   GaAs microwave circuits with submicron geometries;
4.   X-ray mask manufacture;
5.   Experimental submicron lithography.
These applications use the high-resolution capability and flexibility of the electron beam process without the need for high throughput rates.

Ion-beam Lithography   Focused ion-beam lithography is relatively immature, requiring many years of development before it is competitive with other high-resolution techniques. However, this technique offers some advantages over the electron beam approach. The heavier ions suffer far less scattering in the resist and can supply much higher energy to the resist, resulting in greater resist sensitivity and faster writing time. One difficulty to be overcome is the effect of damage that the energetic ions may impart to the underlying semiconductor. Again, this problem may be alleviated by using a double-layer resist technology to prevent the ions from penetrating to the semiconductor surface. This technique can give spot sizes down to 0.1μm.

Deep-UV Optical Steppers   The semiconductor industry is becoming increasingly interested in using the enhanced resolution capability of laser-driven deep-UV steppers. We have already shown how reducing the wavelength of the incident radiation can improve the resolution of the system. Excimer lasers that emit an intense pulse in the 150 - 350nm region are possible sources for such equipment.   However, at present, their reliability is suspect and more reliable solid-state alternatives such as Nd:YAG are being studied. Resist developments are also required to make full use of this technique. The novolac resists currently have too high an absorption at deep-UV wavelengths, necessitating the use of multilevel-resist techniques (25). Nonetheless, deep-UV optical steppers offer higher throughput than either electron beam or Ion-beam technologies, and the equipment cost is far less.

X-Ray Steppers The use of X-rays to expose resist has been a theoretically attractive approach for many years. The short wavelengths (2 - 10Å) practically remove diffraction effects, masks can be used in a noncontacting mode, and backscattering or reflection from the substrate is virtually zero. These properties allow very small lines with vertical walls to be resolved in thick resists. Other attractive features of the process are that resist thickness variations do not affect line width and dust particles will not impair the features, because they are transparent to X-rays. We thus have a most desirable technology, but three major impediments to successful operation remain to be solved:
1.   Reliable x-ray sources;
2.   Cost-effective mask-making processes;
3.   Effective x-ray sensitive resists (26).
    The resolution limits of this approach are set by the quality of the mask. Because there is no image reduction, electron beam lithography must be used to produce the reticle. In practice, this limits the minimum resolvable dimension to about 0.5μm, which does not yet allow the full potential of X-ray lithography to be realised.
    Despite this lengthy list of problems, X-ray lithography is considered to be one of the leading contenders for producing <0.5μm geometry semiconductor devices and circuits.   Considerable sums of money are being invested to solve the problems highlighted in this section, and good progress has already been made.

7   Facilities of the Future

The yields of the lithographic techniques described in Section 5.1.6 are strongly dependent on the environment in which the machines are placed, and people limit the ultimate cleanliness that can be achieved in any clean room facility. This statement is certainly true, as between 500-1000 particles per minute are shed by processing personnel, despite the elaborate clean room procedures adopted in all fabrication areas. The ultimate clean room has long been the dream of the manufacturing manager, and developments since the mid-1980s give hope that processing areas isolated from the intervention of humans, and hence from the particles they generate, will become a reality.

The technology most widely evaluated at present is the standard mechanical interfaces (SMIF) isolation concept. The SMIF approach keeps wafer cassettes sealed in a protective box so that they are not affected by the environment around them. They can then be transported between processes through relatively dirty environments without being contaminated by foreign matter. In most facilities that use this concept, human intervention is then necessary for the particular process step. If this step is, for instance, the loading of the wafers into a process boat, the chances are that many tens of thousands of extra particles will be generated in placing the wafer into the boat. This is a most undesirable situation , but it may be overcome by the use of robotics for 'soft' placement.

Such robotic systems in conjunction with SMIF are now being evaluated. Special clean-operation robotic mechanisms have been designed for specific tasks. These robots can pick wafers from the cassette, insert them into the processing equipment, and return the wafers to the cassette ready for the next step when the process is complete. This processing approach only requires clean areas directly over the process equipment and is therefore less expensive in air processing plant. This approach is believed to be a very cost-effective solution to contamination problems.

The average cost of a new manufacturing facility is around $50 million and can reach $100 million for a large facility. This figure contains the clean room construction cost of approximately $1000 per square foot with air-handling costs of roughly $100 power square foot per year. A 20,000 square foot facility would therefore cost in the order of $2 million per annum. A totally robotic SMIF system (if shown that it can adequately perform the tasks) would perhaps require less than 5% of this yearly cost.

The use of robotics may have other, far less obvious financial advantages. There is, for instance, considerable cost associated with dressing in clean room clothes. Processing personnel can spend up to 30 minutes of unproductive time per shift putting on and removing their gowns, and as much as 5 - 10% of clean room space is occupied by changing rooms. This nonproductive use of time and space could be better deployed.

The need to dress in rather uncomfortable clean room clothing has long been recognised as a deterrent to process engineers entering the clean rooms to rectify equipment and process faults. Delays can occur because of this, causing "knock-on" effects in the production line. The processing personnel may also work more effectively if unhindered by the sometimes cumbersome clothing. The robotic clean area would virtually eliminate the need for clean room clothing, and one can easily envisage operators and process engineering staff controlling isolated dust-free environments while dressed in ordinary casual clothing.

The operator "hands-off" approach would lead to a cleaner working environment and a closer operator-process-engineer manager relationship for they would be able to interact more freely, unrestricted by their surroundings. The yearly costs of such operations could also be considerably lower than today's more conventional facilities if the financial arguments can be justified.

8   Conclusions

This section has dealt with the many factors affecting the volume manufacture of GaAs components. Over the past few years the major European GaAs component manufacturers have put expensive facilities into place to meet the future demands of the market place but the manufacturing infrastructure is, in general, some way behind that of the more mature silicon semiconductor industry. The European Gallium Arsenide industry will become a leading world player in this technology if prompt attention is paid to the issues raised in this section. It is therefore vitally necessary that future EC and National Government funding is channelled into these areas through programmes like ESPRIT and EUREKA so that the partnership between Industry and Government can be strengthened, past funding is not wasted, and European Industry can compete on level terms against the ever increasing threat of the Americans and Japanese.

9.   References

1. 'Taguchi Methods and QFD : Hows and Whys for Management"
   N E Ryan (Ed), ASI Press, Dearborn, Michigan, 1988.
2. "Quality Function Deployment - A Practitioner's Approach"
   J L Bossert, ASQC Quality Press, Milwaukee, Wisconsin, 1991.
3. "Quality Function Deployment - Integrating Customer Requirements into Product Design"
   Y Akao, Productivity Press, Cambridge, Massachusetts, 1990
4. "Quality Function Deployment : A Modern Competitive Tool"
   M Zairi, Technical Communications (Publishing) Ltd,
   Letchworth, Hertfordshire, England, 1993
5. "The Silicon Insulated-gate FET" Hofstein S R and Heiman F P
   Proc IEEE, Vol 51 no 9, pp 1190-1202 (1963)
6. "Cost-Size Optima of Monolithic ICs" Murphy B T,
   Proc IEEE. Vol 52 no 12, pp 1537-1545 (1964)
7. "GaAs MMIC Yield Modelling" Norris G B and Barratt C A
   1990 IEEE GaAs IC Symposium Proceedings, pp 317-320
8. I Deviny Private communications
9. "Manufacturing Strategy" T Hill, Macmillan Education Ltd, Houndmills, Basingstoke, Hampshire, England (1985)
10. "SPC and Continuous Improvement", M Owen, IFS Publications, UK 1989
11. "Statistical Process Control for GaAs MMIC Wafer Fabriction", G E Brehm, Military Microwaves, London, pp 375-380, July 1990
12. "Statistical Process Control Implementation for GaAs Integrated Circuit Fabrication"
    K A Salzman, Digest of the GaAs IC Symposium, pp 217 - 220, October 1992
13. "18 Months at Merged Triquint : Fab Yield Unqualified Success"
    Semiconductor International, p 17, January 1993

14. "The Use and Misuse of Statistical Process Control in GaAs MMIC Manufacture"
    D J Warner, C E Lindsay and C L Sansom, Digest of the GaAs IC Symposium, pp 131-133, October 1993
15. "Statistical Quality Control" 6th Edition, E L Grant and R S Leavenworth, McGraw-Hill, 1988
16. "Control Limits - Continuous vs Batch Process", J Weintraub, Microelectronic Manufacturing and Testing, p 10, July 1990
17. PROMIS System Corporation, Toronto, Ontario, Canada M5X 1E3
18. WORKSTREAM is a product of Consilium Inc. Mountain View, California, USA
19. Bolt, Beranek and Newman Inc., Cambridge MA, USA
20. "I-line Steppers Make MMIC Gates at Hughes"
    Semiconductor International, p 14, December 1990
21. "Four More Significant Japanese Advances in Phase Shifting Technology"'
    Semiconductor International, p 16, December 1991
22. "Subhalf-Micron Gate GaAs MESFET Process Using Phase-Shifting Mask Technology", Digest of the GaAs IC Symposium, pp281 - 284, 1991
23. "Ultimate Limits of Lithography", C Morgan et al
    Physics World, pp 28 - 31, November 1992
24. Spicer DF, Proceedings Institute of Physics Meeting on Lithography, London 1988
25. Pol V. et al, Optical Engineering, Vol 24, No 4, p 311, 1987
26. "Electron beam, X-ray and Ion Beam Techniques for Submicrometer Lithographies V", Fay B S and Novak W T, pp632, 146, 1986

5.2   New Devices and Technologies

H Brugger
Daimler Benz Research Centre, D-89081, Ulm, Germany

with contributions from K.H. Bachem (FhG IAF, Freiburg), J. Dickmann (DB AG, Ulm), K. Eberl (MPI FKF, Stuttgart), H. Heinecke (Univ. Ulm), J. Ralston (FhG IAF, Freiburg), and G. Weimann (TU Munich).

Introduction

Electronic systems operating in the millimeter wave frequency range require advanced device structures and new heterostructure material combinations to improve the high frequency performance and to extend the operating frequency to higher values. New concepts and technological tools are necessary to reduce fabrication cost and to allow a higher degree of monolithic integration of mm-wave devices.

This chapter discusses current issues and future prospects of new hetero-structure device layers based on antimonides, phosphorous-based quaternary materials, and indium-based ternary materials on GaAs substrates. New HFET and RTD device applications and novel low-dimensional devices (lateral-field-effect structures) are discussed. Recent progress on GaAs-related epitaxial regrowth techniques (MBE, MOMBE) on patterned wafers (dry-etched surfaces, selectively implanted layers, masked dielectric area) are reported. Regrowth is a promising way for monolithic device integration and for the fabrication of novel structures. The article reports about technologies and device relevant structures which are compatible with a fabrication on the basis of a GaAs substrate processing technology.

Novel Heterostructures and New Devices

Antimonide-Based Layered Materials

III-V compound semiconductor heterostructures containing antimony have recently become of tremendous technological interest due to their demonstrated application in a wide variety of high-performance electronic and optoelectronic devices, including ultra-high frequency resonant tunneling diodes (RTDs) [1], high-transconductance heterostructure field-effect transistors (HFETs) [2], complementary FETs [3], heterostructure bipolar transistors (HBTs) [4],

344

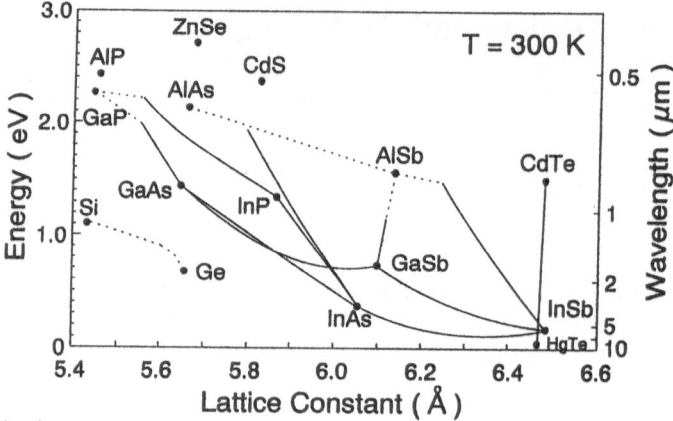

Fig. 1. Energy gap (in eV and μm) at 300 K as a function of lattice constant for various compound semiconductors.

high-efficiency mid-infrared photodiodes [5], [6], mid-infrared MQW diode lasers [7] and long-wavelength infrared detectors [8]. Demonstrated and proposed system applications of such devices include high-performance microwave integrated circuits, optical communication employing fluoride-based fibers, laser radars exploiting atmospheric transmission windows, remote sensing of atmospheric gases, and infrared imaging.

In the following, a brief overview of electronic device structures for ultra-high frequency applications grown on GaAs substrates is given, in particular for HFETs and RTDs. Fig. 1 shows the well-known bandgap versus lattice constant "road map" for III/V-semiconductors. The compounds GaSb, AlSb and InAs form a nearly lattice-matched system, with the subsystem InAs/AlSb having the largest conduction-band offset (1.35 eV) of all fcc III/V-heterostructures. However, the lattice misfit of these compounds with respect to a GaAs substrate is of the order of 7%. To overcome this limitation and to accommodate the lattice mismatch, thick and relaxed buffer layers of AlSb and/or GaSb are used. For the epitaxial growth of layered structures, molecular beam epitaxy (MBE) is currently the most widely used technique, due to the lack of a stable, high-purity Sb precursor for chemical vapor deposition.

HFET Devices. InAs as a current channel material is an interesting candidate for high-speed HFETs. A theoretical study by Cappy et al. [9] attributes potentially better performance of InAs than GaInAs for gate lengths of 1 μm or less. Published results [2], [3] indicate the enormous potential of InAs/AlSb HFETs arising from the very large conduction-band offset and the capability to achieve room-temperature two-dimensional electron mobilities exceeding 20 000 cm^2/Vs together with electron sheet concentrations of the order of 4×10^{12} cm^{-2}. Fig. 2

(a)

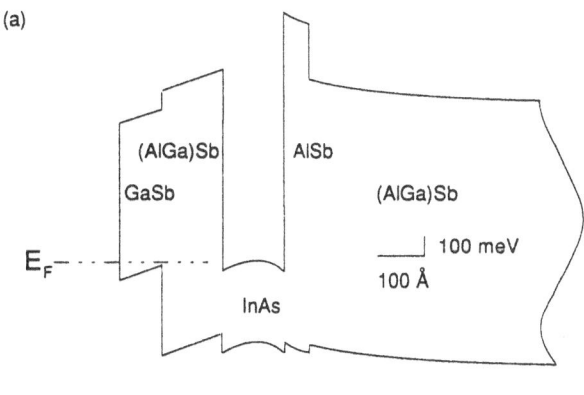

(b)

Ohmic Contact   Schottky Gate   Ohmic Contact

AuGe/Au		Au		AuGe/Au	
	GaSb cap				100 Å
	Al(0.5)Ga(0.5)Sb				150 Å
	InAs channel				150 Å
	AlSb				60 Å
	Al(0.5)Ga(0.5)Sb				2000 Å
	AlSb buffer				2.8 µm
	semi - insulating GaAs substrate				

Fig. 2. Energy-band diagram (a) and schematic layer sequence of an InAs channel HFET (from ref. [10]).

shows a schematic layer structure of an InAs HFET with (AlGa)Sb and AlSb barriers together with the corresponding band diagram.

InAs and GaSb form a type-III heterostructure, i.e. the conduction band minimum of InAs lies below the top of the GaSb valence band maximum. A promising complementary field effect logic circuit has been proposed using a vertically stacked GaSb p-type and an InAs n-type HFET structure [11]. Electron and hole wave function overlap is achieved by a thin wide band gap AlSb barrier layer.

346

A critical drawback of binary InAs channel layers is the small bandgap energy (0.36 eV), which limits both the electrical isolation capabilities and breakdown voltage/current drive, the latter mainly due to impact ionization. Much work remains to be done to optimize the epitaxial growth, device design and device fabrication in order to explore the high frequency and high power limits of such devices.

Resonant Tunneling Devices. Research activities on resonant tunneling are mainly directed for ultra-high frequency electronic device applications, because it is one of the few solid state transport phenomena that can provide a fast negative-differential resistance at room temperature and strong non-linear current/voltage

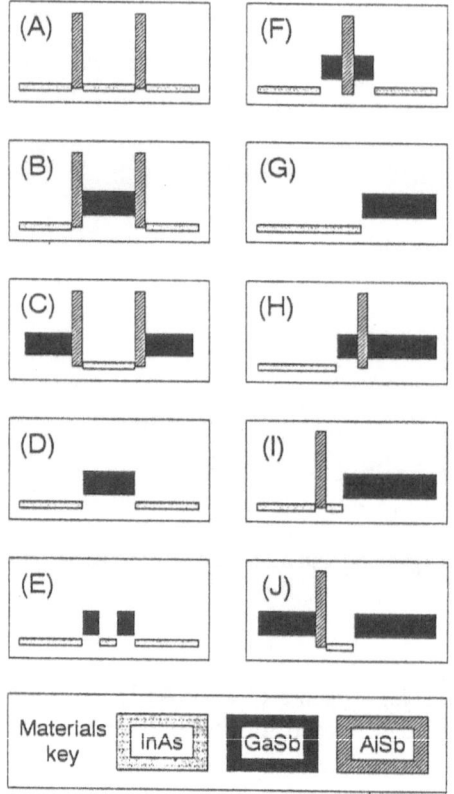

Fig. 3. Schematic energy band diagram for ten different tunnel structures realized in the InAs/GaSb/AlSb material system. Energy gaps for each material are shaded, so that the top (bottom) of each shaded box represents the conduction (valence) band edge in each layer (from ref. [16]).

behavior. Electronic applications are in the field of millimeter wave generating devices [1], [12] and high-frequency detector devices [13]-[15].

The large conduction-band offset between InAs and AlSb and the staggered type-III band alignment between InAs and GaSb are utilized in InAs/AlSb/GaSb resonant tunneling devices. Fig. 3 shows the bandstructure of ten different tunneling structures realized using different InAs, AlSb and GaSb layer sequences. Peak-to-valley current ratios of 30 and 85 at room temperature and liquid nitrogen temperature, respectively, have already been demonstrated [17].

Compared to GaAs/AlAs RTDs,

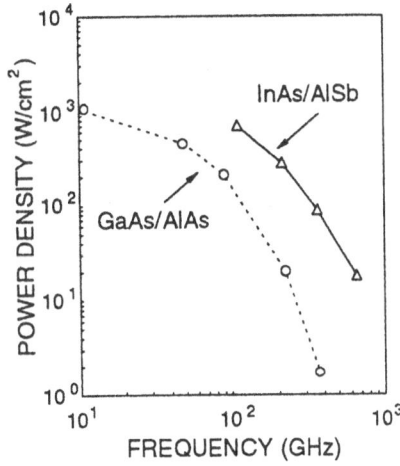

Fig. 4. Experimental room-temperature oscillator power results for an InAs/AlSb RTD (solid line) and GaAs/AlAs (dashed line) (from ref. [1]).

the InAs/AlSb material system has been demonstrated to have several advantages, including higher available current densities and operation up to substantially higher frequencies. In Fig. 4 the measured power performance of ultra-high frequency RTD-oscillators are shown for standard AlAs/GaAs heterostructures and for InAs/AlSb diodes. The InAs/AlSb RTD has shown oscillations up to 712 GHz [1]. The generated power densities are up to 50 times higher than those of GaAs/AlAs RTDs at an oscillation frequency of 360 GHz.

AlGaInP/InGaAs-Heterostructures

The quarternary $(Al_xGa_{1-x})_yIn_{1-y}P$ material allows the fabrication of a semiconductor with the highest direct bandgap energy of all the III/V compounds lattice-matched to the commonly available, high quality GaAs substrate [18]. With $y = 0.51$, $x < 0.7$ the band gap is direct with an energy of $E_G = 2.3 \, eV$ (0.54 μm) and a lattice constant equal to that of GaAs [19]. Such alloys are potentially useful for visible light emitting diodes and visible lasers for optical storage applications. Laser structures made with this material system emit at wavelengths shorter than 0.66 μm, which is useful for optoelectronic applications, e.g. as light sources in optical information processing systems and optical fiber communication systems using plastic fibers [20]-[22]. Additionally, high direct band gap semiconductors may be useful for high temperature and/or high power applications and offer a large potential for low noise microwave- and millimeter wave applications [23].

Microwave transistor devices, like HFETs and HBTs are fabricated traditionally on layered structures composed of AlGaAs, InGaAs and GaAs material. This combination of materials is distinguished by a variety of remarkable features. First, the band-offsets are arranged in such a way that HFETs as well as HBTs can be fabricated from stacks of layers made up from these materials. Second, heterostructures can be grown in extraordinary high quality by using the well developed MBE technique. The AlGaAs/InGaAs/GaAs heterostructure system has become the most widely used material system for GaAs-based microwave devices.

However, the dominating position of this system is being challenged for the first time by the introduction of GaInP [23] and carbon doped GaAs [24] into HBT structures approximately three years ago. It has now been demonstrated by several groups [25]-[28] that MOVPE-grown layered structures using GaInP barrier and carbon doped GaAs base layers are definitely better suited for the fabrication of HBT devices than the conventional AlGaAs/GaAs structures; GaInP is a very promising layer material because it provides a large valence band discontinuity and allows devices to be fabricated in a much easier way.

HFET Devices. The most attractive semiconductor heterostructure for HFET applications is the combination of $(Al_xGa_{1-x})_yIn_{1-y}P$ and $In_zGa_{1-z}As$. The use of $(Al_xGa_{1-x})_yIn_{1-y}P$ as a wide band gap supply material and $In_zGa_{1-z}As$ as an active current channel material allows the realization of a high conduction band discontinuity (ΔE_c) at the hetero-interface. Assuming ΔE_c is about 40 % of the total bandgap difference between the two compounds, values of $\Delta E_c > 0.5$ eV can be achieved just by tailoring the compositions x, y and z as shown in Fig. 5.

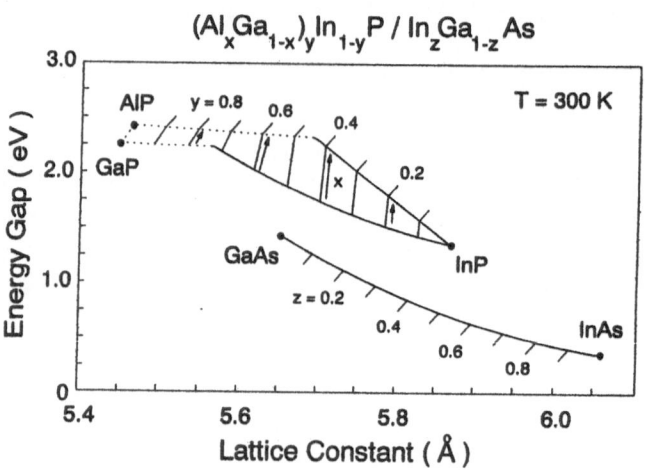

Fig. 5. Band gap energy versus lattice constant for the quarternary material system $(Al_xGa_{1-x})_yIn_{1-y}P/In_zGa_{1-z}As$.

A device epitaxial layer structure sequence is shown in Fig. 6. Two-dimensional electron-gas (2DEG) carrier concentrations up to 4×10^{12} cm^{-2} in combination with excellent room temperature and liquid nitrogen temperature mobility values have been measured [29], [30]. These values are significantly higher in comparison with transport data from AlGaAs/InGaAs HFET-structures on GaAs substrates. Much higher saturation currents and improved noise performance are expected for these devices. Additionally, the higher values of the band gap energy and the valence band discontinuity lead to an

GaAs Cap Layer	40nm
$(Al_{0.7}Ga_{0.3})_{0.51}In_{0.49}P$ $N_D = 3 \times 10^{18}$cm^{-3}	30nm
$(Al_{0.7}Ga_{0.3})_{0.51}In_{0.49}P$	3nm
$In_{0.2}Ga_{0.8}As$	12nm
GaAs buffer	
s.i. GaAs substrate	

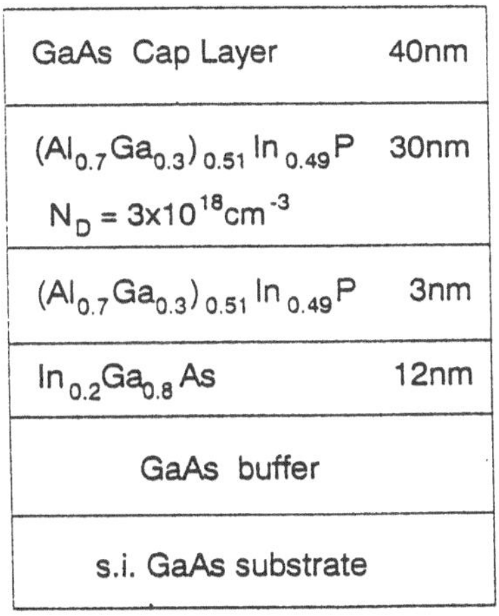

Fig. 6. Schematic layer structure of an AlGaInP/InGaAs HFET device.

improvement of the breakdown and gate leakage current characteristics relative to the AlGaAs/InGaAs material system.

For submicron gate-length devices very high drain/source breakdown voltages in excess of 18 V have been obtained. These breakdown voltages are about 6 V higher than on devices made with AlGaAs/GaAs. For a non-optimized layered structure and a gate length of $L_G = 0.35$ μm, a power gain cut-off frequency of $f_{max} = 120$ GHz has been achieved [29]. First attractive low noise performance data have been published for passivated and packaged devices. At 50 GHz a noise figure of $N_F = 1.2$ dB with an associated gain of $G_{ass} = 5.8$ dB has been measured [23].

Power Applications. A new type of HFET device for power applications was recently presented [29], [31]. For this structure an AlGaInP layer was used as a barrier layer and the conventional Schottky gate was replaced by a heavily carbon doped p$^+$-GaAs junction. The combination of both supports a very simple processing scheme and results in a high gate-drain breakdown voltage (see Fig. 7). Although only a simple direct printing photolithography and wet etching techniques were used, f_t and f_{max} values of 60 GHz and 140 GHz, respectively, have already been realized [32] as shown in Fig. 8.

These first results demonstrate the potential of AlGaInP/InGaAs/GaAs hetero-structures especially for power devices for which high current capability and high breakdown voltages are necessary. The development of GaAs-based HFET devices has obviously not reached its ultimate state. The new materials offer new degrees of freedom for developing improved devices.

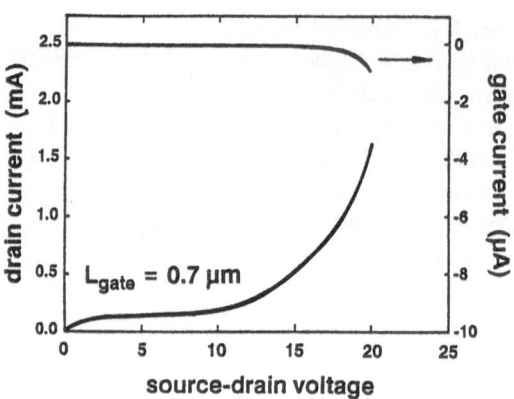

Fig. 7. Breakdown characteristics of an AlGaAsP/InGaAs HFET with a p⁺-GaAs gate structure and a gate length of $L_G = 0.7$ μm (from ref. [32]).

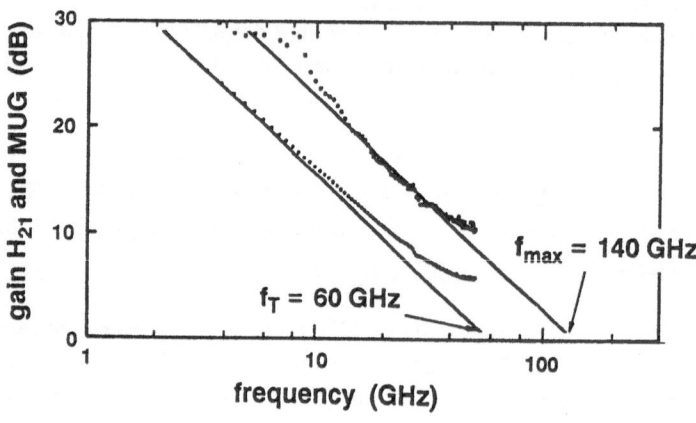

Fig. 8. RF figures of merit of an AlGaAsP/InGaAs HFET with a p⁺-GaAs gate contact (gate length: $L_G = 0.3$ μm, gate width: $W_G = 240$ μm) (from ref. [32]).

Technological Aspects. Beside the physical advantages there exist also some practical aspects which make this material system an attractive candidate for future activities. First, the material is based on GaAs and therefore can be used and fabricated on the basis of mature GaAs processing technology. Second, the AlGaInP material can be selectively etched (dry and wet chemically) with respect to AlGaAs and GaAs. Third, by using GaInP as a supply layer material there are no problems with Al-related effects, e.g. oxidation and DX-centers. These process features make device fabrication much more reliable. These are important facts to be taken into account concerning future mass fabrication of single devices and millimeter-wave modules.

InGaAs/AlInAs Heterostructures on GaAs-Substrates

Strain-relaxed epitaxial InGaAs and InAlAs layers on high quality GaAs substrates are useful for the fabrication of superior high speed and optoelectronic devices [33]-[36]. The higher achievable electron concentration and mobility values improve high frequency performance and noise figure in InGaAs/AlInAs HFET devices in comparison with GaAs/AlGaAs or pseudomorphic InGaAs/ AlGaAs heterostructures. Relaxed InGaAs buffer layers also provide access to an expanded band gap window and allow the use of the strain field as an additional design parameter.

Thick strain relaxed InGaAs layers with a constant indium concentration exhibit surface defect densities which are typically in the range of $10^9 \, cm^{-2}$ [37]. Recently, a siginificant improvement has been achieved by applying linearly graded or step graded buffer layers [33], [34], [38]-[43]. Dislocation densities as low as $10^4 \, cm^{-2}$ can be realized this way. The main problem of the graded buffer layers is the total thickness of several µm and the remaining surface roughness which shows a characteristic structure along the <110> crystal axes [33]. Prepared with optimized growth conditions, the layers have an optically flat surface with a microscopic "cross hatched" structure with step heights of the order of 10 nm.

InGaAs/AlInAs heterostructures lattice matched on InP substrates provide high electron mobilities above $10\,000 \, cm^2/Vs$ at room temperature [44] and 2DEG carrier densities up to $4 \times 10^{12} \, cm^{-2}$. Similar electron densities and 300 K mobilities of about $10\,000 \, cm^2/Vs$ were measured on InGaAs/AlInAs hetero-structures on relaxed InGaAs buffer layers (GaAs substrates) [34]. At 77 K the electron mobility in heterostructures on relaxed InGaAs buffer layers is about $45\,000 \, cm^2/Vs$ compared to $71\,000 \, cm^2/Vs$ on InP [34], [44]. The reduced mobility, especially at low temperatures, is expected to be mainly due to misfit dislocation defects in the relaxed buffer layer.

Additional growth research activities are necessary for a further optimization of the surface quality of relaxed buffer layers. A lower dislocation density and a smoother surface is necessary to make virtual substrates with high quality available for bandstructure engineering, which can be used for a larger range of electronic and optoelectronic features compared with the well established GaAs-based heterostructures.

Novel Device Applications

Field-Effect Transistor with Lateral Gate Electrodes. Low-dimensional transport properties are being intensively investigated for future electronic and optoelectronic device applications. In a one-dimensional electron gas (1DEG) system, a drastic suppression of the electron scattering rate was predicted theoretically [45]. Therefore carrier mobilities in quantum wires should be larger than the values observed in a 2DEG HFET. However, inelastic carrier scattering effects by optical phonons or intersubband transitions reduce the mobility significantly. Therefore, the expected advantage is difficult to realize in a quantum wire electronic device for room temperature application with hot carriers under normal operating conditions.

Novel field-effect transistor structures with narrow electron and hole channels have already been proposed and fabricated by focused-ion-beam [46]–[48], trench etching [49], selective isolation implantation [50], [51], and focused-laser-beam [52] techniques. Device operation has been demonstrated up to 300 K. The current channel conductivity is tuned by a lateral electric field (in-plane gate electrodes) rather than by using a vertical electrical field as in standard HFETs.

An interesting feature of this new type of device is its fully planar structure with an inherent low gate capacitance due to the very small gate area. A device layout with triangular shaped and closely spaced gate electrodes is shown in Fig. 9. Using a thin MBE-grown and by a δ-doped quantum well structure with a

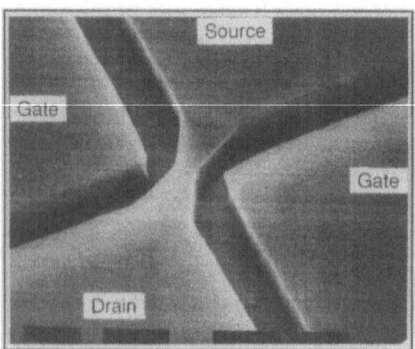

low-capacitive lateral gate electrodes
one-dimensional current transport
selectively implanted isolation barriers

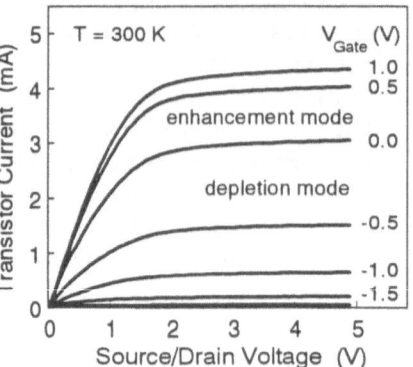

room temperature operation
small output conductance
multi-channel array

Fig. 9. SEM photograph and room temperature operating current/voltage characteristics of a novel field-effect transistor device with a narrow current channel and lateral gate electrodes [53].

high density 2DEG, narrow channels were fabricated by a selective isolation implantation technique [51]. Highest current and transconductance values of 0.6 mA and 0.25 mS, respectively, were achieved in single-channel devices [53]. The transistor current is enhanced by small matrix arrays of narrow channels. The I/V characteristics of a multi-channel FET is also shown in Fig. 9. The device can be enhanced and depleted by applying positive and negative gate voltages (V_G), respectively. Perfect pinch-off is achieved for $V_G < 2\,V$.

A detailed analysis has to be carried out in order to investigate the high frequency performance of this new type of device and to explore the potential for high speed and low noise applications.

Photodetector with Internal Gain. The above mentioned narrow current channels (quantum wires) are formed by carrier confinement in two directions. Vertically, abrupt heterointerfaces form a thin conductive layer (2DEG). Laterally, the 2DEG is confined by two narrow and closely spaced potential barriers. This causes lateral depletion zones at the wire boundaries. The flowing current is limited by the geometrical distance between the barrier regions reduced by the lateral depletion lengths. If above-bandgap light is absorbed, the device current is enhanced due to the reduction of depletion zones.

This effect can be used for a very sensitive light detecting device with internal gain [54]. Fig. 10 shows measured 300 K detector characteristics. Sensitivity values above 10^5 A/W for low power levels (< 1 nW) and a large dynamic range are measured.

This new type of detector has some advantages over conventional photo-FETs. First, the fully planar and lateral gate electrode structure is useful for normal incident light detection. Second, the fabrication of large area detector arrays is straightforward. Further investigations must be carried out to explore the sensitivity limits and to get information about the bandwidth of such devices.

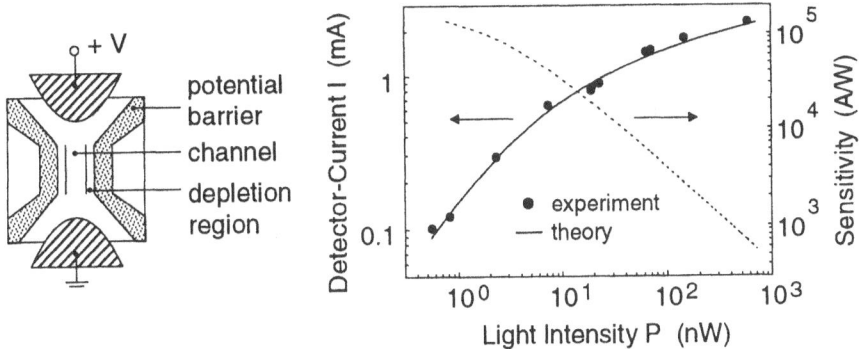

Fig. 10. Room temperature output current and sensitivity behavior of a novel photo-detector device with quantum wire light absorbing regions [54].

Epitaxial Regrowth and Selective Growth Techniques

Molecular Beam Epitaxial (MBE) Regrowth on Dry Etched Surfaces

Epitaxial overgrowth on non-planar structures has been in use for more than a decade, e.g. for buried heterostructure lasers [55] or low dimensional structures, i.e. quantum wire lasers [56]. Liquid and vapor phase epitaxy (LPE, MOVPE) were generally used on wet etched growth surfaces, with the crystalline quality and cleanliness necessary for epitaxial regrowth. The increasing demand on lateral size reduction and control of regrown layer thickness in atomic dimensions makes the combination of dry etching techniques, e.g. RIE or CAIBE, and MBE a viable tool for the growth of novel devices and the direct synthesis of low dimensional structures, i.e. quantum wires and quantum dots.

RIE allows reproducible and material selective etching of growth surfaces, yielding structures with vertical sidewalls in sub-μm dimensions. Etch damage can be kept low, so that two-dimensional, atomic layer by layer growth with MBE is possible, without 3-dimensional island growth due to nucleation at surface contaminants. Epitaxially grown GaAs and AlGaAs layers on GaAs substrates, which were etched with $SiCl_4$, show unchanged morphology, good electrical transport properties and good photoluminescence yield in comparison with layers grown on conventionally cleaned growth surfaces [57].

The MBE growth process, on the other hand, with the formation of non-growth surfaces, such as the (111)B plane, and its orientation and material dependent growth rates, allows a variety of structures with selected boundary facets to be fabricated. The shape of final overgrown structures depends on the vertical and lateral dimensions of the initial surface structures.

Growth on High Ridges: Quantum Wires. The kinetics of MBE growth are determined by sticking coefficients and surface migration lengths of adatoms, both depending on crystallographic orientation, growth temperature, III/V ratio, species of arsenic (As_2 or As_4) and group III-adatoms. Surface steps on patterned substrates influence migration lengths. High steps additionally incur shadowing of molecular beams with subsequent trench formation. The two effects result in non-planar and facetted growth.

High steps or ridges (≥ 500 nm) result in orientation dependent trench formation and facetted ridge overgrowth. The width of the ridges on the initial structured growth surface determines orientation and extent of the boundary facets. On high and narrow ridges the epitaxial structures are terminated by perfect crystallographic planes, which yield quasi one-dimensional structures with practically no surface roughness. Direct growth of quantum wires is thus possible, with lateral size variations comparable to the inherently excellent control of the vertical dimensions.

Fig. 11 shows the cross section of a 0.7 μm wide and 1 μm high ridge, overgrown with 1.5 μm GaAs. There are perfectly smooth boundary planes, namely

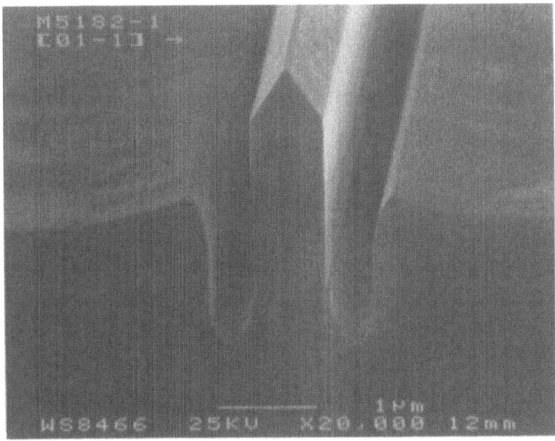

Fig. 11. SEM picture of a 1 μm high ridge in [011]-direction, overgrown with 1.5 μm GaAs (from ref. [58]).

vertical (01-1) side-walls and (111)B planes, resulting in a sharply peaked triangle. Such triangular structures serve as the basis for the direct "self-adjusted" growth of one-dimensional quantum wires. These quantum wires show a high cathodoluminescence yield, comparable to that of quantum wells grown simultaneously on unstructured parts of the wafer [58].

Surface roughness, emanating from masking and etching irregularities of the initial structures, is thus eliminated. Sub-μm structures with lateral size variations on the nanometer or atomic scale can be grown.

Overgrowth of Low Steps: Buried Structures. MBE regrowth over shallow structures, i.e. ridges or steps of less than 100 nm height can be realized without voids, which yields to quasi-planar layers and the fabrication of buried structures. Shadowing of the molecular beams is negligible. The two-dimensional epitaxial growth proceeds on (100) growth surfaces.

Fig. 12 shows an epitaxial structure consisting of a 0.88 μm thick $Al_{0.35}Ga_{0.65}As$ layer followed by alternating layers of AlGaAs and AlAs grown in an aperture of 5 μm surrounded by a 50 nm step. The AlAs and $Al_{0.1}Ga_{0.9}As$ layers form a high reflectivity distributed Bragg reflector (DBR) with 18 pairs of λ/4 layers of 63 nm and 74 nm thickness, respectively, in a vertical cavity surface emitting laser (VCSEL). The initial n-AlGaAs confinement layer, being much thicker than the current blocking layer or step height, planarizes the growth surface and allows undistorted growth of the DBR-mirror. Laser resonators with such regrown top mirrors show unchanged reflectivities of more than 99.7 %, identical to those grown in a single MBE run [59]. The incorporation of these 50 nm thick current blocking layers by epitaxial regrowth serves to reduce current spreading in VCSELs.

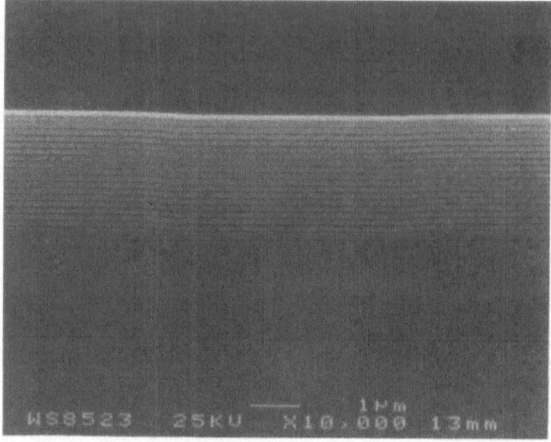

Fig. 12. Planarizing growth of an AlGaAs cladding layer and an 18-pair AlAs/-Al$_{0.1}$Ga$_{0.9}$As high reflecting distributed Bragg reflector. Step height prior to regrowth was 50 nm (from ref. [59]).

MBE-Regrowth on Selectively Implanted Layers

Ion implantation is firmly established in III/V-technology, either to create doped layers or to produce resistive regions for device isolation. In the last decade III/V MBE has been established as a standard technique for the fabrication of high quality heterostructure device layers, which are grown in one epitaxial run for, e.g. HFET and quantum well laser device applications.

The combination of a selective implantation process for lateral patterning and a following large-area MBE overgrowth is a promising way for a monolithic integration of devices with different vertical layer sequences on one chip and allows the design of novel millimeter wave integrated circuits (MMICs) [60]. It enables the parasitic losses and the fabrication cost of MMICs to be reduced.

Fig. 13 shows schematically a monolithic combination of a Schottky-Diode and a HFET device layer sequence. The HFET layered structure is overgrown by a second MBE run after a selective isolation implantation process into the diode layered structure. In doped GaAs material, highly resistive behavior can be achieved by proton or boron ion implantation. This is a consequence of the damage-induced compensation of carriers due to ion bombardment. However, long-term stability of high resistance behavior is only observed for temperatures well below 500°C.

Oxygen ions have also been shown to isolate highly n-type doped GaAs material. However, the isolation characteristics are thermally stable up to temperatures above 750°C after a rapid thermal anneal as shown in Fig. 14 [61].

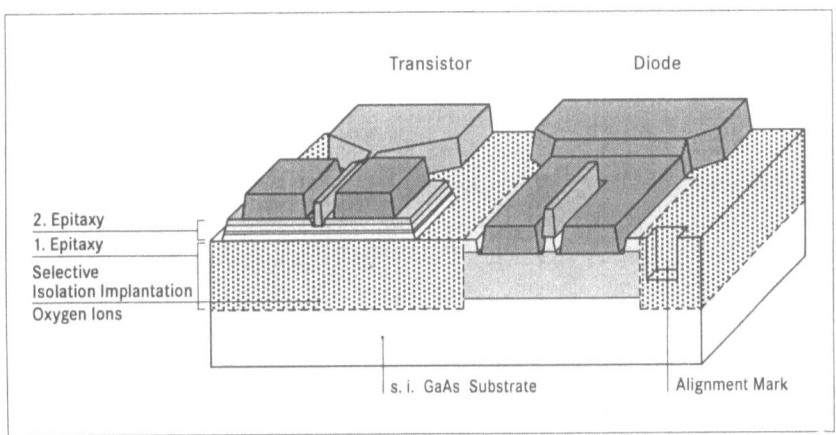

Fig. 13. Schematics of a device layered structure for a monolithic integrated Schottky-Diode/HFET-MMIC.

This behavior allows MBE regrowth of additional layers without any degradation of the electrical isolation characteristics of the implanted buried material. There is strong evidence for a chemically induced carrier compensation [62], [63].

The overgrown MBE layers on the selectively implanted wafers show nearly the same surface morphology and a low defect density in comparison with conventionally grown samples on cleaned substrate wafers. A strong photo-luminescence response is observed from the quantum well region of regrown pseudomorphic (PM) InGaAs/AlGaAs HFET layers.

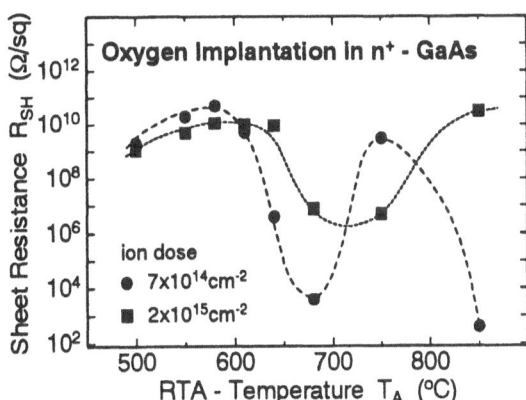

Fig. 14. Oxygen isolation characteristics vs. annealing temperature of highly doped GaAs ($n = 5 \times 10^{18}$ cm^{-3}, $R_{SH} = 17$ Ω/sq).

The 2DEG transport properties of regrown PM HFET layers were investigated by Hall effect measurements. Mobility values above $6\,500\ \text{cm}^2/\text{Vs}$ and carrier densities close to $2 \times 10^{12}\ \text{cm}^{-2}$ were obtained at $300\ \text{K}$ [61]. These values are close to measured transport data from reference samples on standard substrates.

Metal-Organic Molecular Beam Epitaxy (MOMBE)

MOMBE is a rather new growth technique which opens new routes for novel device structures and production technologies. The method enables a much more efficient use of the toxic starting materials compared to metal-organic vapour phase epitaxy (MOVPE). This is an important aspect directed towards the increasing safety requirements in an industrial environment. Presently there are several strong fields of application for the MOMBE process:

- The unique results from carbon doping enable new generations of heterojunction bipolar transistors (HBTs) which require highly doped and very thin base layers. Due to its low diffusion coefficient carbon, is an excellent candidate for p-type doping [64].

- Materials containing phosphorus are another strong area for MOMBE. The method allows sharp transitions to be fabricated between As- and P-containing layers. Due to the molecular beam nature and the fact that the sources are outside of the reaction chamber, system memory effects can be significantly suppressed. State of the art device structures have been demonstrated using this growth technology [65].

- The most outstanding performance of MOMBE can be allocated to the field of future device integration, where epitaxial growth on structured surfaces is

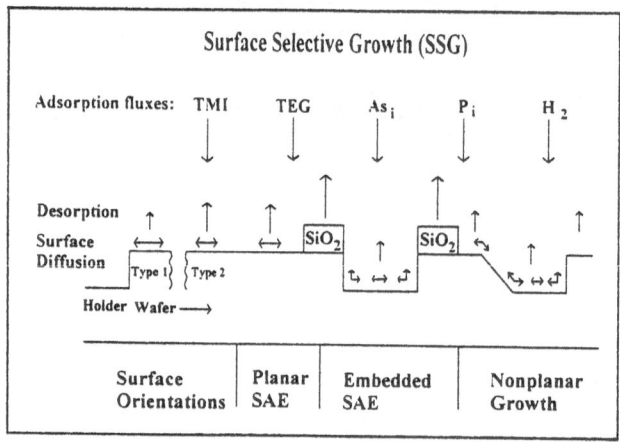

Fig. 15. Surface selective growth conditions in MOMBE.

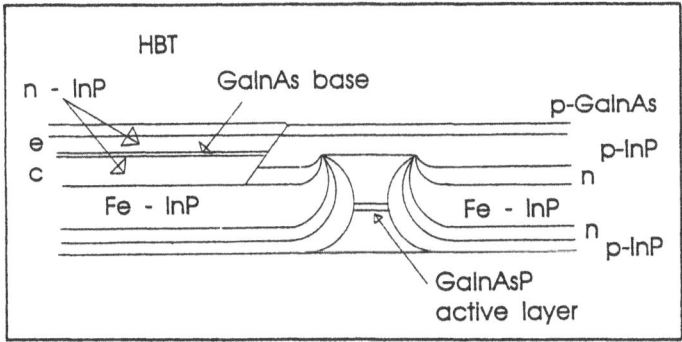

Fig. 16. Schematic cross section of a HBT-laser integration. The laser structure was grown by LP MOCVD, the transistor was selectively regrown by MOMBE (from ref. [68]).

required. In that field the surface selective growth (SSG) in MOMBE/CBE opens new routes for the realization of novel monolithic integrated device concepts. The SSG technique can be divided into different growth modes as shown schematically in Fig. 15:

a) Orientation dependent growth
b) Planar selective growth
c) Embedded selective area growth
d) Growth on non-planar surfaces.

The achieved perfection in selective area epitaxy (SAE) (a und b), where the crystal growth is localized using dielectric masks, allows for the realization of lateral coupling and integration of different device structures. The major driving force to implement MOMBE into integrated device concepts is based on the fact that a wide aspect ratio range of the dielectric mask can be used without disturbing the growth result. This enables uniform localized layered structures with a homogeneous thickness and material composition [66]. This has already been demonstrated on device structures, e.g. laser-waveguide [67] and laser-HBT integration where a lasing threshold as low as 160 μA was obtained (see Fig. 16).

As a young growth technology MOMBE is implemented into demanding device concepts. However, there are still open research areas, e.g. in doping and alternative materials for improved process control and better safety standards.

Acknowledgement. The fruitful discussions and expert technical help of Th. Hackbarth, U. Meiners, H. Müssig, E. Schlosser, and C. Wölk are gratefully acknowledged. The author also thanks H. Dämbkes for his continuous encouragement and support at Daimler-Benz AG.

References

[1] E.R. Brown, J.R. Söderström, C.D. Parker, L.J. Mahoney, K.M. Molvar, and T.C. McGill, Applied Physical Letters 58 (1991), 2291

[2] J.D. Werking, C.R. Bolognesi, L.D. Chang, C. Nguyen, E.L. Hu, and H. Kroemer, IEEE Electronic Device Letters 13 (1992), 164

[3] K.F. Longenbach, R. Beresford, and W.I. Wang, IEEE Transactions on Electronic Devices 37 (1990), 2265

[4] J.J. Pekarik, H. Kroemer, and J.H. English, Journal of Vacuum Science and Technology B10 (1992), 1032

[5] D.H. Chow, R.H. Miles, J.N. Schulman, D.A. Collins, and T.C. McGill, Semiconductor Science and Technolology 6 (1991), C47

[6] J.W. Sulhoff, J.L. Zyskind, C.A. Burrus, J.C. DeWinter, M.A. Pollack, and J.C. Centanni, Applied Optics 31 (1992), 3398

[7] H.K. Choi and S.J. Eglash, Applied Physical Letters 61 (1992), 1154

[8] C. Mailhiot and D.L. Smith, Journal of Vacuum Science and Technology A7 (1989), 445

[9] A. Cappy, B. Carnez, R. Fauquembergues, G. Salmer, and E. Constant, IEEE Transactions on Electron Devices ED-27 (1980), 2158

[10] K. Yoh, T. Moriuchi, and M. Inoue, IEEE Electron Device Letters 11 (1990), 526

[11] T.X. Zhu, W.J. Ooms, J.K. Abrokwah, C.L. Shurboff, and H. Goronkin, European Patent Application EP 0521 700 A1 (1992)

[12] R. Bouregba, D. Lippens, L. Palmateer, E. Böckenhoff, M. Bogey, J.L. Destombes, and A. Lecluse, Electronic Letters 26 (1990), 1804

[13] J.M. Gering, T.J. Rudnick, P.D. Coleman, IEEE-MTT 36 (1988), 1145

[14] I. Mehdi, J.R. East, and G.I. Haddad, IEEE-MTT 39 (1991), 1876

[15] B. Landgraf, H. Brugger, A. Trasser, and H. Schumacher, Proc. of the 20th Int. Symp. on Galliumarsenide and Related Compounds, Freiburg (1993), Institute of Physical Conference Series, in press

[16] D.A. Collins, D.H. Chow, E.T. Yu, D.Z.Y. Ting, J.R. Söderström, Y. Rajakarunanayake, and T.C. McGill, Proc. of the NATO Advanced Research Workshop (1990), El Escarial (Spain)

[17] S. Tehrani, J. Shen, H. Goronkin, G. Kramer, M. Hoogstra, and T.X. Zhu, Proc. of the 20th Int. Symp. on Galliumarsenide and Related Compounds, Freiburg (1993), Institute of Physical Conference Series, in press

[18] M.O. Watanabe and Y. Ohba, Applied Physics Letters 50 (1987), 14

[19] G.B. Stringfellow, Journal of Crystal Growth 55 (1981), 42

[20] S. Naritsuka, Y. Nishikawa, H. Sugawara, M. Ishikawa, and Y. Kokubun, Journal of Electrical Materials 20 (1991), 9

[21] H. Tanaka, Y. Kawamura, S. Nojima, K. Wakita, and H. Ashai, Journal of Applied Physics 61 (1987), 5

[22] M. Korn, T. Körfer, A. Forchel, and P. Roentgen, Electronics Letters 26, (1990), 9

361

[23] M. Takikawa, K. Joshin, Electron Device Letters ED-L14 (1993), 406

[24] H. Kawai, T. Kobayashi, F. Nakamura, and K. Taira, Electronic Letters 25 (1989), 609

[25] B.T. Cunningham, M.A. Haase, M.J. McCollum, J.E. Baker, and G.E. Stillman, Applied Physics Letters 54 (1989), 1905

[26] F. Alexandre, J.L. Benchimol, J. Dangla, C. Dubon-Chevalier, and V. Amarger, Electronic Letters 26 (1990), 1753

[27] P. Zwicknagel, U. Schaper, L. Schleicher, H. Siweris, and K.H. Bachem, Extended Abstracts of the IEEE-MTT Workshop on Heterostructure Technology, Günzburg (1991)

[28] H. Leier, A. Marten, K.H. Bachem, W. Pletschen, and P. Tasker, Electronic Letters 29 (1993), 868

[29] J. Dickmann, M. Berg, Th. Hackbarth, R. Deufel, H. Daembkes, F. Scholz, and M. Moser, Proc. of the 14th Biennial Conf. on Advanced Concepts in High Speed Semiconductor Devices and Circuits (1993), Cornell University, Ithaca (U.S.A.), in press

[30] K.H. Bachem, D. Fekete, W. Pletschen, W. Rothemund, and K. Winkler, Journal of Crystal Growth 24 (1992), 817

[31] K.H. Bachem, W. Pletschen, K. Winkler, J. Fleissner, C. Hoffmann, and P.J.Tasker, Proc. of the 20th Int. Symp. on Galliumarsenide and Related Compounds, Freiburg (1993), Inst. of Physical Conference Series, in press

[32] W. Pletschen, K.H. Bachem, P.J. Tasker, and K. Winkler, Materials Science and Engineering (1993), in press

[33] J.C. Harmand, T. Matsuno, and K. Inoue, Japanese Journal of Applied Physics 28 (1989), L1101

[34] K. Inoue, J.C. Harmand, and T. Matsuno, Journal of Crystal Growth 111 (1991), 313

[35] J. Chen, J.M. Fernandez, and H.H. Wieder, Applied Physics Letters 61 (1992), 1116

[36] J.C.P. Chang, J. Chen, J.M. Fernandez, H.H. Wieder, and K.L. Kavanagh, Applied Physics Letters 60 (1992), 1129

[37] D.I. Westwood, D.A. Woolf, A. Vila, A. Cornet, and J.R. Morante, Journal of Applied Physics 74 (1993), 1731

[38] V. Krishnamoorthy, Y.W. Lin, and R.M. Park, Journal of Applied Physics 72 (1992), 1752

[39] J. Chen, J.M. Fernandes, and H.H. Wieder, Semiconductor Science and Technology 8 (1993), 315

[40] M.J. Eckenstedt, T.G. Anderson, and S.M. Wang, Physical Review B48 (1993-II), 5289

[41] J.C.P. Chang, T.P. Chin, C.W. Tu, and K.L. Kavanagh, Applied Physics Letters 63 (1993), 500

[42] F.K. LeGoues, B.S. Meyerson, J.F. Morar, and P.D. Kirchner, Journal of Applied Physics 72 (1992), 1752

[43] V. Krishnamoorthy, P. Ribas, and R.M. Park, Applied Physical Letters 58 (1991), 2000

[44] E. Tournie, L. Tapfer, T. Bever, and K. Ploog, Journal of Applied Physics 71 (1992), 1790

[45] H. Sakaki, Japanese Journal of Applied Physics 19 (1980), L735

[46] A.D. Wieck and K. Ploog, Applied Physics Letters 56 (1990), 928; Applied Physics Letters 61 (1992), 1048

[47] Y. Hirayama, Applied Physics Letters 61 (1992), 1667

[48] J.A. Adams, L.M. Templeton, E.V. Kornelson, and S.P. McAlister, Proc. of the Int. Semic. Devices Research Symp., Charlottesville (1991), 309

[49] J. Nieder, A.D. Wieck, A. Grambow, H. Lage, D. Heitmann, K. v.Klitzing, and K. Ploog, Applied Physics Letters 57 (1990), 2695

[50] U. Meiners, H. Brugger, B.E. Maile, C. Wölk, and F. Koch, Proceedings of the 6th Int. Conf. on Modulated Semiconductor Structures, Garmisch-Partenkirchen (1993), Solid State Electronics, in press

[51] U. Meiners, B.E. Maile, and H. Brugger, to be published

[52] P. Baumgartner, K. Brunner, G. Abstreiter, G. Böhm, G. Tränkle, and G. Weimann, Applied Physics Letters (1994), in press

[53] U. Meiners and H. Brugger, to be published

[54] H. Brugger, U. Meiners, and E. Schlosser, German Patent Application P 43 26 754.8 (1993)

[55] M. Hirao, S. Tsuji, K. Mizuishi, A. Doi, M. Nakamura, Journal of Optical Communication 1 (1980), 10

[56] E. Kapon, D.W. Hwang, R. Bhat, Physical Review Letters 63 (1989), 430

[57] M. Walther, T. Röhr, G. Böhm, G.Tränkle, G. Weimann, Journal of Crystal Growth 127 (1993), 1045

[58] M. Walther, T.Röhr, H.Kratzer, G.Böhm, W.Klein, G.Tränkle, G.Weimann, Proc. of the 20th Int. Symp. on Galliumarsenide and Related Compounds, Freiburg (1993), Institute of Physical Conference Series, in press

[59] T. Röhr, M. Walther, S. Rochus, G. Böhm, W. Klein, G. Tränkle, G. Weimann, to be published in Materials Science and Engineering B

[60] H. Brugger, C. Wölk, and H. Müssig, Materials Research Society MRS Fall Meeting, Boston, 1992, MRS Symposium Proceedings, ed. by D.C. Houghton et al., vol. 281 (1993), 281

[61] H. Müssig, C. Wölk, H. Brugger, and A. Forchel, Proc. of the 20th Int. Symp. on Galliumarsenide and Related Compounds, Freiburg (1993), Institute of Physical Conference Series, in press

[62] R.D. Schnell, S. Gisdakis, H.Ch. Alt, Applied Physics Letters 59 (1991), 668

[63] H. Müssig and H. Brugger, to be published

[64] C.R. Abernathy, F. Ren, S.J. Pearton, T.R. Fullowan, R.K. Montgomery, P.W. Wisk, J.R. Lothian, P.R. Smith, and R.N. Nottenburg, Journal of Crystal Growth 120 (1992), 234

[65] W.T. Tsang, Journal of Crystal Growth 120 (1992), 1

[66] H. Heinecke, A. Milde, B. Baur, and R. Matz, Semiconductor Science and Technology 8 (1993), 1023

[67] H. Heinecke, Proc. of the 4th Int. Conf. on CBE (1993), Nara (Japan), Journal of Crystal Growth, in press

[68] X. An, H. Temkin, A. Feygenson, R.A. Hamm, M.A. Cotta, R.A. Logan, D. Coblentz, and R.D. Yadvisch, Electronic Letters, in press

5.3   Nanoelectronics

S P Beaumont
University of Glasgow, Glasgow, UK

1. Introduction

It does need a dramatic leap of faith to believe that integrated circuits based on 0.1μm design rules will be commercially available in the early part of the next century. Prototype fully-scaled 0.1μm gate length MOS devices and processes have already been demonstrated[1] and are currently the focus of research by major international manufacturers. GaAs- and InP based High Electron Mobility Transistors (HEMTs) were shrunk to these dimensions some years ago, and circuits based on 0.1μm gate length devices have recently demonstrated not only useful performance at very high frequencies (close to 100GHz), but also significant performance advantages at low frequencies (<10GHz)[2] compared with larger devices.

The study of electronic and optoelectronic devices with critical dimensions of the order of 0.1μm and below has come to be called *Nanoelectronics*. Nanoelectronics research explores semiconductor devices at the limits of manufacturing technology, and seeks answers to the following questions:

- What *are* the constraints on further shrinkage of electronic devices?

Ten years ago it was confidently predicted that silicon devices would never shrink to gate lengths less than one micron for technological and physical reasons. Yet half-micron devices are already in commercial production, as are quarter-micron HEMTs. Technology exists to make transistors with 10nm gate lengths and to control the vertical dimensions of transistors to atomic precision. This provides an opportunity to determine whether or not there are fundamental limitations to miniaturisation of devices, perhaps because the random distribution of dopants introduces randomness in device characteristics.

- Are very small devices useful?

Although it may be possible to make a perfectly respectable transistor with a 30nm gate length, for example, is such a device any better than a larger and more easily manufactured device? Or simply smaller, and therefore potentially capable of higher levels of integration provided the interconnect problem can be dealt with?

- Can very small devices exploit phenomena leading either to improved performance in conventional transistors or a new basis for device operation?

It has been predicted that very short transistors should be faster than long-gate devices because electrons can travel through them without scattering. It has yet to be convincingly demonstrated that this ballistic behaviour has a measureable effect on the terminal properties of a transistor. But are there circumstances in which these effects can be put to good use? Can scattering be reduced in devices where the electron is constrained to move only in one direction, thus enhancing its mobility.

It has also been suggested that nanometre-scale devices might exploit the wave-like character of the electron, or its discrete charge. For example, a wire of the order of the electron wavelength in width should act as a waveguide. By analogy with microwave and optical devices, is it possible to make a switching device which relies on electron interference for its operation? Might it be possible to make a memory element so small that one bit of information can be stored on a single electron?

In optoelectronic devices, confinement of carriers in more than one dimension alters the density of states distribution in such a way as to make the light-emitting properties of bulk semiconductor more 'atom-like'. The emission spectrum of lasers made from wires or dots of semiconductor material should therefore be narrower than a bulk or quantum well laser, and might require considerably less current for lasing action. Can these advantages be demonstrated?

- Can the technology emerging from such research be usefully applied to the engineering of larger devices, or even more widely in other branches of technology?

The accuracy and precision of semiconductor nanotechnology could improve the yield and manufacturability of conventional device processes. And the opportunity to engineer molecular and biological materials using nanometre-scale fabrication techniques developed for semiconductor devices is expected to have applications in chemical sensing and medicine.

All these issues have been studied by ESPRIT and ESPRIT Basic Research projects. NANOFET and MONOFAST studied short-gate III-V transistors and the use of precision fabrication technologies to improve yield and manufacturability. NANSDEV, LATMIC and LATMIC II studied quantum transport devices and the optical properties of nanometre-scale semiconductor structures, the latter topic being pursued further by the NANOPT consortium. The following sections review the current state of the art in Nanoelectronics, drawing on the results of ESPRIT projects and other research as appropriate. In the space available it has not been possible to cover the field in detail or even to mention every relevant topic so the interested reader is referred to the more extensive treatments that can be found elsewhere[3, 4].

For a variety of reasons - not least the wide availability of suitable substrates, the relative ease with which nanometre-scale structures can be fabricated and the physical properties of the material - nanoelectronics has developed much more rapidly in the III-V compounds, and in the (Al,Ga) As alloy system in particular, than in silicon. All ESPRIT nanoelectronics project were focused on these materials. This article therefore concentrates on III-V nanoelectronics. There is nonetheless considerable interest in the the physics of nanometre-scale devices and structures in epitaxial silicon and silicon-germanium, especially as improved epitaxial growth techniques have made available material of very high mobility.

2. Semiconductor Nanotechnology

III-V-based Nanoelectronics is based on a number of enabling technologies which together allow the fabrication of structures with nanometre dimensions in both vertical and lateral directions. These are:

- Epitaxial growth by MBE or MOCVD
- Very high resolution lithography, usually based on Electron Beam Lithography (EBL)
- Pattern transfer often using dry etching, ion implantation or directly by focussed ion beam patterning.

In all areas, European research groups have made significant contributions to the development of technology and its application to nanoelectronic devices.

2.1 Epitaxial Growth

Molecular Beam Epitaxy (MBE) and Metal-Organic Vapour Phase Epitaxy (MOVPE) are both capable of growing epitaxial III-V layers with almost atomic control. The lower growth temperatures, the ultra-clean vacuum environment and the availability of in-situ diagnostic tools have tended to favour MBE as a source of high quality substrates for nanoelectronics work, and in particular for the growth of very high mobility two-dimensional electron- or hole-gases (2DEGs or 2DHGs). In these layers the dopants are remote from the mobile carriers, leading to much reduced Coulomb scattering, and the carriers themselves are confined in thin sheets against a heterojunction barrier. Carefully grown 2DEGs can have mobilities as high as 10^7cm^2/Vs and much effort has been spent on the development of methods of growing consistently high mobility material. One key factor is the growth of very high purity undoped GaAs in which the carriers will be confined. During the NANSDEV project the use of As$_2$ rather than As$_4$ was shown to produce GaAs with background impurity concentrations of 2×10^{13} cm^{-3} and peak mobilities of 400,000 cm^2/Vs, a record for GaAs grown by any expitaxial technique[5].

Apart from their importance as sources of substrate material, MBE and MOCVD are being used to fabricate nanometre-scale laterally patterned substrates directly. The traditional methods of nanofabrication, involving lithography and etching of

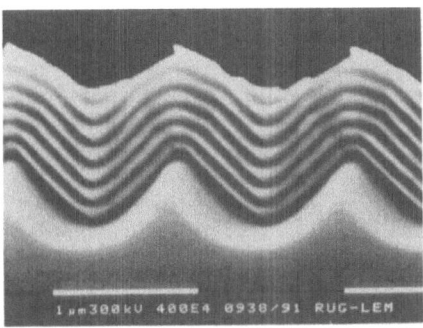

Fig. 1. Quantum wires grown in V-grooves by the migration of Ga during MOVPE overgrowth (couresy University of Gent).

pre-grown layers, are potentially damaging, time-consuming and perhaps ultimately of inadequate resolution. Therefore consideration is being given to the direct growth of nanostructures, either on substrates pre-patterned with large topographic 'clues' which control lateral growth, or by exploiting the kinetics of growth on non-stand-ard crystal planes to form nanometre structures by 'natural' means.

The former method was studied in some detail by the NANSDEV consortium. If quantum well layers are grown by MOVPE on substrates into which grooves have previously been etched, the rapid migration of GaAs on (111) planes causes a thick-ening of the quantum well at the apex of the groove. This thickened region forms a 'quantum wire' which can be used to confine and guide electrons. Indeed, the first realisation of a 'quantum wire' laser was claimed for a device fabricated using such wire structures[6]. Figure 1 shows a stack of quantum wires grown by MOCVD in V-grooves by the University of Gent as part of the NANSDEV project. The quantum wire region is clearly visible at the bottom of each 'V'. The same migration process gives rise to an accumulation of Ga-rich material during the growth of the AlGaAs cladding layers. This is clearly seen in the sample shown in Figure 2, from which the

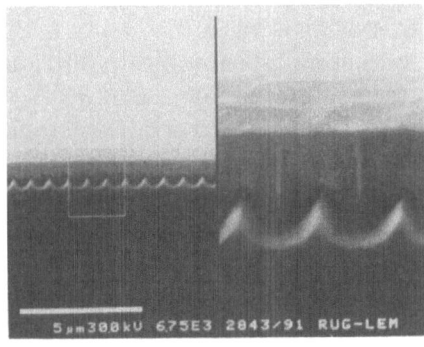

Fig. 2. Vertical quantum wells growing from the base of overgrown grooves as a result of migration of GaAs during MOCVD overgrowth (University of Gent)

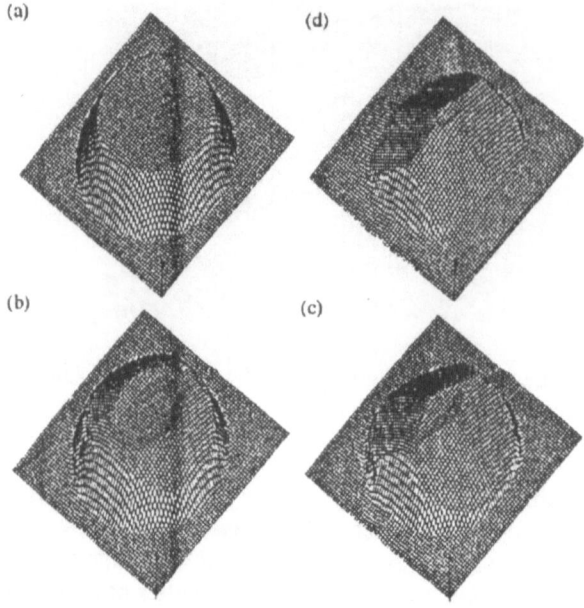

Fig. 3. Simulation of MBE overgrowth on etched GaAs cylinders (a) As-etched structure (b)-(d) Evolution of facetted shape as overgrowth proceeds

conformable quantum wells have been omitted[7]. A vertical quantum well is formed down which carriers are channeled during the operation of the quantum wire laser. Self-organised growth therefore not only forms the wires, but also enhances carrier injection!

NANSDEV also studied the overgrowth of etched stripes and cylinders using MBE[8]. We showed that, for certain orientations of the stripes with respect to crystallographic axes, smooth facets would form spontaneously to produce triangular shaped wires[9]. These same facets formed as overgrowth of cylinders progressed. This faceting could be realistically modelled using simple assumptions of growth kinetics: a typical result is shown in Figure 3.

Further discoveries under investigation by a new ESPRIT project, NANOPT, have shown that faceted wires form spontaneously during MBE growth on specially oriented substrates without any pre-fabrication of surface texture. For example, growth on (311) surfaces produces oriented wires with nanometric dimensions[10, 11]. Other routes to the direct growth of nano-structures include the use of zeolite cages to act as templates for the formation of ultrasmall particles of GaAs[12]: this approach is under investigation in the ESPRIT SOLDES project. Provided such structures can be shown to offer useful device-related properties, these growth techniques offer a route to mass production of substrates embedded with nanometre-scale features without resort to lithography.

2.2 Nano-Lithography

Although direct growth may be used in the long term, most experimental nanoelectronic devices are currently made by conventional lithography. Optical lithography does not have sufficient resolution to fabricate 100nm features although its performance continues to improve with the development of short-wavelength optics and novel masking techniques. Optical holography, however, has been used frequently to make periodic structures such as gratings with ~100nm linewidth for nanoelectronic devices . For research and low-throughput production, electron beam lithography offers easiest access to sub-100nm dimensions. Most research laboratories can undertake e-beam lithography simply by converting a scanning electron microscope using home-built or commercial hardware/software packages. For those requiring dedicated e-beam machines for research or production, systems are now available offering resolution down to 20nm over 150mm substrates. Leica-Cambridge Ltd, an EC-based company, is one of two suppliers of such equipment worldwide. Further improvement of these machines is likely to allow routine fabrication of structures on 10nm design rules.

The versatility of e-beam lithography is in part due to the availability of suitable resists. Most are based on simple polymers which are easily coated onto a wide range of substrates and can be used either for lift-off or as etch masks. PMMA (poly-methyl methacrylate) is the highest resolution e-beam resist and the resist of choice for most experimental work done on the smallest fabricated structures. The resolution achieved appears to be determined by a number of intrinsic and extrinsic factors including electron scattering, adhesion and other aspects of the physics and chemistry of the exposure and development process. For some time the resolution limit of this material was commonly believed to be 10nm, but recently, a number of groups have demonstrated that the resolution can be improved by careful processing. In particular, work done at the University of Cambridge (a member of the LATMIC II consortium) has shown that very brief development under ultrasonic agitation reduces the effects of resist swelling and allows features just 5nm in size to be defined[13]. Similarly work done at Glasgow has shown that the gap between adjacent structures can also be as small as 5nm if the pattern is designed so that adhesion is not lost. Figure 4 is an example of a pair of Au-Pd contacts separated by a 5nm gap lifted off from a PMMA resist pattern written at 100kV. The results represent an important breakthrough because 5nm lithography appears to be necessary for novel devices operating at room temperature.

Focused ion beam technology can be used to fabricate nanometre-scale structures directly, without the need for resist. Focused ion beam machines generate a sub-100nm probe using a liquid metal ion source and a series of lenses. The probe is scanned under computer control to write a pattern. Using heavy ions (e.g. Gallium), direct etching of a substrate can be performed. Using dopant ions such as Be or Si it is possible to control a doping profile laterally and write p-n junctions directly. Quantum transport devices have been made in this way, using lateral p-n junctions to confine electrons into narrow wires. Implantation of dopants and non-dopants may

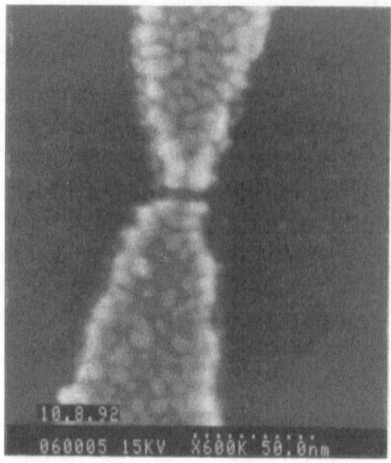

Fig. 4. 5nm gap between Au/Pd electrodes fabricated by electron beam lithography in PMMA resist followed by lift-off.

be used to to enhance the intermixing of quantum well material during annealing so that a substrate with a laterally varying bandgap can be produced. The LATMIC and NANOPT consortia have used this technique to make quantum wires for optical experiments[14, 15]. At the current state of the art the structures fabricated by ion implantation techniques are insufficiently smooth to show good quantum properties.

2.3 Processing

The translation of a nanometre-scale pattern from resist into a useful structures is as critical as the initial lithography. Liftoff methods have a long history of success in forming metallic structures, though patterns having minimum dimensions much smaller than the grain size of many useful metals do not translate into smooth, continuous structures. Liftoff is also limited to patterning films which are comparable in thickness to the resist which, for very high resolution patterns, may be only a few 10's of nanometres.

For the fabrication of semiconductor nanostructures, dry etching has been a popular route. The resolution of this technique is at least as high as the lithography step, and dot structures in the 5-7nm diameter range have been demonstrated in GaAs using silicon tetrachloride reactive ion etching (RIE)[13]. In contrast to liftoff the aspect ratio of an etched nanostructure can be very high as a result of self-masking, where the etching structure becomes passivated on its sidewalls by the formation of a protective layer, such as oxide. Thus a thin resist film can be used to mask a structure several hundred nanometres in height. Alternatively, selective etch techniques allow very thin layers of semiconductor to be removed with almost atomic layer accuracy - in a sense, the reverse of the MBE process. For example, selective reactive ion etching using Freon-12 will, under appropriate conditions, etch GaAs 4000 times faster than AlGaAs. Consequently, two monolayers of AlGaAs inserted in a GaAs epilayer are sufficient to arrest etching at a depth determined precisely by the substrate growth process. This method was used in the MONOFAST project to control the depth of the gate recess in high performance MESFETs[16]. Reactive ion etching techniques therefore offer three-dimensional control to nanometre accuracy using thin masks compatible with nanometre-scale lithography.

Dry etching techniques are not free from drawbacks, and one of the most important of these is the damage done to the semiconductor during etching. Nanoelectronic devices are particularly sensitive to dry etch damage because, for a variety of reasons, their active regions are relatively close to the etched surface and associated damage region. Consequently, the properties of semiconductor nanostructures can be used as probes of damage. For example, the conductance-width characteristics of dry etched n^+ GaAs wires and the luminescence-width characteristics of quantum wires etched into quantum well substrates have both been used to assess damage effects. During the NANSDEV project such measurements were used to probe the mechanisms of defect production on the sidewalls of etched nanostructures, leading to the development of a semi-empirical model for the accumulation of damage during etching. This enables the damage distribution in a conducting device to be predicted[17]. Similarly, sensitive probes for damage allow low-damage processes to be developed. Low damage is usually found to be associated with very low ion bombardment energy. Processes have now been developed to work in the 10-20eV range with excellent lateral resolution, good aspect ratio capability through self-masking of the sidewalls, and undetectable levels of damage[18].

2.4 Conclusions

Molecular Beam Epitaxy, Electron Beam Lithography and Dry Etching in combination provide a versatile foundation for the fabrication of nanometre-scale structures. Interest in nanoelectronics has stimulated growth of the highest quality material, the definition of sub-10nm size structures and the etching of semiconductors with undetectable damage. These developments alone have already had considerable impact on the fabrication of conventional electronic devices, as we will discuss in the next section.

3. Short Gate Transistors

The fabrication of short gate transistors is one of the most obvious applications of semiconductor nanotechnology. The current gain cutoff frequency (f_T) of a transistor is inversely related to gate length: a 100nm gate length transistor fabricated in GaAs should have an f_T greater than 100GHz. Shorter gate length devices, or transistors fabricated on material having a higher electron saturation velocity, will have even higher cutoff frequencies. Provided parasitic elements such as gate resistance and gate-drain capacitance are also scaled properly, these devices should allow integrated circuits to be operated in the 60-100GHz range, where new commercial applications are expected to emerge within the decade, at even higher frequencies for scientific applications and offer advantages even at low frequencies.

The miniaturisation of III-V transistors has been the subject of academic research for about ten years. One of the early quests of this research was evidence for so-called overshoot, in which electrons are able to attain velocities significantly greater than their saturation velocity because the time taken to transit the device is less than the time taken to lose kinetic energy through scattering. Despite many theoretical predictions of overshoot, and a number of papers which purport to have measured it, its existence remains to be proved by direct measurement. Even when gate lengths are shrunk down to 30nm, a variety of transistors - MESFETs, MOSFETs and HEMTs - exhibit perfectly normal dc behaviour provided they are properly scaled[19].

Optimisation of short-gate pseudomorphic HEMTs was one of the goals of the NANOFET project and devices with gate lengths down to 30nm were fabricated[20]. Scaling of a transistor requires not only a reduction in gate length, but also a reduction in channel thickness and an increase in doping density to counteract the effects of surface depletion and maintain adequate device current. Steps must be taken to confine electrons within the channel in the presence of strong two-dimensional electric fields which inject carriers into the substrate. These requirements have been met through the use of nanoelectronic technology. Substrates grown by epitaxial techniques (MBE or MOCVD) incorporate heterojunctions to provide electron confinement and modulation doping for high sheet electron concentration and mobility. ∂-doping minimises the distance between gate and channel for the highest mutual conductance. E-beam lithography is used for gate definition, using multi-layer resist processes to fabricate gates with mushroom cross-sections with the lowest possible electrical resistance. This combination of processes results, for example, in HEMTs with f_T's approaching 300GHz. [21].

III-V MESFETs and HEMTs normally have recessed gates to minimise the effects of series resistance and surface charge on device performance. Obviously the dimensions of the recess play an important role in determining the electrical characteristics of the device. In most industrial processes, recessing is carried out by wet etching techniques. When applied to short gate devices, wet recessing gives poor uniformity, yet a simple study shows that the recess depth must be controlled within a few nanometres to maintain the best performance. One solution is to adopt selective dry etching methods, so that the etch depth is set by an etch stop layer grown in with great accuracy by epitaxial techniques. In this way target etch depths are easily achieved, and over-etching can be used to adjust the offset between the edge of the gate metal and the sidewall of the recess. Such a process was used very successfully in the 'MONOFAST' project, as reported in the chapter by Ladbrooke and elsewhere[16]. Here, GaAs MESFETs rather than HEMTs were adopted for simplicity. Because all the physical parameters of the transistors, including the gate recess depth and offset, were well characterised, discrete devices and MMICs met their forecast performance in a single fabrication run. Moreover, the noise and associated gain of these MESFETs were comparable to those of commercial pseudomorphic HEMTs, confirming that the selective etch introduced negligible damage, and showing how nanotechnology coupled with a good understanding of device physics can improve the performance of classical transistors. Finally, by characterising the tolerance of each technological step it was possible to predict the functional yield of the demonstrator circuits - the first time such a capability has been properly demonstrated.

Clearly, as device dimensions shrink in all directions to achieve high performance yet dimensional variations remain limited to a few atomic diameters, device and circuit yields will be low unless the very best nanoelectronic technology is adopted and the circuit is designed to be tolerant to physical fluctuations. The approach demonstrated by the 'MONOFAST' programme could help to achieve economic yields of monolithic circuits operating in the millimetre- and sub-millimetre wavebands. But to do so will require further improvements in our understanding of the physics of short-gate transistors made in complex, poorly characterised materials systems such as (In,Ga,Al,As) and high precision processing technologies for these materials which, at present, are in their infancy. In addition, we will require a more detailed understanding of the sources of variations in FET-type devices. Some of this information is coming from basic research into quantum transport and single electronics.

4. Quantum Transport

4.1 Point contacts and quantum wires

What factors are likely to set the physical limit to the further miniaturisation of transistors? There are some obvious extrinsic effects, particularly those associated with parasitic reactances and resistances, notably the gate resistance. Intrinsically, if the transistor is to be properly scaled to retain good electrical characteristics, then the distance between channel and gate shrinks to such an extent that there will be no room to incorporate the dopants. In addition, the number of dopants contributing charge to the channel will become so small that statistical fluctuations at different points on the gate will lead to soft cutoff characteristics. And in devices which are scaled in width to increase packing density, there will be statistical differences between different devices which will therefore not have the same threshold voltage, or channel current at a set gate bias.

It is fair comment that so far, statistical fluctuations in doping appear to be irrelevant to the behaviour of ultrasmall transistors. Such factors are, however, playing a significant role in the behaviour of quantum transport devices.

The possibility of making devices in which electrons are not only unscattered in their transit from one terminal to another, but remain phase coherent, has been the subject of intense study. Such devices should not only be fast, but the current through them should be controllable by the same interference effects that allow photonic devices such as Mach-Zehnder interferometers to act as optical switches. For quantum-coherent devices to work, clearly the electron must not suffer scattering, implying that the device should be very short and/or be operated at low temperatures. Also, to get good switching behaviour, the electron transport should all occur in a single lateral mode. This latter constraint is the equivalent in optics or in microwave devices of working with single-mode waveguides. A further effect of operating a device in a single one-dimensional lateral mode could also be a reduction in electron scattering and an improvement in mobility: the basic idea is that the lateral confinement places a severe restriction on the directions in which the electron can be scattered[22]. One-dimensional structures may therefore offer a way of 'modifying' the transport properties of semiconductor materials.

The most convincing evidence of lateral quantisation in electron transport devices has come from very short wires fabricated by squeezing electrons in high quality two-dimensional electron gas layers between closely spaced gates. Figure 5 is a schematic diagram showing how this is done. Measurements of the resistance of such devices as a function of gate voltage were made simultaneously by van Wees et

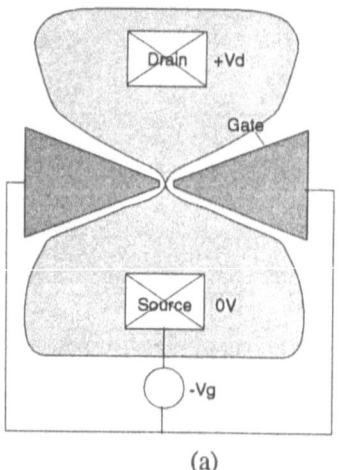

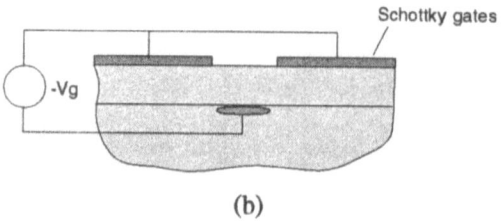

(b)

Fig. 5. Schematic of quantum point contact: (a) Layout, showing electron sheet defined by gates (b) section through gap between gates where point contact is defined.

(a)

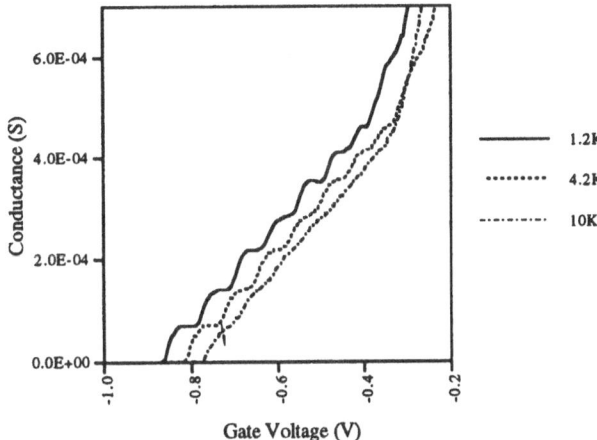

Fig. 6. Quantisation of conductance of quantum point voltage as additional lateral modes are permitted in the gap defined by the applied gate voltage. This device was fabricated on shallow 2DEG material on which quantisation can be detected up to at least 10K.

al in the Netherlands[23], and Wharam et al in Cambridge[24]. The experiment is equivalent to measuring the conductance of a transistor at different gate voltages: but remarkably, in this structure, the channel conductance does not pinch off smoothly - instead, it is quantised in steps of $2e^2/h$, where e is the electronic charge and h is Planck's constant (Figure 6). The origin of these steps is the lateral quantisation in the channel. As the gate voltage decreases, the channel narrows and fewer lateral electron modes are allowed in its width. It can be shown that a lateral mode contributes a fixed unit of conductance, so as each mode cuts off, the conductance of the device falls by one such unit until the channel is so narrow that no modes are allowed, and the conductance falls rapidly to zero.

So the lateral quantisation needed to make quantum-controlled electronic devices has been demonstrated. Moreover, remembering that this device is just a very small transistor twisted through 90°, this result tells us that quantum phenomena could play a role in the behaviour of highly miniaturised HEMTs. Except that these effects, when first discovered, were only significant at temperatures well below that of liquid helium (4.2K) and even at such temperatures, elongated structures - longer wires - showed no evidence of the reduced scattering that had been predicted. Similarly as the voltage drop along the channel of the device was increased from very small values, the effects of quantisation were rapidly washed out. Clearly it was not possible to increase the electron confinement in these almost ideal structures to overcome the effects of thermal and voltage smearing simply by increasing the gate voltage.

Two ESPRIT Basic Research projects began to address the issues of why these devices would only operate at low temperatures and could not be elongated. In the NANSDEV project, we looked at a theory which involves the random spatial fluctuations of donors. Simulations done by Nixon and Davies at Glasgow[25] showed that the random distribution of donors causes considerable fluctuations of width and electron density in elongated quantum wires squeezed between two gates. Figure 7 shows a typical result. From this work it appears that elongated devices would suffer from severe roughness effects unless a high electron density could be maintained as the width decreases. Otherwise as we squeeze a single mode device to increase the energy separation between modes to prevent multi-mode behaviour at elevated temperature, it becomes so rough that scattering destroys the coherent ballistic transport along the channel.

Evidence in support of this theory came from experiments done by Williamson et al[26], who examined the conductance characteristics of a point contact as it was moved electrically from side to side. The presence of fluctuations is clearly visible although the effects are significantly smaller than the simulations predict: this may be due to donor correlation effects.

More recent work has shown how redesign of the substrate can improve the thermal robustness of laterally quantised effects. The problem is to maximise the confining potential whilst minimising the effect on electron density in the channel. This can be done by making the 2DEG as shallow as possible without increasing electron scattering. Layers can now be grown having mobility $\sim 10^6 \text{cm}^2/\text{Vs}$ with the 2DEG

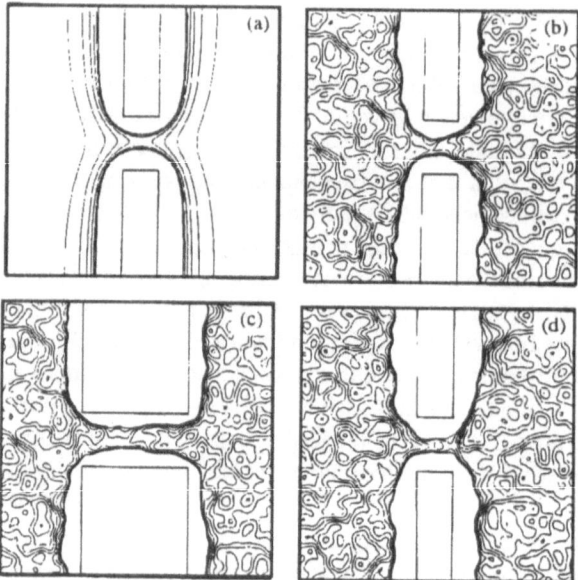

Fig. 7. Simulations of electron density in a quantum wire formed between two gates for different wire widths. The increasing disorder in the wire is due to the random potential imposed by donor atoms in the 2DEG substrate.

just 28nm from the surface. On such material, point contact devices exhibit quantisation effects at temperatures of 10-15K (Figure 6), and in channels up to 500nm in length. In material with smaller effective mass, e.g. InAs, there are claims that quantisation effects are visible at 80K.

In addition to improving the confinement potential, the effects of fluctuations would be minimised if the doping could be eliminated and electrons induced in the channel by the application of a positive gate voltage. This idea has already been used to make high-speed transistors known variously as SISFETs (semiconductor-insulator-semi-conductor FET), MISFETs (metal-insulator-semiconductor FET) or HIGFETs (Heterojunction Insulated Gate FET). Under the auspices of the NANSDEV project and its successor working group QUANTECS, studies have been carried out on 2DEG HIGFET material in which the doping lies in an n^+ layer 100's of nanometres below the channel, the remainder of the structure being totally undoped. To populate the 2DEG, a positive voltage is applied to a surface gate. At the moment, the technology for making standard devices on these layers has not been adequately developed, so non-contact methods of characterisation have to be adopted. These show that the mobility in the 2DEG at 4K is at least $10^6 m^2/Vs$[27]. Moreover, when the surface gates are fabricated into arrays of interdigitated fingers making it possible to split the 2DEG up into quantum wires occupying a large area, C-V measurements at 4K reveal directly the presence of laterally quantised sub-bands[28]. We believe that it is the absence of potential fluctuations due to ionised donors which allows us to see this quantisation so clearly over such a large array of quantum wires. Strong lateral quantisation over large areas could be used in infra-red detectors, sensing transitions between the laterally-confined sub-bands but more sophisticated processing will be necessary before this device potential can be exploited.

4.2 Periodic Devices

Periodic devices have been proposed as micro- and mm-wave power sources and there is some evidence that electron drift through a semiconductor underneath a periodic slow-wave structure gives rise to a gain mechanism[29]. A more exotic idea proposes that a periodic potential imposed on an electron gas would generate a set of minibands within which electrons could execute Bloch oscillations at THz frequencies. This last prediction remains unsubstantiated. Experiments have nonetheless tried to understand the nature and strength of the coupling between periodic surface potentials and electrons in an underlying channel using low temperature quantum transport techniques[30]. The upper insert to Figure 8 shows the resistance of a 2DEG device in which electrons are passing under an array of surface gates spaced on 280nm centres across the direction of charge flow. The resistance is plotted as a function of magnetic field. The oscillations in resistance at low magnetic field are due to the cyclotron motion of the electron interacting with the periodic potential. The amplitude of the oscillations allows us to extract the strength of the periodic potential experienced by the electrons. Unfortunately there is no evidence of mini-band formation, largely

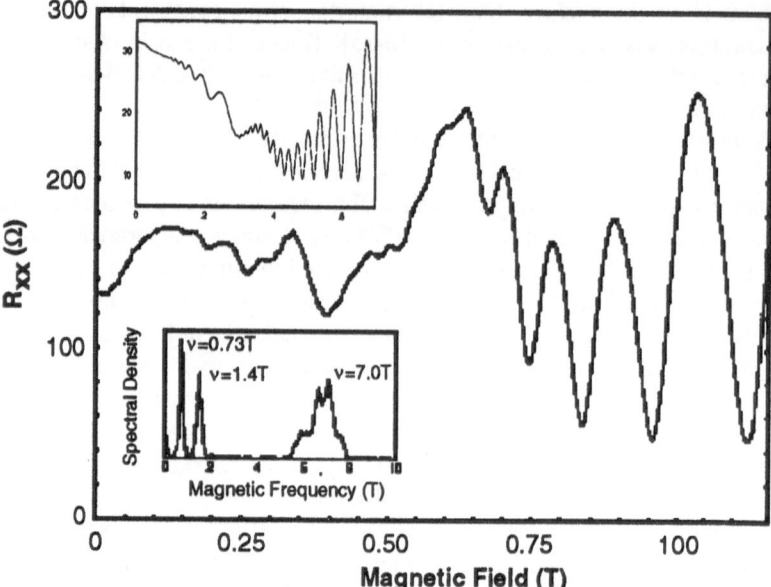

Fig. 8. Magnetoresistance oscillations in lateral surface superlattice device fabricated on a shallow 2DEG. The lower insert shows the power spectrum of the oscillations. The upper insert corresponds to a device fabricated on a deep 2DEG. The richer harmonic content is a result of strain sensed by the shallow 2DEG.

because the period of the superimposed potential is too large. To make it smaller, we also have to scale the depth of the 2DEG, otherwise the influence of surface potentials becomes negligible. The main part of Fig 8 is the magnetoresistance of the 280nm period device replicated on one of the shallow 2DEG substrates referred to previously. Now, instead of smooth sinusoidal oscillations of magnetoresistance, there is a strong harmonic content in the characteristic, confirmed by the power spectrum of the oscillations shown in the lower insert to this Figure. Vertical scaling of the device has considerably altered the interaction between the surface potential and the electrons. In fact the interaction in this sample is believed to be a consequence of strain imposed by the differential contraction between the semiconductor and the metallic gates on cooling. As yet we have not scaled the spacing between the surface gates to look for real superlattice effects.

Further understanding of the transport of electrons in periodic structures can be obtained by studying the microwave absorption of electrons confined by periodic arrays of etched pits or surface gates containing holes at low temperatures[31]. By examining the position and strength of the absorption it is possible to show that at low temperatures electrons move between and around adjacent cells just like billiard balls - the transport is indeed 'ballistic'. By studying the absorption spectra over a range of gate voltages, it is also possible to show how the coupling between adjacent

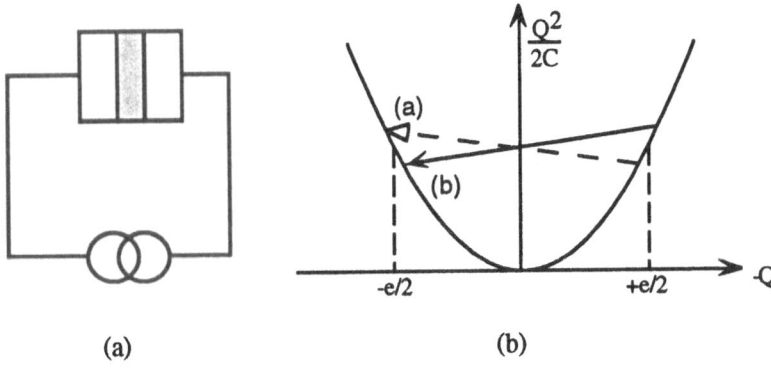

       (a)   (b)

Fig. 9. Illustrating the Coulomb blockade of tunneling through a capacitor (a) due to changing energetic favourability of tunneling process as charge builds up on the capacitor and is then reduced (solid arrow) by a single electron tunneling event.

cells of the periodic array can be modified by the gate voltage. This work has a strong connection with the idea of using periodic arrays as 'cellular automata' in single electron computing.

5. Single electronics

A typical HEMT structure contains on the order 10^{12} electron per cm^2 of substrate. This is equivalent to an average of one electron in every 10x10nm square. Therefore at the current limits of nanofabrication technology one can make devices which contain very small numbers of electrons. We might therefore envisage electronic systems in which computation and storage of data is based on the manipulation of single electrons. Recent research has shown that some of the basic requirements for such a technology may exist.

    Experiments carried out by ESPRIT Basic Research groups and others have shown unequivocally that electrons can be manipulated one-by-one. The basis of the manipulation is the so-called Coulomb blockade of electron tunneling. This effect can best be explained by considering the simple circuit shown in Figure 9(a). Here, a constant current source is charging a capacitor which contains a dielectric so thin that electrons can tunnel through it. Figure 9(b) shows the energy of the capacitor as a function of the charge on its plates. Suppose that the nett charge lies between 0 and e/2. Then if an electron tunnels across the junction - a process shown by the dashed arrow in Figure 9b, the energy of the system must increase. Hence the tunneling process is blockaded until the nett charge exceeds +e/2, when a tunneling event will reduce the stored energy of the system. Once the electron has transferred, the charge will have to build up to +e/2 again, before a second electron can tunnel. The Coulomb blockade of the tunneling process is responsible for the transfer of electrons individually across the junction.

In large capacitors, this process is undetectable. The energy change brought about by a single electron transferring across the capacitor is so small ($e^2/2C$) that thermal fluctuations swamp the system. However, nanotechnology allows us to make very small tunneling junctions with capacitances of the order of 0.1fF (10^{-16}F). Now the process is detectable especially at low temperatures, where the randomising effects of thermal fluctuations is small

Single electronic effects have already been detected in a large number of systems, including metal-insulator-metal junctions, semiconductor quantum dots, metal clusters and molecules[32-35]. In the latter two cases, Coulomb blockade is detectable at room temperature, so here we have a true nanoelectronic effect which, given properly engineered devices, could be used as the basis for the ultimately miniaturised electronics.

So far, most demonstrator devices have been limited to very low temperatures (<4K) for technological reasons, but this is good enough to demonstrate principles which should be extendable to high temperature. Very recently, workers at the Microelectronic Research Laboratory of Cambridge University have shown that single electron effects are detectable in Si-Ge devices at temperatures up to 50K[36].

By combining several tunneling junctions in series, it is possible to make single electron transistors, where a change in gate charge of one electron is sufficient to control the flow of current[37]. Multiple junctions can also be formed into frequency-controlled current sources by pumping a single electron through the device on each cycle of a high frequency clock. The combination of a single electron pump and a single electron electrometer (charge detector) is proposed as a capacitance standard, using the pump to charge an unknown capacitor with a known number of electrons, and the electrometer to measure the voltage across it. One source of inaccuracy in such systems is the charge pump which, on some clock cycles, fails to deliver exactly one electron. The reasons for this are an important area of study, as error rates are currently too large to make a single electron computer a worthwhile proposition.

Some researchers are beginning to consider how single electron systems could be made to compute[38]. Their ideas assume that single electron systems will consist of regular arrays of cells fabricated, perhaps, by a direct growth technique and fed with data only along their edges. By using the polarisation of these cellular arrays it has been shown that simple logic functions (AND, OR and invert) can be implemented.

6. Optical Properties of Nanostructures

Interest in the optical properties of semiconductor nanostructures arises from the modifications to the bulk density of states distribution which occur in confined structures. As carriers are confined first in quantum wells (2D systems), wires (1D systems) and then in dots (0D systems) the density of states changes from a continuum, through a series of steps and overlapping spikes to the set of discrete levels that most of us learned to calculate in elementary 'particle in a box' quantum mechanics. Thus the optical emission and absorption properties become more 'atom like', potentially enhancing the performance of lasers, detectors, optical modulators and optical switching devices. These potential benefits have encouraged a search for evidence of lateral quantisation in the optical properties and an improved understanding of the processes governing the emission of light from fabricated nanostructures.

Etched nanostructures have been subjected to the most intensive study as they are relatively easy to make by a combination of electron beam lithography and wet or dry etching of quantum well starting material (Figure 10). Optically excited carriers are confined vertically by the quantum well and laterally by the sidewalls of the etched structures. Many groups have detected blue shifts of the photoluminescence spectra, additional features in the photoluminescence emission spectra and polarisation dependence of the emission which have all been interpreted as characteristics of lateral confinement and quantisation.

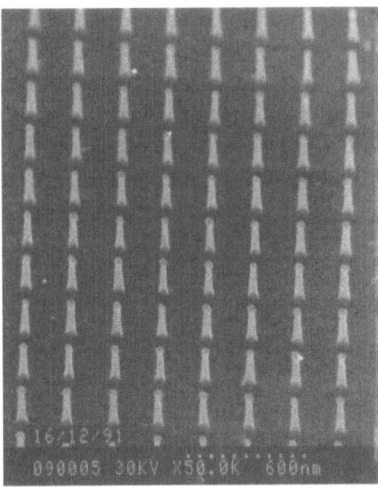

Fig. 10. Array of dry-etched quantum dots approximately 60nm in diameter. The dot mask was written by e-beam lithography in a negative resist, and the dots were etched with silicon tetrachloride RIE into underlying quantum well material.

However the optical emission efficiency of etched devices declines rapidly as the size of the quantum wire or dot is reduced. This has been interpreted in terms of exciton diffusion to surface states where recombination is non-radiative, and also by a model put forward by NANSDEV consortium involving quenching of luminescence by a phonon-blockade mechanism[39]. Detection of hot phonon signals from quantum dots, and changes in rate of cooling of hot electrons in photoexcited quantum dots lend weight to this theory[40]. Attemps to overgrow etched structures by MOCVD carried out by the LATMIC consortium have led to improved luminescence in very small dots[41]. Similar effects have also been obtained by annealing etched devices under overgrowth conditions without depositing a coating layer.

More attention is being paid to the optical properties of 'naturally' fabricated quantum and dots as discussed above. The use of lithography and etching is rightly believed to introduce unwanted defects and is far too slow to contemplate for production. The demonstration of the V-groove quantum wire laser has also stimulated this trend and has confirmed that room temperature operation needs quantum wires which are too small to fabricate easily by conventional means.

7. Conclusions

The stimulus of nanoelectronics research has led to the development and improvement of fabrication technology with much wider applicability. High resolution electron beam lithography, epitaxial growth technology and low damage dry etching all find applications in the fabrication of larger-scale electronic and optoelectronic devices, and further afield in non-semiconductor technologies.

The commercial production of GaAs and Si-based transistors on $0.1\mu m$ design rules is highly probable with benefits to performance as well as the level of integration . Some of the mechanisms likely to affect adversely the performance of conventional devices, such as quantisation and fluctuation effects, are detectable but appear to have a negligible effect even in 50nm-scale devices under normal ambient operating conditions. However at very low temperatures and in certain devices, they may begin to play a role. Thus in cryogenically cooled low noise front ends, or in circuits integrating semiconductor devices with superconductors, some of these phenomena may have a significant effect on device performance. By the same token, devices exploiting these phenomena are proving very difficult to develop because of their high thermal and voltage fragility. New techniques for fabricating smaller structures are needed to give greater confinement whether the applications are in electronics or optoelectronics.

The principles of single electronics have been demonstrated and prototype devices exhibiting single electron control have been fabricated. Again the temperature of operation is at present too low for widespread applications and new fabrication concepts are needed.

Quantumwires and dots show evidence of lateral quantisation in their optical characteristics, and a low-threshold quantum wire laser has been demonstrated, though more understanding of the optical properties of nanostructures is needed.

The future of Nanoelectronics probably lies in the exploitation of conventional devices and the development of 'natural' nanostructures for use a substrates for optical and single electronic devices and systems. A great deal about the properties of prototype structures and devices remains to be understood, giving much scope for further study under ESPRIT and other programmes.

Acknowledgements

I should like to thank all my colleagues at the Nanoelectronics Research Centre and members of the NANSDEV, QUANTECS, MONOFAST and SETRON consortia for their contributions to this article and in particular for their past and present comradeship.

References

1. Reeves, C.M., et al., Design and characterization of compact 100nm-scale silicon metal-oxide field-effect transistors. Journal of Vacuum Science and Technology B, 1992. **10**(6): p. 2917-2921.

2. Rosenbaum, S.E., et al., A 2-GHz three stage AlInAs-GaInAs-InP HEMT MMIC low-noise amplifier. IEEE Microwave and Guided Wave Letters, 1993. **3**(8): p. 265-267.

3. Weisbuch, C. and B. Vinter, Quantum semiconductor structures - fundamentals and applications. 1991, San Diego: Academic Press.

4. Davies, J.H. and A.R. Long, ed. Physics of Nanostructures. P. Osborne. 1992, Institute of Physics: Bristol.

5. Stanley, C.R., et al., $4 \times 10^5 cm^2 V^{-1} s^{-1}$ peak electron mobilities in GaAs grown by solid source MBE with As_2. Journal of Crystal Growth, 1991. **111**(1-4): p. 14-19.

6. Kapon, E., D.M. Huang, and R. Bhat, Stimulated emission in semiconductor quantum wire heterostructures. Phys. Rev. Lett., 1989. **63**(4): p. 430-433.

7. Vermeire, G., et al., Anisotropic photoluminescence behaviour of vertical

AlGaAs structures grown on gratings. J. Cryst. Growth, 1992. **124**(1-4): p. 513-518.

8.   Böckenhoff, E. and H. Benisty, Evolution of 3D growth patterns on nonplanar substrates. J. Crystal Growth, 1991. **114**: p. 619-632.

9.   Benisty, H., E. Böckenhoff, and A. Talneau, Evidence for facets with <210> azimuth in molecular beam epitaxial growth on patterned GaAs(001) substrates. Applied Physics Letters, 1992. **60**(16): p. 1987-1989.

10.   Notzel, R. and K. Ploog, Direct synthesis of GaAs quantum-wire structures by molecular-beam epitaxy on (311) surfaces. Journal of Vacuum Science & Technology A, 1992. **10**(4): p. 617-22.

11.   Notzel, R., L. Daweritz, and K. Ploog, Surface structure of high- and low-index GaAs surfaces: a direct formation of quantum-dot and quantum-wire structures. Journal of Crystal Growth, 1993. **127**: p. 1-4.

12.   Tomiya, S., et al. Germanium loaded zeolite Y: preparation and characterization. in MRS meeting. 1992. Boston: Materials Research Society.

13.   Chen, W. and H. Ahmed, Fabrication of 5-7nm wide etched lines in silicon using 100keV electron-beam lithography and polymethyl methacrylate resist. Appl. Phys. Lett., 1993. **62**(3): p. 1499-1501.

14.   Vieu, C., et al., Optical characterization of selectively intermixed GaAs/AlGaAs quantum wells by Ga^+ masked implantation. J. Applied Physics, 1991. **70**(3): p. 1444-1450.

15.   Prins, F.E., et al., Intermixed GaAs/AlGaAs quantum wires and the influence of implantation species on the steepness of the lateral potential. Journal of Applied Physics, 1993. **73**(5): p. 2376-80.

16.   Cameron, N.I., et al., Selectively dry gate recessed GaAs MESFETs, HEMTs and MMICs. J. Vac. Sci. Technol. B, 1993. **To be published**

17.   Rahman, M., et al., Model for conductance in dry-etch damaged n-GaAs structures. Applied Physics Letters, 1992. **61**(19): p. 2335-2337.

18.   Murad, S.K., C.D.W. Wilkinson, and S.P. Beaumont. Selective and Nonselective RIE of GaAs and $Al_xGa_{1-x}As$ in $SiCl_4$ plasma. in Microcircuit Engineering '93. 1993. Maastricht: North Holland.

19.   Adams, J.A., et al., Short-channel effects and drain-induced barrier lowering in nanometer-scale GaAs MESFET's. IEEE Transactions on Electron Devices, 1993. **40**(6): p. 1047-1052.

20.   Aniel, F., et al., Gate length electric parameter dependences of ultra-submicron delta doped pseudomorphic HEMTs. Electronics Lett, 1993. **29**(17): p. 1570-1571.

21.   Thayne, I., et al. Comparison of 80nm-200nm gate length $Al_{0.25}GaAs/GaAs/Al_{0.25}GaAs, Al_{0.3}GaAs/In_{0.15}GaAs/GaAs$ and $In_{0.52}AlAs/In_{0.65}GaAs/InP$ HEMTs. in International Electron Devices Meeting. 1993. San Francisco: IEEE.

385

22.   Sakaki, H., Quantum wires, quantum boxes and related structures - potentials and structural requirements. Surface Science, 1992. **267**(1-3): p. 623-629.

23.   Wees, B.J.v., et al., Quantized conductance of magnetoelectric sub-bands in ballistic point contacts. Phys. Rev. B, 1988. **38**(5): p. 3625-3627.

24.   Wharam, D.A., et al., One dimensional transport and the quantisation of the ballistic resistance. J. Phys. C, 1988. **21**(8).

25.   Nixon, J.A. and J.H. Davies, Potential fluctuations in heterostructure devices. Physical Review B, 1990. **41**(11): p. 7929-7932.

26.   Williamson, J.G., et al., The quantum point contact as a local probe of the electrostatic potential. Phys. Rev. B. Condensed Matter, 1990. **42**(12): p. 7675-7678.

27.   Drexler, H., et al., Transport studies and infrared spectroscopy on AlAs/GaAs MIS heterojunctions without dopants in the barrer. Semiconductor Science and Technology, 1992. **7**: p. 1008-1013.

28.   Drexler, H., et al., One-dimensional electron channels in the quantum limit. submitted to Physical Review B, 1993. .

29.   Thompson, J.J., et al., Gallium arsenide solid-state travelling wave amplifier at 8GHz. Electronics Lett., 1991. **27**(6): p. 516-518.

30.   Cusco, R., et al., Anharmonic periodic modulation in lateral surface superlattices. Surface Science, 1993. (To be published).

31.   Kotthaus, J.P., et al., Field effect confined electron dots, in Localization and Confinement of Electrons in Semiconductors, F. Kuchar, H. Heinrich, and G. Bauer, Editors. 1990, Springer-Verlag: Heidelberg.

32.   Wilkins, R., E. Ben-Jacob, and R.C. Jaklevic, Scanning-tunnelling microscope observations of Coulomb blockade and oxide polarization in small metal droplets. Phys. Rev. Lett., 1989. **63**: p. 801.

33.   Kouwenhoven, L.P., et al., Single electron charging effects in semiconductor quantum dots. Z. für Physik B, 1991. **85**(3): p. 367-373.

34.   Barner, J.B. and S.T. Ruggiero, Observation of the incremental charging of Ag particles by single electrons. Phys. Rev. Lett., 1987. **59**: p. 807.

35.   Kuzmin, L.S., et al., Single-electron charging effects in one-dimensional arrays of ultrasmall tunnel junctions. Phys. Rev. Lett., 1989. **62**(2539).

36.   Paul, D.J., et al., Coulomb blockade in silicon based structures at temperatures up to 50K. Applied Physics Letters, 1993. **63**(5): p. 631-2.

37.   Jin, Y., et al., Single electron effect transistor - fabrication and observation. Microelectronic Eng., 1992. **17**(1-4): p. 513-516.

38.   simple logic functionsLent, C.S., et al., Quantum cellular automata. Nanotechnology, 1993. **4**(1): p. 49-57.

39.   Benisty, H., C.M. Sotomayor-Torres, and C. Weisbuch, Physical Review B, 1991. **42**: p. 8947.

40.   Wang, P.D., et al., Photoluminescence intensity and multiple phonon Raman scattering in quantum dots: Evidence of the bottleneck effect. Surface Science, 1993. **To be published**

41.   Izrael, A., et al., Microfabrication and optical study of reactive ion etched InGaAsP/InP and GaAs/GaAlAs quantum wires. Appl. Phys. Lett., 1990. **56**(9): p. 830-832.

Appendix - List of ESPRIT projects on GaAs

232   Compound Semiconductor Materials and Integrated Circuits - I

522   Compound Semiconductor Materials and Integrated Circuits - II

830   Packages for High Speed Digital GaAs Integrated Circuits

843   Compound Semiconductor Integrated Circuits

927   Basic Technologies for GaInAs MISFETs

971   Technology for GaAs-GaAlAs Bipolar ICs

1128   Large-diameter Semi-insulating GaAs Substrates Suitable for VLSI Circuits

1270   Advanced Processing Technology for GaAs Modulation-doped Transistors and Lasers

2035   Advanced GaInAs-based Devices for High Speed Integrated Circuits (GIANTS)

5003   Multiwafer Planetary MOVPE Reactor (PLANET)

5018   GaAs Monolithic Analogue Circuits for Microwave Communication Systems (COSMIC)

5031   Metal Organic Research for Semiconductor Epitaxy (MORSE)

5032   Advanced Integrated Millimetre-wave Sub-assemblies (AIMS)

5052   Monolithic Integration Beyond 26.5 GHz (MONOFAST)

6016   Components for Large-Signal 60 GHz GaAs ICs (CLASSIC)

6050   Manufacturable Power MMICs for Microwave System Applications (MANPOWER)

Index of Contributors

Springer-Verlag
and the Environment

We at Springer-Verlag firmly believe that an international science publisher has a special obligation to the environment, and our corporate policies consistently reflect this conviction.

We also expect our business partners – paper mills, printers, packaging manufacturers, etc. – to commit themselves to using environmentally friendly materials and production processes.

The paper in this book is made from low- or no-chlorine pulp and is acid free, in conformance with international standards for paper permanency.